T0327413

Geotechnical Problem Solving

Geotechnical Problem Solving

John C. Lommler

Principal Geotechnical Engineer, USA

A John Wiley & Sons, Ltd., Publication

This edition first published 2012
© 2012 John Wiley & Sons, Ltd

Registered office
John Wiley & Sons Ltd, The Atrium, Southern Gate, Chichester, West Sussex, PO19 8SQ, United Kingdom

For details of our global editorial offices, for customer services and for information about how to apply
for permission to reuse the copyright material in this book please see our website at www.wiley.com.

The right of the author to be identified as the author of this work has been asserted in accordance with the
Copyright, Designs and Patents Act 1988.

All rights reserved. No part of this publication may be reproduced, stored in a retrieval system, or
transmitted, in any form or by any means, electronic, mechanical, photocopying, recording or otherwise,
except as permitted by the UK Copyright, Designs and Patents Act 1988, without the prior permission of
the publisher.

Wiley also publishes its books in a variety of electronic formats. Some content that appears in print may
not be available in electronic books.

Designations used by companies to distinguish their products are often claimed as trademarks. All brand
names and product names used in this book are trade names, service marks, trademarks or registered
trademarks of their respective owners. The publisher is not associated with any product or vendor
mentioned in this book. This publication is designed to provide accurate and authoritative information in
regard to the subject matter covered. It is sold on the understanding that the publisher is not engaged in
rendering professional services. If professional advice or other expert assistance is required, the services
of a competent professional should be sought.

Library of Congress Cataloging-in-Publication Data

Lommler, John C.
   Geotechnical problem solving / John C. Lommler.
      p. cm.
   Includes bibliographical references and index.
   ISBN 978-1-119-99297-4 (hardback)
 1. Engineering geology.   2. Soil mechanics.   3. Soil-structure interaction.   I. Title.
   TA705.L64 2012
   624.1′51–dc23

                                                                2011044002

A catalogue record for this book is available from the British Library.

Set in 10/12.5pt Palatino by Aptara Inc., New Delhi, India

Contents

Preface

7 December 2007, 6:19 am

For several years, my friend Ralph Peck has been gently encouraging me to participate more actively in the geotechnical engineering community and to write a book. A few weeks ago he told me about starting to write his famous book, *Soil Mechanics in Engineering Practice*. The day he started to write was December 7, 1941, better known as "Pearl Harbor Day." Talk about hard times to write a book! Ralph told me it took seven years to write that first book and that it was impossible, or nearly impossible, to finish, but worth the effort.

I suggested to Ralph that at 61 years old, I am past my prime for writing, and that in 1948 at the age of 36, when "Terzaghi and Peck" was published, he was in his prime. Ralph pointed out that although he was 36, Terzaghi was 65 years old, and that the old gentleman wrote papers and books until the day he died at age 80.

Although Ralph is too much of a gentleman to say it directly, he does suggest by comparison to his generation, that my generation of geotechnical engineers took the money, kept information proprietary, and did not share our experience with the engineering community at large. I guess (alright I know) that Dr. Peck is right, and so this book is my first serious attempt at sharing with you the practicing engineer.

For his example, and his encouragement, we all owe a great debt of gratitude to Ralph B. Peck, thanks Ralph.

John C. Lommler, Ph.D., P.E.
Sandia Park, New Mexico

Preface

1

General Topics

1.1

How to Use this Book

I want you the reader to be a good, if not great, problem solver. Problem solving is what engineers do, and it represents your value to society. When a client or employer pays for an engineer's services, they are purchasing a solution to their problems. Often this process is called designing, investigating or analyzing, but in the end it all comes down to solving a problem or a series of problems.

Engineering problems involving geotechnical issues are difficult to solve, primarily because geotechnical parameters are difficult to measure, difficult to characterize and difficult to analyze. Some of the geotechnical difficulty comes from spatial variability in a large volume of soil on a building site. Some of the geotechnical problem is due to correlating field and/or laboratory measurements to the soil parameters required for analysis, and some of the problem is associated with limitations of analysis methods. I want to help you figure out how to solve geotechnical problems, and I want you to enjoy the problem-solving process.

I want to be your personal mentor, and if you have a mentor, I want to help them mentor you. If you are a student, I want you to start thinking about what is required to become a practicing engineer and to start now to develop the problem-solving tools you need.

Right from the start, I want you to accept the fact that you will never be able to include all of the physical processes involved in natural systems in your model of reality. You have to simplify real-world problems by use of models that have a few essential parameters, or, to use mathematical terminology, you need to limit the number of variables included in your models (equations). Later, in Section 1.2, I will discuss and explain the phenomenon of increasing complexity. Let's just say for now that you will need to know how to adjust the number of variables considered in your problem-solving efforts to fit the needs of your particular problem.

Speaking of adjustment of the number of variables considered in your engineering problems brings me to a rather thorny engineering management problem that

Geotechnical Problem Solving, First Edition. John C. Lommler.
© 2012 John Wiley & Sons, Ltd. Published 2012 by John Wiley & Sons, Ltd.

frequently arises between problem-solving engineers and project managers. Before starting to work on the solution of an engineering problem, there needs to be agreement between the engineer and the manager on the level of detail and complexity to be included in the planned analyses. If the project manager thinks that the problem at hand is a simple issue, and you perform a highly complex analysis without informing the project manager, he or she will be unpleasantly surprised. There is going to be an issue over your charging excessive analysis hours to the project manager's budget. Fights over man-hour budgets for engineering analysis tasks versus the actual number of man-hours expended to solve the problem are quite common in consulting engineering practice these days. Matching the complexity of an engineering investigation and analysis to the requirements of the problem is called applying the "graded approach ." I will give you more information about the graded approach to problem solving in Section 1.3.4, and don't worry, we will include a discussion of how to handle surprises requiring more work than was initially anticipated.

Unlike most engineering/technical books that you have used, the presentation given here is conversational and personal from me to you. I want to give you practical advice on solving geotechnical problems and give you keys to the use of material that may not have been included in your college work. This "advanced" geotechnical material may be familiar to you, or it may be new; in either case, I want to help you understand the underlying assumptions and limitations of various geotechnical problem-solving techniques. You may not agree with my preferred choices of analytical methods. I would be surprised and a bit suspicious if you did agree with me on everything presented. It is alright to disagree, but we have to agree to base our disagreements on logic and interpretation of physical principles, not on arbitrary preferences. You may conclude that the available data and problem requirements need a more intensive analysis than I suggest is required. That's fine; if you need to do more detailed analysis work to feel comfortable with the solution, it's your choice. Just be ready to defend your man-hour charges with your boss, project manager, or client, or come in early or stay late and do the extra work on your own time. Having confidence in your solution to an engineering problem gives a sense of self-satisfaction. Remember that increasing your problem-solving skills increases your personal worth. Please do not think of extra work as giving "the company" something for nothing, consider it as money in your personal problem-solving account. It is your engineering career, not theirs.

During the early part of my career, I was a structural engineer. It was common in an earlier time for geotechnical engineer s to start their careers as structural engineers. My friend Ralph Peck was a famous geotechnical engineer who started as a structural engineer. He had a Ph.D. in structural engineering and no formal degrees in geotechnical engineering. I converted to geotechnical engineering during my graduate studies to help me understand how settlement-induced load redistributions in a structure could lead to structural failure. By the time my Masters degree studies were completed, I was hooked on geotechnical engineering. When I started working as a consulting geotechnical engineer, it quickly became quite clear to me that my structural engineering work had not been a waste. Knowledge of structural engineering helped me communicate with my structural clients because I knew what they needed from their geotechnical consultant. I have included material in this book to help geotechnical engineers understand what structural engineers need for their work, and I've tried to clarify geotechnical engineering topics to structural engineers so they can better communicate their project requirements to geotechnical engineers.

I have included what I consider to be important topics on selection and interpretation of soil laboratory tests, on analyses of shallow and deep foundations, retaining structures, slope stability, behavior of unsaturated soils including collapsible and expansive soils, and geotechnical Load and Resistance Factor Design(LRFD) topics, to name a few. I want you the reader to develop problem-solving tools for each of these geotechnical problems. We will start you with simpler standard practice approaches in each article, and work up to the advanced material. I'll give you examples of standard practice analysis methods including their assumptions and limitations. I don't want you to pick a standard practice approach for solving your problem if it doesn't fit the requirements of *your* problem! Advanced problem-solving methods are often required to deal with problems that have additional complexity. I do not believe that so-called advanced geotechnical material is only for Ph.Ds. I am convinced that if you graduated from college with a degree in engineering (or science and mathematics), you *can* use all of the material covered in this book.

In each section of this book, I will give suggested references and include a "Further Reading" section that provides materials for your study and consideration. At the end of each section, I will include a list of the references discussed in the section. I hope you will forgive me, but I do not like to repeat figures and equations from other books. Some equations and discussions are essential and I cannot avoid repeating them, but hopefully with added insights. Over the years I have grown weary of seeing the same material repeated and repeated over and over. I will refer you to the books where these materials are covered, and I hope you will take the opportunity to grow your geotechnical reference library.

It is my goal to help you understand the "how" and "why" of each topic, and to give you tools to use to solve problems that are not always included in text books, but are present in the real world. My advice is that you do not need a geotechnical engineering "cookbook." What you need is an understanding of geotechnical principles so that you can use them as tools to solve your engineering problems.

1.2

You Have to See it to *Solve* it

1.2.1 Introduction to Problem Solving

A question that I am asked by students and 60 year old engineers alike goes something like this, "Why is geotechnical engineering and engineering in general so difficult? When are the codes and requirements going to be simplified like they were in the good old days?" Give me a chance, and I'll answer these questions, but I need to build my case for the answer.

Did you ever notice that there is one person in the group that almost always disagrees with your opinion, conclusion, report, presentation, problem solution, etc? Sometimes they start by saying, "I'm just a Devil's Advocate here, but . . .." Personally, I don't want to give the Devil or his attorneys the credit for this phenomena, I believe that skepticism is a natural human trait that at least 10 to 15% of your students, clients, or associates will possess at any given moment. No matter how hard you try to convince these people that you have considered all of the important problem variables, they always seem to come up with new variables for consideration. How do they always manage to complicate your work?

I have come to believe (but cannot prove) that all problems in engineering have at least 10 to 15 variables that could be measured, analyzed, and used in their solution. The ideal or perfect solution to a given problem is a function that considers the impacts and interactions of all of these 10 to 15 variables. This hypothetical perfect solution considers all of the theoretical complexity involved. The engineering profession accepts two to three of these variables as the primary variables required in analysis and design. Analysis based on these two or three variables is referred to as standard practice, see Figure 1.2.1 below.

Geotechnical Problem Solving, First Edition. John C. Lommler.
© 2012 John Wiley & Sons, Ltd. Published 2012 by John Wiley & Sons, Ltd.

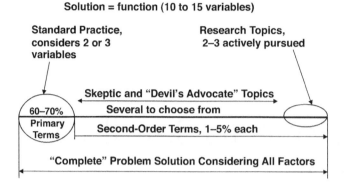

Figure 1.2.1   Engineering standard practice versus complete problem solution

After removing the primary two to three variables from the solution set, the re-maining seven to 13 variables are not considered to be standard practice , and I will call them second-order term s. These second-order terms may not often have a great impact on any given problem, *but* sometimes they do have significance. When second-order terms are important to the solution of an experienced engineer's problem, and when he or she has a feel for the magnitude of the impact of these terms is referred to as "experience" or "engineering judgment."

Thinking of an engineering problem as a number line of issues or variables, such as Figure 1.2.1, helps us to see an aerial view of the problem landscape. On the left end of the scale, say from one to three are the engineering design issues that make up standard practice. Let's assume that standard practice includes 60 to 70% of the weighted factors of significance . The 30 to 40% that standard practice is off the mark is covered by standard factors such as factors of safety, load factors, resistance factors, and so on. Presumably, standard practice suggests that if you are 60% correct and use a factor of safety of 3.0, or if you are 70% correct and use a factor of safety of 2.5, you should safely bound your problem.

On the right end of the scale in Figure 1.2.1, three issues from the remaining seven to 13 are commonly selected engineering research issues. These second-order terms on the right end of the scale are the habitation of university researchers and professors seeking research funding. If you want to get the problem solved, the design completed, stay in budget and meet your client's schedule, then you need to stay with the "engineering standard practice analyses" end of the problem scale. If you want a research topic, and you don't want to cover the same old ground of standard practice variables, then you need to select a second-order variable for study that may prove to be more significant than is currently understood. And finally, if you want to argue with others at conferences and monthly technical meetings, as many skeptics do (and you know who you are), please feel free to bring up issues from the portion of the problem scale that is not considered by standard practice or by current mainline researchers.

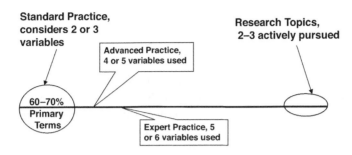

Figure 1.2.2 Advanced practice, expert practice, and research topics

1.2.2 Advanced and Expert Practice

Imagine that a client like a national laboratory, government agency, or high-tech project owner wants you to analyze something out of the ordinary. The measurements, analysis tools and advanced theories required involve something that is not covered by standard design practice. This type of engineering work is often referred to as advanced practice or expert practice. You can see from Figure 1.2.2 that standard practice is bounded by advanced and expert practice. These high-tech practices may be the standard of practice in some other parts of the United States or in other continents outside North America, such as in Europe or Asia. The point is that the definition of standard practice varies from place to place, and it varies with time.

After a natural disaster like Hurricane Katrina in New Orleans on 29 August 2005 or a major failure like the I-35W bridge collapse in Minneapolis on 1 August 2007, standard practice often expands rapidly to deal with issues uncovered during forensic analyses.

1.2.3 Theories Can Be Wrong. . .

You may say, "Standard practice or the accepted notion of what variables are important can be wrong. What about Albert Einstein and his proof that the standard idea of ether as the element that fills space was wrong?" You have a point, just like those skeptics I'm talking about.

Albert Einstein is famous for his attack on the concept of ether as the substance that fills space and transmits light. The concept of ether was an accepted principle of physics in the nineteenth century. Ralph Peck and I read the Einstein biography by Walter Isaacson (Isaacson, 2007) shortly after it came out in 2007. Ralph's eyesight was failing at the time, so he had a reader named Nida read the book to him. I recommended the book to my mother, a retired accounting professor from Kent State University. Nida and my mother commented on the parts of the Isaacson book that illustrated how human and fragile Dr. Einstein was in his personal and family relationships.

Ralph and I were drawn to the sections of Isaacson's book that illustrated how Einstein solved problems and developed theories. Ralph commented that Einstein solved problems by visualizing the solution. Einstein used analogies such as an elevator, a train, or a spacecraft traveling at the speed of light to frame problems and suggest their solutions. Ralph also pointed out that Einstein's physics journal papers were short and directly to the point, a trait that Ralph admired and strived to accomplish in his own work. I was taken by the contrast of Einstein's early work where he visualized problems, and his later work where he focused on mathematical formulations. Einstein accomplished a large amount of highly significant work when he relied on visual models or analogies to help guide his theories, and after 30+ years of work he never came up with a solution to the Unified Field Theory problem, although he formulated endless equations that appeared promising but failed.

Theories may be wrong, but we have to prove them wrong. Clearly seeing the issues involved seems to be the best guide to working through a complex problem.

1.2.4 Seeing is Better than Not Seeing . . .

My point in all of this is that you need a simple, visual model to understand and solve engineering problems. Focusing on two or three primary variables, as is the custom in standard design practice helps clarify the solution model because most people can see things in two or three dimensions. Clarifying the complex problem of changing engineering practice was my intention in showing "Problem Solutions" and "Standard Practice" and numerous potential variables in Figure 1.2.1 as a one-dimensional number line.

For those few of you who are experts and can see problems in five- or six- dimensional space, and have the mathematical tools to solve equations in these spaces, consider your selves blessed. It's up to you to reconfigure these complex problems in ways that are easy for the rest of us to see. *We have to see a problem clearly to solve the problem.*

1.2.5 Why is Standard Engineering Practice Changing?

The concept of seeing problems clearly brings me to a question that seems to be on many engineers' minds these days. Why does engineering practice seem to be becoming more and more complicated? To help you see this problem and predict the answer to this question, please refer to Figure 1.2.3.

As time goes on, more and more clients become involved in designs that require solutions to complex problems and require application of advanced engineering practice. Infrastructure failures, such as the failure of levees in New Orleans and the collapse of the World Trade Center towers in New York, raises questions about the

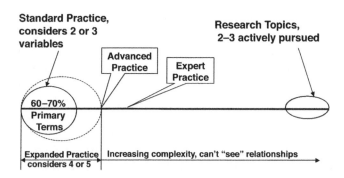

Figure 1.2.3   Growth of standard practice requirements

variables and principles used in standard engineering design practice. These problems become topics for funded university research. As university research into these advanced topics is published and made available to the engineering community through technical journals and conference proceedings, more and more engineers and clients want to work at a higher level. Computer programs are written that incorporate advanced topics, and everyone wants to have the latest and greatest computer modeling tools. The result of all of this pressure to do more is that standard practice expands to cover advanced practice, as illustrated in Figure 1.2.3. The amount of engineering analysis work increases, the cost and effort increases, and the complexity increases as we consider first three, then four, and then five or more variables in the problem solution. When asked if engineering practice will ever get easier, like in the good old days, my answer is *no*!

1.2.6 An Example of Increasing Complexity of Standard Practice . . .

As an example of how the engineering design process becomes more complex, consider paving design, which is an interaction problem between traffic, pavement, aggregate base, soil subgrade , drainage and weather. Let's say that a local university professor gets funding from the State Department of Transportation to do research on the standard problem of paving design. Say, for instance, that paving in their State is failing prematurely. Standard practice indicates that performance of paving depends on climate, traffic volume, wheel loads, and paving-base-subgrade strength. The premature paving failures could be related to excessive traffic, overweight vehicles, severe winters, or inadequate strength of a portion of the paving section.

But then one day an engineer from the Department of Transportation drives down a roadway in the State and notices that rainfall water is soaking into and seeping out of the paving. As he drives along, the Department engineer starts to think that maybe water in the paving section is the real problem behind premature paving failures.

After the engineer convinces his colleagues, the Department puts out a supplemental request for paving research on the topic of the effect of water in the paving section. The local university professor researching paving performance submits a proposal to further his study to include the affects of water in the paving section. His new paving proposal is accepted and research into the permeability of pavement and the affects of water on paving is funded, conducted, reported, and guess what, it does affect the life span of the paving section. As a result of findings of this research, the State Department of Transportation requires that two more variables related to paving permeability and paving saturation's affect on paving life be included in all standard paving designs. Referring to Figure 1.2.3 above, note that addition of a fifth and a sixth variable into the paving standard design practice increases the amount of work required and as a result increases the design complexity. In this way, the scope of standard practice increases and the work required increases, but likely the budget to do this additional work is not increased. From this analogy, I suggest that increasing complexity is to be expected. As time goes on, the size of the circle representing standard practice in Figure 1.2.3 will become larger and larger, and the ability of most engineers to see the impact and interaction of the design variables will be clouded. An engineer I know told me that he is working with an EPA standard contaminate transport model that has 23 input parameters. How's that for increased complexity!

1.2.7 Helping You See . . .

This book is designed to help you see each geotechnical problem clearly, so you can solve your problem. The discussions of each geotechnical problem are presented in a sequence from basic principles of standard practice to more complicated advanced practice issues. For those of you interested in advanced topics, a reference section is included with each problem discussion. These references to advanced material are included for those who have the need to dig into the body of available information.

To help you see concepts, analogies will be used. Don't worry that we're already introducing complexity; the definition used here for an analogy is a simple example that is similar to or has the same physical principles as the more complex problem. These simple analogies give you a visual key to help see the concept in your mind. An example of a simple analogy is the concept of frictional resistance in a granular soil, which is like a sliding wooden block on a table. At low strains the soil's frictional resistance or friction angle is higher, which is like static friction of the block on the table. At higher strains the soil's frictional resistance or friction angle is lower, which is like sliding friction of the block on the table.

We are going to discuss problem solutions from simple to complex, but how do you know which problem solution method to use on your problem. How do you decide when to use a simple, quick method or a highly complex, sophisticated method? How do you adjust the number of problem variables to fit your clients' scope and budget and still solve the problem? The answer is you need to understand and use

Figure 1.2.4   Would you expect to find expansive soil in a nearby boring?

the "graded approach." What is the graded approach? It's just a case of good old common sense, check out Section 1.3.4 for the answer to this question.

Please remember two things: (1) geotechnical engineering is not physics, we use constitutive equations that are based on tests that are designed to solve specific problems. Our tests and equations are *not* "laws of physics;" (2) whether it is geotechnical engineering, physics, accounting or everyday life . . . to *solve* it you have to *see* it. What do you see in Figure 1.2.4?

Reference

Isaacson, W. (2007) *Einstein, His Life and Universe*, Simon & Schuster, New York, 675 pages.

1.3

My Approach to Modern Geotechnical Engineering Practice – An Overview

1.3.1 Introduction

This book is intended to bridge the gap between geotechnical material covered in university civil engineering course work and the geotechnical topics required for practicing civil, structural, and geotechnical engineers to solve real world problems. Over the past decade or so there has been a tug of war between competing groups for the size and content of curricula included in undergraduate civil engineering programs. Several groups, the American Society of Civil Engineers comes to mind, have adopted programs requiring a Masters degree or an additional 30 credit hours of specialized training beyond the bachelor's degree to qualify for a professional engineering license. Will these additional training requirements discourage potential engineering students from entering the profession? Maybe it will. The fact remains that additional knowledge and skills beyond the bachelor's degree and even beyond a Professional Engineer's License are required to become a fully competent senior engineer.

For a long time, I've had the idea for a book to explain practical geotechnical problem solving to practicing engineers. My basic thought was to discuss alternate analysis methods and approaches. I'm not embarrassed to admit that I have struggled for years with many geotechnical topics, conflicts between competing theories of soil mechanics, and issues of increasing complexity, as discussed in Section 1.2. Having a mentor to help you through these geotechnical topics is a good thing, a very good thing.

Geotechnical Problem Solving, First Edition. John C. Lommler.
© 2012 John Wiley & Sons, Ltd. Published 2012 by John Wiley & Sons, Ltd.

In the later part of my career, my mentor was Ralph Peck. Ralph Peck was one of the founding fathers of soil mechanics and foundation engineering (now called geotechnical engineering or geo-engineering), and I was honored to have him as a friend and mentor. This book was gently suggested by Ralph. He also suggested that I keep working and forget about retirement. Since engineering mentors are in short supply world-wide, it is my hope that this book will help mentor you in geotechnical engineering.

As you have probably figured out by now, this book is not a text book (although I hope that advanced students may find it useful), nor is it a book of standard cook book solutions that don't match your design problem. It is my intent that this book will give you the why and how; that is, it will give you the understanding of the principles involved so that you can solve your own project problems. I will include some real world problems that illustrate my approach to geotechnical problem solving, and I will discuss alternate approaches and even theories that apparently conflict with one another so that you understand the different positions taken by experts.

Ralph Peck said, "The highest calling of an engineer is to make engineering as clear as possible." That is my intent.

1.3.2 Summary of Problem-Solving Approach

We have several choices of analytical techniques to solve the many classes of geotechnical problems. Each geotechnical analysis involves many levels of assumptions that are used to reduce a very complex situation to a solvable model of the real problem. As described in Section 1.2, there will be at least one person in the group that will disagree with your selection of field testing, laboratory testing, analysis procedure, computer program selection, analysis results, conclusions, and report format. Your only defense against opposing opinions is to have a logical, well-defined, thought-out approach to the problem, often referred to as a problem solution plan. If you can clearly define why you did something, a defense to opposing opinions can be constructed. If you just did something because "that is the way we always do it," or if you grabbed an equation out of a textbook because it was convenient and available, you are in for serious difficulties. Maybe not today, maybe not tomorrow, but some day you will be asked the question, "Why?" My experience has been that it is easier to answer your own questioning of why when you have time to think, rather than to deal with someone else (like a government QA auditor) pushing you later. Besides, exploring the "why" to any engineering problem is just plain fun. It's kind of like solving a mystery. I like fun, how about you?

1.3.3 Geotechnical Overview and Approach

The first thing you need to do when solving a geotechnical problem (or any problem for that matter) is to slow down enough to think. You have to start with your best understanding of what the problem really is. Both your client and your boss (if you have one) will tell you what the problem is all about. You should listen carefully and take notes, but do not jump to the conclusion that they have fully or even properly characterized the problem. Consider your client's and your boss's direction as data or input into the problem characterization process. The first thing I do is collect information in the most open-minded manner than I can muster. This information often includes:

- A site location plan or map.
- A drawing showing the facility or building on the site, hopefully with site topography (i.e., showing existing and planned ground surface elevations).
- Any available architectural or engineering drawings and plans.
- Site photographs.
- Historical aerial photographs, hopefully a series of photographs from well before site development to just before construction and to date.
- A geologic map of the site and surrounding area.
- Geologic publications for the surrounding area and a site or project geologic reconnaissance report if available.

After reviewing all of the above information, I start to develop a theory of what types of soil and bedrock problems may be (or have been) encountered on the site. Then I look up at the sign on my office wall that says, "You have to see it to solve it," and I let out a short gasp of air, sighing about the distance to the site, my lack of available time, my overly full schedule, and I remember that seeing the site is often the most important piece of a geotechnical engineer's work. Ralph Peck told me many stories of how Karl Terzaghi (The Father of Soil Mechanics) used extensive site visits often lasting one or two weeks to identify the critical pieces of information required to solve difficult geotechnical problems.

Ralph told me that Karl Terzaghi always keenly studied the geology of every site, he frequently visited his project sites and studied the landscape for signs of earlier landforms that would give him clues to the magnitude of the soil's pre-consolidation pressures, the nature of groundwater on the site, and existing shearing stresses that might have affected soils at his project site. For those of you who are interested in Terzaghi's methods of site investigation, I recommend the book *From Theory to Practice in Soil Mechanics, Selections from the Writings of Karl Terzaghi,* copyright 1960 by John Wiley & Sons, Inc. I found a used copy on the internet that was originally owned by Robert L. McNeill.

OK, now that you have done a thorough office study of available information, it is time to visit the site. When you get to the site, step back, climb up on the high ground and take in the landscape. Make notes and take photographs of everything you see at the site. Sometimes you are investigating a site for a future planned facility, and other times you may be investigating an existing damaged facility. For those latter cases, don't allow the client or his attorney to take you on "the tour" of their damaged building without doing your own investigation of site geology and a reconnaissance of the surrounding area. An example illustrates this point.

Worked example

On the way to the site, I heard that we were going to a very expensive house that had severe floor settlement and numerous cracks in the walls. I heard that the construction inspection reports indicated that all fill under the house was compacted to 95 to 100% of modified Proctor maximum density. I heard that one engineer expected that settlement was due to soft clays and that another engineer was convinced that expansive soils were present. I also heard a theory that roof drainage soaked foundation soils due to poor site drainage and that foundation support soils had collapsed.

When we arrived at the site, the driver pulled up to the front of the house and everyone bolted for the front door, except me. I went in the opposite direction. I walked out about 1000 feet perpendicular to the front of the house, turned around and surveyed the terrain. What I saw were outcropping layers of sandstone and layers of shale. The rock layers were upturned, with the sandstone forming ridges that ran approximately perpendicular to the front of the house. The shale material was highly weathered and buried beneath clayey soil between the sandstone ridges, see Figure 1.3.1.

Figure 1.3.1 Upturned sandstone with highly weathered shale

Following the sandstone ridges toward the house, it was clear that one ridge ran under the left end of the house, another ridge ran under the center of the house, and a third ridge ran under the right end of the house. It seemed to me to be a perfect geology for a differential settlement problem, see Figure 1.3.2.

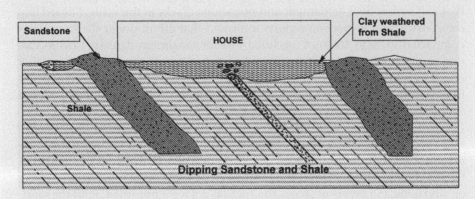

Figure 1.3.2 Ridges of sandstone and weathered shale under house

Based on field observations, I had my proposed theory of how the house settled, and now I was ready to join the others in touring the damaged structure. After reviewing room after room with settled floors and cracked walls, I started to become confused, because not all of the observed crack damage fit my proposed "theory of the case." After the tour, I asked a junior engineer to map all of the crack damage on a building plan drawing. The client gave me a copy of the original geotechnical report, and it showed soft, wet, and moderate to high plasticity clay in some borings to more than 20 feet, and refusal on sandstone at three feet in other borings. Checking the geotechnical report's boring location plan showed that borings with shallow sandstone lined up with my ridges of sandstone, and the borings with soft wet clay lined up with the shale zones located between ridges of sandstone. Settlement of the house's floors generally matched this pattern of little or no settlement where sandstone was indicated to be shallow. One immediate problem was the presence of exterior wall cracks indicating that the front and back walls were settling more on their ends than in the middle. But sandstone was located at the ends, so the ends should not have settled more than the middle unless there was a problem with my geologic model.

To resolve this problem with my theory, we dug test pits adjacent to the ends of the front and back wall of the house. Soil under the house footings was probed and determined to be very soft and wet. Further investigation disclosed that soils beneath the house footings were over-excavated and replaced with three feet of compacted fill. Although compaction tests indicated that the fill was compacted to greater than 95% of modified Proctor maximum density, the contractor blended off-site select granular fill soil with expansive clay and clay shale from the site. The 50–50 blend was done to save money. The expansive clay and clay shale mixed into the fill beneath the footings did not become wetted and softened until well after the house was built. Upon wetting, the expansive clay and clay shale lost strength turning to a muddy consistency resulting in settlement of the house footings.

After all of the competing engineers' theories of "what happened" to this house were reviewed based on detailed findings, it was apparent that they all had part of the problem correct. Expansive soils were present, and there had been collapse upon wetting, and so on. The common issue with all of the competing theories was that they all had most of the issues correctly characterized, but they missed important parts of the problem.

1.3.4 What Do I Do? Answer: The Graded Approach

There are several competing methods of solving each class of geotechnical problem. Some of these solution methods apparently conflict with other published methods. Some of these methods require much more field and laboratory testing data and analysis work to develop a problem solution, which of course cost more money. How do you select the right method for your problem?

I've found that many engineers are ready and raring to go, ready to dive into a geotechnical problem, but they are unsure how to start work or which solution techniques to use. Often they wait for the principal engineer to outline the approach for them, and then when the problem is laid out for them, they start to work. As a young engineer told me last month, "You better start writing down what you know because you'll be dead before you give it all to us, and then what will we do!" I hope she is wrong about my eminent demise, but she has a point. Older engineers do need to pass the information along.

I hate to break it to our younger readers, but money is involved in the geotechnical investigation and design process. It would be wonderful if we could use all of our high tech tools on each and every project. But no, it's true, the person or the agency with the money controls the work scope of the project. The engineer does not control the project unless he controls the money. This is not necessarily a bad thing. The client knows what he has to spend, and he has some idea of what he wants or needs. It is the engineer's role to guide and discuss options with the client until a reasonable accommodation has been made. You might think that large United States government national laboratories want the highest technology that money can buy applied to their projects. Not true. They want and need value as much or more than most clients because of the federal scrutiny involved with their budgets. A Los Alamos National Laboratory engineer once told me when referring to my proposed geotechnical work scope, "We don't need a super computer on this one. A small laptop will do!"

The point of this LANL engineer's comment was that the scope of work including the extent of field investigation, the complexity of laboratory testing, and the engineering analyzes had to fit the size and intent of the project. The building he was referring to was a large metal building used to store landscaping equipment: lawn mowers and tractors. The final cost of the investigation for this landscaping equipment storage building with a break room for the landscape workers was less than the quality assurance budget on a geotechnical investigation for a LANL facility where the primary mission of the lab was conducted. Federal government agencies often refer to this notion of the work scope fitting the project size and project importance as the "graded approach." By this they mean that small, less important projects require lesser work scopes, and that large, important projects require more costly, larger work scopes.

To the best of my knowledge, the United States Department of Energy developed the concept of the graded approach. I went on the internet to the US Department of Energy (DOE), Brookhaven National Laboratory (BNL) website. BNL conducts research in the physical, biomedical, and environmental sciences, as well as in energy technologies and national security. They have a definition of the graded approach on their website as:

A process for determining that the appropriate level of analysis, controls, documentation, and actions necessary are commensurate with an item's or activity's potential to

- Create an environmental, safety, or health hazard;
- Incur a monetary loss due to damage, or to repair/rework/scrap costs;
- Reduce the availability of a facility or equipment;
- Adversely affect the program objective or degrade data quality;
- Unfavorably impact the public's perception of the BNL/DOE mission.

I am always impressed at how DOE documents can be read without clearly understanding what they mean. I believe the operative word in the DOEs definition of graded approach can be summarized simply as the appropriate level of analysis. What is appropriate analysis for safety evaluation of a nuclear reactor is likely overkill for analyses of a parking lot subgrade.

Having read the above material, you may be fooled into thinking that the purpose, size and type of structure (a nuclear facility, a large hospital, or a landscaping shed) being scoped for a geotechnical investigation are the only factors that control the cost and extent of the geotechnical work. Not so. Several other important factors have to be added for selection of an appropriate level of analysis, including: complexity of the structure, complexity of the site stratigraphy and geology, requirements of the people in charge, and needs of the public.

Even if a project is small and seemingly unimportant, a small project's loading and geometry can still be complex. This presents a dilemma for the geotechnical engineer: how to simplify the complexities or how to deal with these complexities within a limited budget. Geology should be easy for geotechnical engineers to understand. If the geology and the site stratigraphy are highly complex, no matter what type of structure is planned, the complex site requires a greater scope and number of tests, and quite likely more complex analyses to obtain a solution with similar reliability to a simple site with uniform stratigraphy and geology. But what does "the public" have to do with the geotechnical work scope?

The answer to this question about the impact of "the public" on a geotechnical work scope may have an infinite number of answers, and if not infinite at least a very, very large number of answers. I will give two examples that come to mind:

1. Simple levees and hurricane Katrina – Levees have always been considered to be simple geotechnical structures, being much less important than dams. When I first started my geotechnical career, I couldn't wait to work on dams. The bigger the better. Levees were just a pile of dirt pushed up along a stream . . . not too interesting. When a dam fails, like the Vaiont Dam or the Teton Dam, now there's a real disaster. The 305-foot high earthen Teton dam structure failed, washing out more than 1500 homes, killed 11 people and over 18 000 livestock animals, and caused $322 million in damages. The 858-foot high Vaiont Concrete Dam in Italy failed when a massive landslide located above the dam dropped debris into the reservoir splashing out a flood wave that drowned 2000 people. After the flood the concrete Vaiont Dam remained intact.

But then hurricane Katrina came along and it changed the public's perception. Small, apparently simple earthen levee structures less than 15 feet high failed, resulting in the death of 1253 people, damage to thousands of homes, and causing approximately $28 billion in damages. Even though engineers and politicians may have considered the New Orleans levees as simple structures in the past, the public now considers them as very important. It is not just the size of the structure; it is the consequences of failure that count. Changes to the design and analysis practices of levees are still on-going, years after the disaster occurred.

2. A simple, single-span bridge investigated by a typical state Department of Transportation (DOT) for a bid-build project may require two to four borings, depending on the width and span of the structure. The embankment, retaining walls and bridge foundations may be analyzed and designed by correlations to standard penetration test (SPT) blow count supplemented by grain-size, plasticity index (PI) and water content laboratory testing used to classify the soils encountered. The DOT is probably not in a big hurry to finish the geotechnical work, and they would

rather save money (read initial cost) than time. The same bridge investigated by a design-build team may be a much different matter. Design-build teams nearly always have to develop an aggressive schedule for completion of a transportation project, and they have to provide performance assurances like "no more than 2 inches of embankment settlement in five years after completion of the project." To make accurate predictions of consolidation settlement rates of embankments built on soft saturated clay deposits, and to monitor progress of preloading settlements to provide assurance that the 2-inch limit will not be exceeded, the design-build team has incentives to do more field testing, more detailed laboratory testing, much more geotechnical analyses using advanced techniques, and to install a sophisticated monitoring system. The monitoring system in the field is used to provide weekly updates on the progress of embankment vertical settlements and lateral deformations (excessive lateral deformations could damage nearby structures or critical utility infrastructure). Data from field monitoring is used to update the settlement model providing revised estimates of time required to preconsolidate foundation soils. Design-build geotechnical investigations for the single-span bridge with its approach embankments described above may easily expand to include 36 borings and 36 piezocone tests not two to four typical bridge borings. The increased cost and expanded complexity of the design-build team's geotechnical scope of work to fulfill their contractual schedule and performance requirements may be five to ten times greater than the typical geotechnical investigation for a design-bid-build project. From a cost-benefit perspective, the extra geotechnical investigation and analysis costs are well worth it to the design-build team managers.

1.3.5 Geotechnical Investigations for $15 per Foot

At our Albuquerque monthly geotechnical meeting, we call it the Albuquerque Soil Mechanics Series, I recently discussed how the scope of a geotechnical investigation varies with the client, the project complexity, and so on, in a similar fashion to the discussion presented above. After the meeting, one of our older members pulled me aside and reminded me of geotechnical practice in the 1960s and early 1970s when we did things a bit differently. In those days a client would call up and ask for five borings for a new office building. We would discuss building height, location, and structural type and then I would suggest boring depths. Let's say we agreed upon 30-foot deep borings. So right on the phone I would say 150 feet of borings and the resulting report will cost about $2300. The client would say that sounds good, and we would schedule the drilling rig, often for the same afternoon or the next morning. If we found a big surprise at the site during drilling, I would drive to a pay phone booth (no cell phones in those days) and call the client to discuss a major change in project scope and budget. Often, he or she would ask me to stop drilling and move to a second site, or verbally authorize me to do more drilling and sampling work costing more money.

Assuming we could get the lab testing done in a day or two, I would often deliver the report to the client's office with an invoice that Friday or at the latest the following Monday. I would classify soil samples in the field during drilling and start a draft of the report. After laboratory testing was completed, I checked my visual, manual classifications, and used consolidation and shear strength data to perform analyses that were reviewed and checked by one of our principal engineers. My draft report was severely critiqued and revised by one of these same principal engineers and they would sit with me and discuss their comments in detail.

You may ask, "What about a proposal and a signed contract? What about utility clearances? What about a health and safety plan? What about archeological clearances?"

We didn't write a proposal, our agreements with the clients were verbal or with a hand shake after thoroughly discussing their problem and the likely required scope of work. We looked for utilities, but not infrequently we drilled through them! We never heard of a tailgate safety meeting before starting work. We didn't think about site contamination or archeological concerns, and sometimes we drilled through hydrocarbon contamination that was so strong we could smell it on the samples and then mentioned it to the client. We drilled through Indian mounds and historical Indian battle sites that were significant archeological discoveries that resulted in project delays of two to three years and even resulted in relocation of some projects to completely new sites. So it is clear that standard practice was different in those days, and many of our early mistakes resulted in the utility, environmental, and archeological clearance procedures that are common today.

Clearances notwithstanding, one thing was very different in those days. Without a contract, what did the client say if our invoice was more than the number quoted to them over the telephone? Most often, the client did not complain, because they discussed and understood the proposed work scope, they knew about and understood changed conditions, and they trusted their geotechnical engineer's expert advice. They wrote us a check upon delivery of the report and receipt of our invoice. The few times that they did question our billing, they were simply asking for clarification of our invoice and why additional field testing, laboratory testing, or engineering analyses were required. Trust and communication were the bonds of geotechnical practice in those days.

What happened to change the former days of client–consultant trust to present day client–consultant mistrust, contract disputes and lawsuits? That's a good question, and I'm afraid it would take a book on that topic to answer this question. Many engineers are quick to blame lawyers and greedy owners, but to quote my first mentor Neil Mason, "It takes two to tango." That is, it takes two parties to dance, and it takes at least two parties who both believe they are right, but both are partly wrong, to cause a dispute and the resulting legal battle. I still believe that effective communication with your client is the best way to avoid problems.

We in the geotechnical profession were quick to adopt canned reports that all looked the same, and sometimes by accident they were the same! We were hungry for more and more profits. We were too busy to mentor our young engineers. We were eager to get in on the "environmental business" work. We were eager to limit risks by using slick "lawyer-created" phrases in our reports, and we were too busy to closely study the soil and geology on the site to see if standard practice analysis and reports were sufficient. I could go on, but the bottom line is that the entire society changed, resulting in a climate that could not sustain "the good old days." Enough! The good old days are gone. Let's get on with solving the geotechnical practice problems of today.

1.3.6 Geotechnical Problems

What kinds of problems do geotechnical engineers have to solve and how do we solve them? We will use the "graded approach" to present each topic. First we will discuss the basic material used in standard geotechnical engineering practice, and then with increasing complexity, we will discuss advanced material to find out the "how" and "why" of the topic. Chapter 2 starts with discussions of soil classification, and stresses and strains in soils, and progresses through the geotechnical catalog of topics. You may wonder why I put off discussing foundation bearing capacity until later in Chapter 3. Check out my explanation below.

1.3.7 Bearing Capacity

Often the first thing that structural engineers look for in a geotechnical report is the foundation allowable bearing pressure, or bearing capacity. Why is that? It is actually quite simple, and quite pragmatic. Structural engineers use the allowable bearing capacity to size spread footings that support their structures. In many design guides, structural engineers are directed to use service loadings and the allowable bearing pressure to size their footings. Then they take factored building loads with the design footing area to recalculate an increased soil bearing pressure. They use this increased "factored" soil bearing pressure to design the footing concrete and steel reinforcing bars using concrete load and resistance factor design (LRFD) principles. I'll give you more on foundation design and use of geotechnical LRFD procedures in later chapters, but for now I would like to consider allowable bearing capacity of foundations. Does it really make sense?

The analysis of shallow and deep foundations is a soil-structure interaction problem. Most soils and geotechnical textbooks consider bearing capacity of shallow foundations as one of the first three topics discussed. Why did I put off discussing bearing capacity of foundations until the middle of Chapter 3? My answer is, the issue of foundation bearing capacity for shallow and deep footings or piles may be a common geotechnical analysis, but it is one of the most complex geotechnical issues, due to the nature of soils. Foundation bearing capacity analyses involve problems of limit equilibrium, settlement, soil-structure interaction, and the definition of foundation failure.

Bearing capacity is commonly described as involving strength and stability of a foundation, but it is more complicated than allowable bearing pressures. What is the limiting state that you are applying a factor of safety against? Is it really a bearing capacity failure? In over 40 years of geotechnical practice, I can recall seeing only three foundation bearing capacity failures, and reading about three or four major foundation failures where the foundations punched into the ground and the building rolled over, resulting in a massive structural failure.

Failure of the Fargo grain elevators on 12 June 1955 comes to mind as a classic example of a bearing capacity failure, see Figure 1.3.3 below from *Collapse of Fargo Grain Elevator*, by Nordlund and Deere, ASCE SMFD Journal, March 1970, pages 585 to 607. Loading on the Fargo grain silo foundations was reported to range from 4.15 to 5.47 kips per square foot with an average value on the total area of 4.75 kips per square foot. The upper approximately 20 feet of the soil profile at the site was predominately silty clay and stratified clay, silt and sand with undrained shear strengths ranging from approximately 1.6 to 2.4 kips per square foot. Below 20 feet there was a stratum of dark gray varved clay with undrained shear strengths ranging from approximately 1.0 to 1.6 kips per square foot with natural water contents of approximately 70%, which was well above the plastic limit of the clay, given as 37%. In 1970, the failure loading on the silo foundation was calculated to be 4.11 kips per square foot, which was well below the actual silo loading. Nordlund and Deere concluded that this was a "classic example of a full-scale bearing capacity failure"; see failed silos below in Figure 1.3.3.

If a bearing capacity failure is so rare, what are we worried about? Good question! You might say, "See, the foundations designed using the bearing capacity equation work fine!" Although catastrophic foundation bearing capacity failures are rare and are apparently limited to heavily loaded foundations on deep soft clay deposits, I have seen and read reported case histories of hundreds if not thousands of foundation failures where the foundation settled excessively without experiencing a classical bearing capacity failure. These foundations that failed by excessive settlement were sized based on the allowable bearing capacity. How could they have failed if their bearing pressure was OK?

In some cases, these foundation settlement failures are reported as punching shear failures where the foundation pushed into loose granular soils until confining stresses generated by the footing reached equilibrium with the soil mass below. To quote my early mentor, Neil Mason, "Where can the foundation go?" By this he meant that if the footing didn't roll over, where could go to fail except down. If the footing went down, compacting soil beneath it until coming into equilibrium, the supported structure may or may not experience excessive crack damage. But then when is a crack defined as a structural failure? Who is designated as the "person-in-charge" of defining crack damage that reaches the threshold of failure? Whether wall cracking or sloping floors

Figure 1.3.3 Fargo grain elevator bearing capacity failure. © 1970 The American Society of Civil Engineers. Reproduced with permission from ASCE

represent structure failure is (most often) "in the eye of the beholder." We have to carefully define the critical functions of a structure to have sufficient data to define foundation performance as acceptable or unacceptable. The point is that settlement and differential settlement limits are most often used to define satisfactory or unsatisfactory performance of a foundation, not some arbitrary definition of inadequate soil shear strength linked to applied foundation stresses.

To perform a thorough analysis of a building foundation, we need to answer several questions. If allowable bearing pressure is a specified foundation design requirement, how do we determine if the foundation has enough strength to resist applied loads? What soil parameters are used to characterize the soil strength and the foundation–soil interaction strength? How is foundation failure defined? How do we link foundation settlements to loading and report the results as an allowable bearing pressure for use by structural engineers or architects? What loads should be considered in our analyses? How are loads factored to include variability and unknowns (i.e., uncertainty)? Are loads factored or are soil strength terms factored, or are both factored, that is, how do we consider LRFD foundation analysis? These are important questions in modern geotechnical practice that need to be answered.

1.3.8 Summary

To appropriately solve a geotechnical problem, you have to have a problem solution plan. Your plan may have to be changed with changing conditions and investigation findings that conflict with your model of the site. That's OK, at least you have an initial plan to change and were not just "shooting from the hip." Two apparently identical projects may have very different problem solution plans if the client, the public, or the risks dictate a larger, more complex investigation. Don't get in a rut. Don't just photocopy last week's geotechnical report and change a few names to customize it. Please take the time to think about what you are doing and why you are doing it. Write your reports from scratch each time, and I guarantee that repeated writing practice will make you a better and a much faster writer. Being organized and having a plan is not only a good idea, it can give you a great feeling, the feeling that you will likely get a better answer and have a better understanding than the "shoot from the hip crowd."

1.3.9 Additional Material

For additional material describing Karl Terzaghi's problem solution approach, please refer to the case history of the Republic Steel Ore Yards in Cleveland, Ohio. Material describing the Republic Steel Ore Yards case history is included in Ralph Peck's Biography, *The Essence of the Man*, published 2006, page 44. I also recommend that you read the Terzaghi Lectures in ASCE Special Publication No. 1, published by ASCE, 1974, and Karl Terzaghi's report of 1 December 1948 in *From Theory to Practice in Soil Mechanics*, publisher John Wiley & Sons, 1960, pp. 299–337.

Further Reading

Bjerrum, L., Casagrande, A., Peck, R.B., and Skempton, A.W. (1960) *From Theory to Practice in Soil Mechanics, Selections from the Writings of Karl Terzaghi*, John Wiley & Sons, New York and London, 425 pages.

Dunnicliff, J. and Peck-Young, N., 2006, *Ralph B. Peck, Educator and Engineer, The Essence of the Man*, BiTech Publishers Ltd., Vancouver, Canada, 350 pages.

Goodman, R.E. (1999) *Karl Terzaghi, The Engineer as Artist*, ASCE Press, Reston, Virginia, USA, 340 pages.

Nordlund, R.L. and Deere, D.U. (1970) *Collapse of Fargo Grain Elevator*, American Society of Civil Engineers, Soil Mechanics and Foundation Division Journal, pp. 585–607.

Peck, R.B. and Raamot, T. (1964) *Foundation Behavior of Iron Ore Storage Yards*, American Society of Civil Engineers, Journal of Soil Mechanics and Foundations Division, Vol. 90, No. SM3, May 1964, reproduced in ASCE publication Terzaghi Lectures 1963–1972, pp. 3–40.

1.4

Mistakes or Errors

1.4.1 Mistakes

Developed early during my engineering career, I've become a mistake catcher. It has become part of my nature to look for mistakes everywhere: work, home, my fellow workers, my wife, and myself. Soon after we were married, my wife told me to "stop checking what I do!" I told her that I even check myself. She didn't care, and told me to "stop it!" So I learned early on that checking laboratory data, checking calculations, and checking myself was OK, but that checking was not a good idea for others (and if you do check others it is best to be discrete.)

Engineering practice depends on working as mistake free as we can humanly work. If members of our engineering team make mistakes, we need to have a system to catch their mistakes as often as possible. Catching laboratory data mistakes, although not a simple matter, can be systematic if your laboratory has a thorough quality assurance and quality control (i.e., QA/QC) system in place. Mistakes in engineering calculations can be caught by having a peer reviewer check the logic and methods used, and by having another engineer check the equations' terms and the numerical calculations (i.e., a number check). Some of these laboratory and calculation mistakes are simple issues and some of these mistakes are more complex issues to resolve, but they are all solved by standard engineering practice procedures. I would like to mention two other kinds of mistakes that I will call observation mistakes and logic mistakes. Let me give you a few examples.

> **Working example**
>
> Last summer my wife and I had an invasion of small ants around our kitchen sink. Not wanting to poison the kitchen area, we tried all kinds of traps and baits to catch and discourage the ants (a word here, ants are never discouraged,

Geotechnical Problem Solving, First Edition. John C. Lommler.
© 2012 John Wiley & Sons, Ltd. Published 2012 by John Wiley & Sons, Ltd.

because to them it is war!). One morning just after breakfast, I observed 10 to 12 dead ants on a sheet of paper towel that I had left on the kitchen counter. Eureka I thought. What a great discovery! I had found that this variety of ants when walking across a plain paper towel became dehydrated and subsequently died! What a great, environmentally sensitive way to rid our kitchen of these pesky ants.

I rushed back to our home office to share my discovery with SueAnn. "I found some dead ants on a paper towel in the kitchen," I started to say, but before I could continue, she said, "Oh yes, I killed those ants with my finger, wiped them off on the paper towel, and forgot to throw it into the trash can."

I was crestfallen. Both my observation and my conclusions were completely wrong! My first mistake was an observation mistake. I glanced at the small ants and didn't detect that they were crushed. There is a good reason why Sherlock Holmes used a magnifying glass to inspect the evidence. If I had closely examined the dead ants with a magnifying glass or with a microscope, I would have seen that they were crushed and that such physical damage was highly unlikely to occur during a dehydration process.

My second mistake was a logic mistake. I jumped to the conclusion that the towel surface had dehydrated the ant bodies. What I should have done was stop my thinking process at an intermediate step. Before jumping to a conclusion, I should have proposed a theory of the case, leaving doubt that some data was missing. This is nothing more than a simple application of the scientific method which suggests that you propose an answer which is called a "theory," then you test the theory to see if it stands up to scrutiny and checking. In this case, it is a good thing that I talked to my wife before I told my colleagues at work about my new ant removal device. Checking with SueAnn saved me from humiliation, but then would my colleagues have checked my untested theory or accepted it without question just because I am a PhD and a Principal Geotechnical Engineer?

Working example

Another example comes closer to my geotechnical problem-solving topic because it involves my recent research into the nature of collapsible soils in New Mexico. As the saying goes, my research into collapsible soils is a long story, so for this example I want to focus on one question, "Why do collapsible soils still exist?" By this I mean, if these soils are highly moisture sensitive, and collapse upon exposure to water, why didn't they collapse hundreds of years ago, given that they are out of doors in the rain and snow month after month and year after year possibly for thousands of years. We have wet years even in the desert southwestern United States, so given their long existence how can today's collapsible soils still be collapsible?

My first approach to this question was to study an exposure of well-documented collapsible soil that exists south of Albuquerque, New Mexico. Upon close examination I noticed that some of the layers of these collapsible soils include small fragments of carbon, possibly small pieces of burned wood, see Figure 1.4.1.

Figure 1.4.1   Carbon pieces in collapsible soil exposure

Considering the observed small fragments of carbon in the subject collapsible soils, I theorized that they were the result of forest fires that occurred on the nearby mountains. When the mountain forests were burned by fires, the barren soil surfaces were unprotected from erosion and intense summer monsoon rains violently washed these soils down onto the alluvial fan deposits at the base of the mountains, carrying fragments of carbon with them.

After the large Cerro Grande fire in May 2000 that burned mountain forests in and around Los Alamos, New Mexico, I observed the development of hydrophobic soils on the burned soil surface. Such soils greatly increase rainfall runoff from the burned surfaces because they repel water. Hence the name, hydrophobic: hydro meaning water and phobic meaning fear of or aversion to.

Putting the evidence of carbon fragments in collapsible soils and awareness of the formation of hydrophobic soils after a forest fire together, I started to research the literature on the presence and nature of hydrophobic soils. Soon I came up with a theory that the persistence of collapsible soils in modern day soil deposits was a function of the hydrophobic nature of such soils. I theorized that upon short exposure to wetting, these hydrophobic soils repelled water and did not collapse, but upon long-term wetting, such as lawn watering that could persist for months and reoccur year after year, that the hydrophobic chemicals were broken down and leached out of these soils resulting in collapse of their metastable structure. I tried out this collapsible soil theory on several of my geotechnical friends and colleagues, and they all more or less agreed that it sounded reasonable. I had a good theory. What do you think?

I was not satisfied. The collapsible soils site is located 15 miles from my office, water was readily available, and I even had a friend whose PhD studies were on hydrophobic materials (primarily asphaltic concrete). He would surely know a hydrophobic material when he tested it. All I had to do was call my friend and schedule a date to go test the collapsible soils for hydrophobicity (his word). All I had to do was get off of my backside and step away from my computer. I am as lazy as the next fellow, but then there is that sign hanging on my office wall that says, "You have to see it, to solve it."

My friend Eric and I had a great outing to the collapsible soils exposure site, and we excavated and tested dozens of small test spots for hydrophobicity. Darn that data, it ruined my perfectly good theory. Rather than repel and form a drop of water on the soil surface, as a good hydrophobic material should do, each drop of water sucked into the soil surface immediately upon exposure. No matter how hard I hoped for a positive result, the water drops didn't even pause for a fraction of a second. Check out photographs in Figure 1.4.2, which show a typical test on the collapsible soil versus the same test on a nearby roadway asphalt paving surface.

Figure 1.4.2 Soil is not hydrophobic, asphalt is hydrophobic

Do you see the difference between making a logic mistake of jumping to a conclusion like I did with the ant-killing paper, and using the intermediate step of developing and testing a theory like I did with the unsuccessful hydrophobic soil explanation for the modern day existence of collapsible soils? Oh, by the way, I didn't prove that collapsible soils were never hydrophobic. What I proved was: I am studying an exposure of collapsible soils that are not presently hydrophobic. I did find a better explanation for the presence of metastable structured, collapsible soils that I will discuss later in Chapter 3.

1.4.2 Errors

Now, what about errors? Aren't errors the same as mistakes? No they are not the same. They are quite different. First, I should clarify that the common meaning and usage of words is not always in agreement with the scientific or engineering meaning and usage of words. If you look up the word *error* in the dictionary, you are likely to see that its common meaning is that someone is incorrect or has false knowledge. The dictionary says that the word *error* comes from a Latin word that means wandering or straying. I presume that they are straying from the correct path defined by society.

In science and engineering, errors are generally considered to be the result of measurements or lack of knowledge. Considering measurements, the *error* is the difference between the measured value and the actual, theoretically correct value of the measurement. In geotechnical engineering, errors of many kinds occur because we are trying to characterize a huge volume of soil that may have highly variable properties with a few test borings or test holes that likely don't penetrate the worst soils on the site. Errors in geotechnical work are much more complex than just taking a measurement. We will discuss geotechnical errors and uncertainty in Chapter 6.

How can we visualize the concept of measurement errors? Using my own advice, you have to see it to solve it. OK, how do you visualize errors? I did some research and came up with a better definition of an engineering error: "An error occurs when you try to do something and you cannot achieve the ideal result." There are reasons why you cannot achieve the ideal result every time and the reasons are based in error analysis.

For my visual example of errors, I am going to try to shoot an air rifle ten times at a target located 15 feet from my standing position. If every variable is controlled and I have perfect human mechanics, the ideal result would be that all ten of the copper B-Bs will pass through the same shot hole at the exact center of the target. The result of my first ten shots is shown in Figure 1.4.3

Notice that the ten shots form a pattern of holes that fit into a circle about 1.375 inches in diameter. The center of the pattern circle is located below and to the left of the center of the target. In engineering, we call this type of error a calibration error. Sights of the air rifle are not adjusted for the target distance of 15 feet. The sights are also off a little to the left which is an alignment error. Riflemen call the required adjustment to raise the pattern to the center of the target "sighting in" their rifle. For our purposes, these adjustments are calibrating our measurement device. The miss calibration observed in the low center of the first group of ten shots is an error we call systematic error.

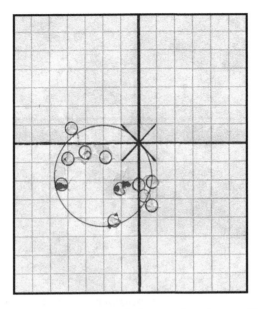

Figure 1.4.3 Ten B-B shots from a standing position

Adjusting the sights, I shoot a second group of ten shots at a new target. The results of my second ten shots are shown in Figure 1.4.4.

Notice that the second ten shots form a pattern of holes that is better aligned with the center of the target, but is still to the left of target center. Adjusting the air rifle's sights to center the pattern of shots is a process of calibrating the air rifle. Several

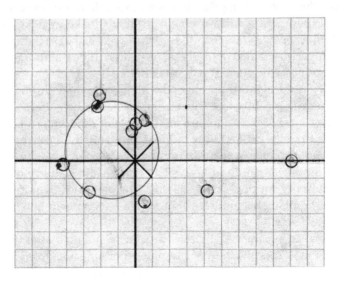

Figure 1.4.4 Second group of ten shots after sight adjustment

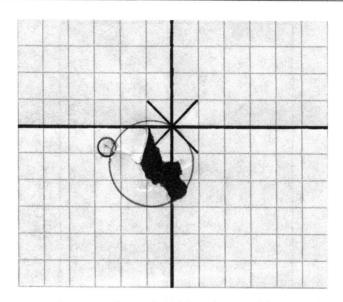

Figure 1.4.5 Third group of shots with reduced random error

additional adjustments will be required to center the pattern of shots with the center of the target.

In Figure 1.4.4 there are two shots that are located to the right of the main pattern. If the air rifle's sights are being calibrated, why did two shots in the second pattern badly miss the center of the target? The answer is human error. I didn't properly squeeze the trigger on those two shots, resulting in deflection to the right, which is called "pulling" the shot. Pulling the shot is an example of human error which can happen in any experiment or measurement. Humans make mistakes, and the best indicator of a human error is data that doesn't make sense, that is, data that doesn't fit the pattern of properly obtained data. Often, off-pattern data points are call "outliers" because they lie outside the general pattern of data.

Why aren't all ten of the B-B shots aligned with one shot hole through the center of the target if the air rife is calibrated? The errors that cause the pattern of shot holes to deviate from the center of the target are an example of random errors. The distance of each shot hole from the center of the target is a measure of the error of the shot.

I can reduce the diameter of the pattern of shots by purchasing a more accurate telescopic sight for the air rifle, and I can sit at a bench and stabilize the air rifle on a sand bag to reduce my support arm's shaking. The pattern shown in Figure 1.4.5 is about 0.75 inches in diameter, and was shot from a seated position with my support arm resting on my knee.

I could purchase an expensive target-grade air rifle, but no matter what I do, I can never eliminate the slight deviation of my pattern of ten shots from an ideal single hole in the center of the target. What this illustrates is that you can never completely eliminate random errors. You can reduce them, but you can't eliminate them. Now what does this result suggest about geotechnical measurements taken in the field or laboratory? Does one number characterize the range of potential deviations of any measurement? How could one number possibly characterize any geotechnical property on a site? The obvious answer is . . . it can't.

There is another error type that can be illustrated by the air rifle experiment. Assume that the air rifle is calibrated and performing well, and you get a pattern of shots centered 3 inches left of target center. This may be explained by a 45 mph wind coming from right to left across the path of the shots. This type of error is an error caused by an observable outside factor (in this case the wind) that skews the test results. This type of error is called an environmental error. If this type of error is identified, adjustments should be made in the measurements to correct for this phenomena, but the required correction will likely be different each time you make a measurement (in my case the wind speed changes every few minutes). If the wind speed maintains a constant 45 mph, I would just aim 3 inches right of center of the target, or I move into an indoor shooting area to avoid wind affects. It is not quite so easy to correct for geotechnical field measurement environmental errors.

1.4.3 Mistakes and Errors – Closing Remarks

A few additional words about mistakes and errors: be on the outlook for mistakes, but don't be too obvious when checking your associates, and please don't rub it in when you find mistakes. Many years ago, I used to try to imitate Robby the Robot from the movie *Forbidden Planet* and television show *Lost in Space* when I found a mistake in my secretary's work. I would go "whoop, whoop, whoop. . .error, error" when I spotted a mistake in her typing. She obviously didn't like my pointing out her mistakes in this juvenile manner, so she took up saying "whoop, whoop, whoop . . ." whenever she caught a mistake in my work. I was a young, rather naïve engineer, but I wasn't stupid, I quit using the Robby the Robot warning immediately!

When it comes to errors, I would like to give you a warning, don't be fooled into thinking that a few tests, borings, or measurements, or even worse that a single measurement can characterize a building site or a soil property.

2

Geotechnical Topics

2.1

Soil Classification – Why Do we Have it?

2.1.1 Introduction to Soil Classification

Do you remember back when you were in college? What did your professor in that required soil mechanics class say about soil classification? He probably said that geotechnical engineers the world over need to communicate with one another about the types of soils they are studying. They need a common soil-language to help other engineers understand what type of soil they have on their site or in their laboratory. As a result, nearly every published geotechnical paper has an introductory section describing the soil's geology, color, grain-size distribution, dry unit weight, Atterberg limits, Unified Soil Classification or AASHTO Soil Classification, water content, percentage organic content, relative density, friction angle, undrained shear strength, and so on.

Below I'll explain what standard classification tests are communicating to experienced geotechnical engineers.

2.1.2 Soil Properties Suggested by Classification Tests

Before we dive into a discussion of soil classification tests, there are a couple of basic questions to consider. When we think about soil what do we imagine it to be? What does the ideal soil look like and what does the ideal water look like? These seem to be simple or trivial questions, but please bear with me for a couple of pages because the answers to these questions significantly impact how you think about soil mechanics.

Geotechnical Problem Solving, First Edition. John C. Lommler.
© 2012 John Wiley & Sons, Ltd. Published 2012 by John Wiley & Sons, Ltd.

Figure 2.1.1   Ideal soil as you imagine it

When most of us visualize a soil, we picture a material similar to the soil shown in Figure 2.1.1. I don't know about you, but my introductory soil mechanics text book included a hand sketch of soil grains with appropriate spaces between grains in Chapter 1 on page 8. Standard textbook figures and photos commonly illustrate quartz grains of sand with open spaces between the sand grains labeled as pores. Some texts skip a picture or sketch of soil and go directly to presentation of a phase diagram where the soil is separated into solid, liquid, and air. Soil characterization problems can arise when real-world soils don't match the ideal soil illustrated in Figure 2.1.1.

What about water, how do you visualize water? I've never seen an introductory soil mechanics text book showing a picture of water or discussing water's basic form. When most people think of water, they think of liquid water or what we sometimes refer to as free water.

Between freezing and boiling temperatures at atmospheric pressure, water is primarily a liquid (don't forget the vapor phase!). On page 4 of my freshman chemistry book (Sienko and Plane, 1961), the characteristic of a liquid state is defined as "retention of volume but not of shape." Liquid water is nearly incompressible, it maintains its volume no matter the shape of its container, it has no characteristic shape, it has very low shear strength, and it evaporates from open containers. Given these definitions of liquid water, which I will use as my definition of free water, I imagine water to be as shown in Figure 2.1.2. That's my hot tub; it has 500 gallons of water with a surface area of 5712 square inches. Assuming water in my hot tub weighs 8.3454 pounds per gallon, the total weight of water in the hot tub is 4172.7 pounds. Calculating the water surface area per pound of water in my hot tub, we get 1.37 square inches per pound of water.

Figure 2.1.2   Water surface in hot tub – free water

Moving into the kitchen, I filled a pitcher with 0.5 gallons of water from the tap, see Figure 2.1.3. The surface of water in the pitcher was examined, see close up photograph of the water surface in Figure 2.1.4.

The surface area of water in the pitcher is 17.72 square inches and the weight of a half gallon of water is 4.173 pounds. Calculating the surface area of water per pound of water in this case we get 4.25 square inches per pound of water.

Figure 2.1.3   Pitcher with 0.5 gallons of water – free water

Figure 2.1.4   Close up of water surface in pitcher with 0.5 gallons of water

OK, both the hot tub and the pitcher's water appear to be free water. Let's check my free water criteria:

1. Liquid water retains its volume, but not its shape, check.
2. Liquid water is nearly incompressible, no easy way to confirm this property.
3. Liquid water has no characteristic shape, check.
4. Liquid water has very low shear strength, check.
5. Liquid water evaporates from an open container, didn't wait to confirm evaporation properties.

Given the nature of water in the hot tub and the pitcher, it appears that water in similar containers with a ratio of surface area per pound of water from about 1 to 5 square inches per pound of water is likely free water. Let's try another case; how about pouring water out of the pitcher into the sink, see Figure 2.1.5.

I photographed several dozen pours from the pitcher, and although no two photographs are identical due to varying light reflections, the basic shape of the stream of the water was the same. Is the flowing water shown in Figure 2.1.5 free water? My suggestion is that the flowing water stream is not free water because it forms a characteristic shape each time. Why is the stream of water different from the static or nearly static water in the hot tub and pitcher? My answer is that hydraulic forces generated by flow out of the pitcher in response to gravity and attractive forces between water molecules generated a water stream with a characteristic shape. I admit that the answer to this question is debatable, but you must admit that each stream of water poured from a pitcher has a very similar shape and my definition of free water is water without a characteristic shape.

Figure 2.1.5 Water stream from pitcher – free water?

Consider one more example; I photographed water falling from a dropper, see Figure 2.1.6.

The first thing that strikes me about the photograph in Figure 2.1.6 is the characteristic shape of falling drops of water. Note the small drop is 0.02 inches in diameter and the larger drop (please ignore the motion blur of the lower larger drop) is about 0.10 to 0.13 inches in diameter. All of the drops are nearly spherical, a characteristic shape! By definition, free water does not have a characteristic shape, and as you see in Figure 2.1.6, drops of water *do have* a characteristic shape. By definition, drops of water cannot be classified as free liquid water.

Looking at a 0.10 inch diameter drop, its volume is 0.000 523 6 cubic inches, weight is 0.000 018 92 pounds, and its surface area is 0.031 42 square inches. Calculating the surface area per pound of water of a 0.10 inch diameter drop gives us 1660.8 square inches per pound of water. This large surface area per pound of water is more than 1000 times that of the free water in the hot tub and nearly 400 times greater than that of the free water in the pitcher. This finding suggests that water having a large surface area relative to its weight (or mass) has molecular attractive forces that predominate over gravity, resulting in a skin or a characteristic surface, as seen on water drops.

What about the interaction of soil and water. Consider the pores in a fine sandy soil with a grain size of 0.006 inches or approximately the size of a No. 100 sieve. Assuming nearly spherical soil grains, the pores between grains of this soil are approximately

Figure 2.1.6   Drops of water – free water, not!

1/5 of the particle size or 0.0012 inches. A soil pore size of 0.0012 inches is much smaller than the small water drop in Figure 2.1.6, which is 0.02 inches in diameter. If the fine sandy soil is partially saturated so water drops partially fill the pores between soil grains, the water present in the fine sandy soil cannot be free water. In fact, the water in partially saturated soils generates what is known as soil suction, resulting in increased strength of partially saturated soils compared to dry or fully saturated samples of the same soil. We will discuss more about soil suction and soil tension in unsaturated soils in Section 2.7.

The only case where water in a soil can be considered to be free water is when a granular soil is fully saturated. In fully saturated soils, the soil grains act like pieces of aggregate in a large body of water. Saturated or not, clayey soils exert complex chemical attractions on water molecules so that there may not be any free water in a clayey soil. In clay soils, pore water's structure is moderately to highly structured in most, if not all, cases.

Now that we have some idea of what our ideal soils look like, let's continue with our discussion of soil properties suggested by standard classification tests.

2.1.2.1 Water Content Test

The water or moisture content test is a measure of the water driven out of a soil sample by heating the soil to 110 °C until the sample has a constant weight, although most soil laboratories assume that overnight heating is sufficient. Often engineers assume that soils with higher moisture contents are weaker and more compressible than soils with lower moisture contents. Obviously this is not the case for dry to damp sandy soils where increases in moisture content result in increases in soil strength, but it is often the case for clayey soils where increases in water content cause decreases in strength and increases in compressibility of samples.

Given a silty, clayey, or organically contaminated soil, increasing moisture content is an indicator of decreasing strength. For granular soils, changes in water content with time are indicators of water line leaks, excessive lawn watering, or wet weather patterns, which may be linked to soil settlements although the soil–water–settlement interactions are often complex. Moisture content in granular soils is often described as dry, damp, moist, or wet. Dry sandy soil is dusty dry with no visible indications of moisture. The color of dry sandy soil is usually light tan to brown or light gray. As the sandy soil becomes damp, the color of the sample is darker, but there is no visual water present. As a sandy soil transitions from damp to moist, the sample color is much darker than the dry state and there are visible signs of water in the sample. A moist sandy soil is still an unsaturated soil. A wet sandy soil is near saturated to saturated, water is clearly visible, in fact water flows out of wet samples if they are vibrated or densified.

By itself, the water content test is not highly useful in characterizing sandy or clayey soil samples without local experience or without markers. In the desert southwestern United States, I know that a sample of silty sand with a moisture content of less than 2% is unusually dry, and with a moisture content of 2 to 5% is dry and likely in a natural state in desert and alluvial fan deposits. A sample of silty sand with a moisture content of 5 to 10% is damp and was somehow wetted. Silty sands with moisture contents over 10% suggest water problems, and silty sands with moisture contents over 20% are very wet or saturated.

For clayey soils changes in moisture content suggest changes in soil consistency from dry to plastic to liquid. This is where the Atterberg limits become very useful.

2.1.2.2 Atterberg Limits

Atterberg limits were developed by Swedish Chemist Albert Atterberg (19 March 1846–4 April 1916) just prior to World War I. Atterberg's limit tests described the effect of increasing water content on the consistency of clayey soils. In Atterberg's 1911 paper, five limit tests were described, although during his research work Atterberg developed many more than five limits. Today two of Atterberg's limits, the plastic and liquid limits, are commonly used and a third, the shrinkage limit, is infrequently used. Atterberg's work defined consistency limits of fine-grained "clayey" soils.

When I think of consistency, I imagine making a cake. You start with dry flower in a mixing bowl. As you add in liquid and eggs the flower starts to clump up in the bowl and is very difficult to stir. As you continue to add liquid the cake batter is easier to stir, but still clumpy, and finally as you stir and add liquid it becomes a viscous liquid that is easy to stir. If you add too much liquid, the cake batter becomes a liquid. Table 2.1.1 below indicates the boundaries of soil consistency defined by Atterberg's shrinkage, plastic, and liquid limits.

Table 2.1.1 Soil consistency defined by the Atterberg limits

Soil Consistency	Description of Soil	Lower and Upper Limits
Solid State	At the lower bound moisture content, soil is very hard to break by hand or brittle-crushes, at the upper limit the soil crumbles.	Lower bound moisture content theoretically zero but practically is 4 to 6% moisture content and the upper bound is at the shrinkage limit (SL)
Semisolid State	Deforms and crumbles at the lower bound and deforms and cracks at the upper bound moisture content	Lower bound at the shrinkage limit (SL) and upper bound at the plastic limit (PL)
Plastic State	Deforms without cracking at lower bound to moldable to very soft "butter" at upper bound moisture content	Lower bound at the plastic limit (PL) and upper bound at the liquid limit (LL)
Liquid State	Flowing soil at lower bound to slurry to muddy water at upper bound water content	Lower bound at the liquid limit, upper bound is highly dispersed soil slurry in water.

I don't want to confuse you, but I have always thought that the Atterberg limits were mislabeled. I suppose it is my Midwestern American version of English, but I think of "limits" as speed limits. To me a limit means a speed that should not be exceeded, that is as an upper bound. If you use my definition of limit, the upper end of the semisolid consistency would be semisolid limit and the upper end of the plastic range would be the plastic limit that is the highest moisture content that still has a plastic consistency. In my system the limits would be upper bounds of consistency. Why did Atterberg define his limits as lower bounds of consistency? Atterberg defined the lower limit of liquid consistency as the liquid limit, the lower limit of plastic consistency as the plastic limit, and the lower bound of semisolid consistency as the shrinkage limit. I have a theory based on the fact that Atterberg was a chemist. Chemists in Atterberg's day did much of their work as wet chemistry. I believe that it would be natural for a nineteenth century chemist from a wet country like Sweden to start from wet soil and work toward drier soil consistencies. He would have to dry out his soils to reach the boundary of each limit, so I imagine since he was working from higher to lower moisture contents it was natural for him to define his limits as lower bounds.

Now that we have Atterberg limits to define consistencies of clayey soils, we have a tool to evaluate the moisture content of a given sample. If a clayey soil's moisture content is at or slightly below the plastic limit, it has a semisolid consistency and is suitable for compaction and densification. If the same soil has a moisture content well above the plastic limit, it has a plastic consistency and will pump and deform if compacted and will not densify, but remold during the compaction process. If the soil's moisture content is near the liquid limit, it is near the upper end of the plastic range and near the lower end of the liquid consistency so its shear strength will be very low. A soil sample with moisture content near the liquid limit is often defined as having zero shear strength by standard geotechnical tests. The problem with this characterization of soil shear strength by comparing the natural moisture content to the liquid limit is founded in the fact that Atterberg's liquid limit is a remolded test having a remolded soil structure. Natural soils can have structure allowing significant shear strength at moisture contents above the liquid limit!

If the clayey soil's moisture content is well below the plastic limit, it is possible that soil wetting may cause softening and collapse of the dry soil's structure, resulting in settlement, or if it is a relatively compact, high-plasticity clayey soil, wetting of the soil could cause swelling. Discussions of soil settling, collapse and swelling upon wetting are included in Sections 2.7 and 3.2.

The plastic range of clayey soil moisture contents is of particular interest to engineers. The range of moisture content from the plastic to the liquid limit is called the plasticity index (PI). If the clay mineral present is a highly active clay called a high plasticity (CH) clay the PI is generally over 30 and the liquid limit is at or over 50. If the clay mineral present in the soil is a low activity clay called a low plasticity (CL) clay the PI is generally less than 15. We have a bit of a problem in the United States because we allow clayey soils with properties nearly equal to CH clays to be classified as CL clays. To avoid a large jump in perceived plasticity from CL to CH clay, having a classification for intermediate plasticity clay would be nice. In the United States, Arthur Casagrande was the engineer who did the original research on standardizing soil classification, and in his early work he defined intermediate plasticity clay between CL and CH. In some countries today, Canada being one I recall, they define a CI, intermediate plasticity clay as a clayey soil with a liquid limit between 30 and 50. Having an intermediate plasticity clay classification helps avoid thinking of a CL clay sample as having low activity clay minerals with little potential for swelling and then getting the surprise when a highly compacted sample on the dry side of optimum moisture content becomes wetted and swells significantly. Personally, when studying test boring logs, I always check the plasticity index (PI) of my CL samples and make sure that they don't have liquid limit values in the mid to high 40s. If they do, I make note of the "high end" CL clayey soils by adding a plus (+) or marking them as CI.

2.1.2.3 Grain Size Distribution – Sieve Tests

In the early days of soil engineering, before Casagrande published the results of his work on soil classification, many engineers thought that the size of soil particles and the distribution of the sizes of soil particles in a sample defined the soil's properties sufficiently for engineering purposes. Today we know that the grain size distributions of soils do not completely characterize their properties, but the grain size distribution contributes to an understanding of the soil properties when considered with other soil tests like the Atterberg limits.

2.1.2.3.1 Gradation – Particle Sizes

Aggregate materials are composed of individual particles or pieces of rock. If rock pieces are fine textured, the aggregate would be a silt or sand. If the rock pieces are coarse textured, the material would be gravel, and if it were very coarse, the material would be cobbles or boulders. For hundreds of years, craftsmen called aggregate materials sand, gravel, or boulders. When a craftsman asked his supplier for masonry sand or pea gravel, the supplier was expected to know what he meant. Engineers and builders in the modern age needed a standard way to describe aggregates, just like auto mechanics working on mass-produced cars needed to know they were getting a standard sized bolt with a standard thread-type.

The solution to this aggregate description problem was to measure the diameter of the aggregate particles and describe them by their diameter size in inches or millimeters. In theory, this sizing of particles concept was a good idea, but in practice it is nearly impossible to measure the diameter of particles less than 1/8 inch without extraordinary, tedious effort. Another problem is that small rock pieces are not spherical, so which diameter dimension do you measure?

This is where the sieve or screen comes in to the picture to provide engineers and technicians with a tool for particle size measurement. By definition, if a particle passes though a sieve's openings, it is smaller than the sieve size, and if it does not pass through or is retained on the sieve, it is larger than the sieve size. Use of sieves *does not* measure the diameter of small pieces of aggregate. What sieves do is approximate a soil sample's particle size distribution. If hundreds of sieves with each successive sieve just one millimeter larger than the sieve below were used to test an aggregate sample, theoretically the sizes of the aggregate particles in the sample could be determined to within one millimeter. This sieve test would be very impractical to perform and the results would be cumbersome.

In practice, aggregate samples are sieved on several standard sieve sizes and the percentage of the sample passing each sieve is calculated and plotted on a curve called a gradation curve or a grain size distribution curve, see Figure 2.1.7.

An aggregate sample consisting primarily of one-sized particles is called uniformly sized or poorly graded. Note that the poorly graded sample 1 in Figure 2.1.7 has a steep, near vertical curve on the gradation plot. A sample having a broad range of particle sizes from fine to coarse grained is called well-graded. The well-graded sample 3 in Figure 2.1.7 has a flatter, relatively uniform slope on the gradation plot. If a sample is made up of only two or three particle sizes, it is called a step graded or gap graded sample, see Figure 2.1.7 sample 2.

Particle size alone does not sufficiently describe the nature of an aggregate sample. Some aggregate samples have rounded and sub-rounded particles, while other samples have sharp, angular particles. Stone and brick masons in Great Britain call rounded sands "soft sand," and they call sands with sharp, angular grains "sharp sand." These tradesmen are quite proficient in blending soft and sharp sands in their mortar mixes to achieve consistencies from creamy architectural mortars to grainy high-strength structural masonry mortars.

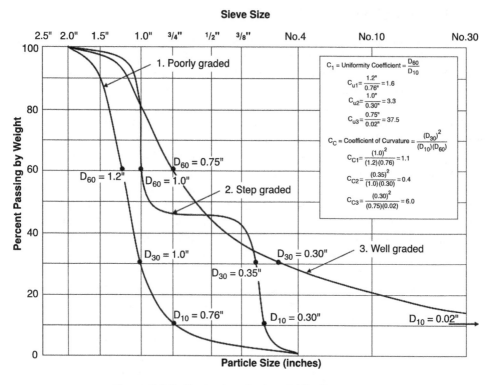

Figure 2.1.7 Typical grain size distribution curves

2.1.2.3.2 Sieve Analyzes

As I introduced above, it is impractical to measure particle sizes, so sieves are used to determine the percentage of different particle sizes present in an aggregate sample. To perform a sieve analysis the aggregate sample is shaken and/or washed through a series of sieves that are stacked with the largest sieve on top and the smallest sieve on the bottom. The sieves used in any particular analysis are commonly determined by agency specifications or engineer requirements. The smallest sieve commonly used for geotechnical purposes is the number 200 sieve (0.074 mm, 0.002 91 inches), which separates fine sands from silt/clay. Practically, the smallest size sieve that could be used for aggregate particle size analysis is the number 400 sieve. This is because water for washing particles does not readily pass through sieves with openings smaller than 0.04 mm (0.0015 inch).

To determine the grain size distribution curve for particle sizes passing the No. 200 sieve, a hydrometer analysis is used. To perform a hydrometer analysis on the fine portion of an aggregate sample, a weighed portion of the fines are split

Figure 2.1.8   Hydrometer test – particle sizes finer than No. 200 sieve

from the main sample and dispersed in water to form a thin slurry or suspension that looks like muddy water. Immediately after the sample is shaken up, a hydrometer is inserted into the suspension to measure its specific gravity, see Figure 2.1.8. With time the suspended particles settle to the bottom of the container, with the large particles settling first and the finer sized particles settling later. By taking hydrometer readings at specified times from initial shaking, the sizes of particles remaining in suspension can be determined by use of Stokes' law. For details of the hydrometer test, the reader is referred to ASTM D 422 or AASHTO T 88.

Fine particles passing the No. 200 sieve are further classified by determining if they are silt- or clay-sized particles. Silt particles under the microscope look like sand particles, and in fact they are just very fine pieces of sand. Clay-sized particles that are very fine silt material also look like sand particles under a microscope. Clay-sized particles that actually consist of clay are much different than silt particles because they are clumps of clay minerals that have unbalanced electrical charges on their surfaces. These unbalanced electrical charges act like small electromagnets and cause clay particles to attract and hold water molecules. This ability to attract water and form a "sticky" surface is typically characterized by the Atterberg limits test described above. Clay can coat and stick on sand and gravel-sized aggregate particles, adversely affecting their performance in construction applications. Recently I've become aware of several poorly performing MSE walls that had small to moderate amounts of high PI clay contamination of their reinforced zone aggregate backfill.

2.1.2.4 Dry Unit Weight

Dry unit weight is equal to the weight of soil solids in a cubic foot of soil. You calculate the dry unit weight from the wet or moist unit weight by the equation:

$$\gamma_{dry} = \frac{\gamma_{wet}}{(1 + w)}$$

where water content w is expressed as a decimal. In general, if a soil sample's dry unit weight has been increased, it has been compacted and its shear strength increased.

Soil unit weight is not formally used in classification systems to classify soils; rather it is used by engineers as an indicator of relative soil properties. For example, two samples of the same soil are compared. The sample with a higher unit weight is expected to have lower void ratio, lower porosity, greater shear strength, lower permeability, and lower compressibility than the lower unit weight sample.

2.1.2.5 Shear Strength and Angle of Internal Friction, Φ

Similar to soil unit weight, soil shear strength and angle of internal friction are not formally used in soil classification systems. Soil shear strength is intended to be used in geotechnical design and analysis, but table and chart values of undrained shear strength and drained angle of internal friction are treated like classification properties. This has caused some confusion in the engineering community, because it appears that you can pick your design shear strength values out of a book without testing soils from your project site. At best, table and chart values of soil shear strength and internal friction angles are useful for preliminary calculations or estimates; they should not be used for final design. For design analyses, testing of representative soil samples from your site are required. I'll discuss soil shear strength in detail in Section 2.3.

2.1.2.6 Unified and AASHTO Soil Classification Systems

Before starting into a discussion of soil classification systems, I have a confession to make. I don't like taking about soil classification . . . it is just plain boring to me. I used to play the clarinet and the piano. My music teacher made me practice endless scales of notes. She told me that I had to learn the craft before I could learn the music, the art. So my advice to you is to consider soil classification as the notes or the language of geotechnical engineering. Classification is a basic communication tool that you have to learn to move on to solving problems. It's not fun, but it is necessary. OK, here we go.

The two main soil classification systems used in the United States are the Unified Soil Classification System (USCS) and the AASHTO Soil Classification System (AASHTO). The USCS is used primarily in general civil and structural engineering practice and the AASHTO soil classification system is used in highway engineering practice. These two soil classification systems look very different, for example high plasticity clay is a CH in USCS and A-7-6 in AASHTO. Although many engineers use these classification systems interchangeably, they are not the same because their design and basic philosophies are quite different.

2.1.2.6.1 The Unified Soil Classification System (USCS)

The basis for our modern USCS system is credited to Arthur Casagrande. Casagrande's work in the 1930s, 1940s, and 1950s was primarily to define basic soil types and identify them by measurable characteristics of grain size distribution and Atterberg limits. The USCS divides soils into three types: coarse grained, fine grained, and highly organic soils. If 50% or less of a soil sample passes a No. 200 sieve then the soil is classified as coarse grained, and if more than 50% of a sample passes a No. 200 sieve then it is classified as fine grained. From that basic split of soil samples into coarse and fine grained soils, using only grain size distribution and Atterberg limits values, all soil samples can be classified into one of 15 soil classifications from well-graded gravel GW to high plasticity clay CH. Highly organic soils or organic fine grained soils can be identified by texture, color, odor, elevated moisture content, or weight loss when heated to elevated temperatures. For details of classification of organic soils such as peat, please refer to ASTM standards such D 4427.

I can't help but believe that Casagrande's division of all non-organic soils into coarse-grained and fine-grained classifications comes from Karl Terzaghi's original papers on soil mechanics, where he described all soils as having characteristics of sand or clay. If you are interested in history, I suggest that you get a copy of ASCEs Geotechnical Special Publication No. 118, Volume One, *A History of Progress, Selected US Papers in Geotechnical Engineering*. In this volume of historical geotechnical papers, check out the reprint on page 22, *"Principles of Soil Mechanics, V – Physical Differences between Sand and Clay, by Charles Terzaghi from Engineering News Record, December 1925."* Who is Charles Terzaghi? Well, I hate to be the one to break it to you, but before, during and after World War I, it was not popular to have a German sounding name in the United States, so they changed Karl to Charles to make Terzaghi's name sound less German. My grandmother changed her name for the same reason even though she was from Bohemia and not from Germany.

2.1.2.6.2 The AASHTO Soil Classification System

Federal and state highway officials were working on building embankments, compacting roadway subgrades and backfilling bridge abutments years before Casagrande started his work at Harvard University or before the Army Air Corps was interested in classifying soils for airfield construction. Throughout the 1920s the United States Bureau of Roads conducted extensive research on dirt roads and tar-sealed dirt roads. Based on the USBR research, Hogentogler and Terzaghi (Hogentogler and Terzaghi, 1929) published a Public Roads Classification System based on the stability of soils on dirt roads. With several revisions, the Public Roads classification system became the current AASHTO system, where roadway subgrade soils are divided into seven groups from A-1 (gravel) through A-7 (clay).

2.1.2.6.3 Differences between USCS and AASHTO

Many engineers think that there is no real difference between USCS and AASHTO classification systems. Are they right? Let's see for ourselves.

Silt and clay soils classifying as fine-grained soils by the USCS have more than half of the soil smaller than the no. 200 sieve. AASHTO silt-clay soils have to have more than 35% passing the no. 200 sieve. The higher plasticity A-7-6 silty-clayey soils have to have a liquid limit of 41 or more and a plasticity index greater than 11 in the AASHTO system. The AASHTO system doesn't differentiate between high plasticity silt and high plasticity clay. The USCS requires high plasticity clay to have a liquid limit greater than 50 and a plasticity index that plots over the A-Line, which means a minimum plasticity index of 22. High plasticity silt is required by USCS to have a liquid limit greater than 50 and a plasticity index that plots below the A-Line, which means a maximum plasticity index of 22 when the liquid limit is 50. From this example you can see that AASHTO doesn't seem to care if the high plasticity soil is silt or clay, and it can classify as a high plasticity soil when 35% of the sample passes a no. 200 sieve. I interpret the differences between USCS and AASHTO classifications to mean that the USCS classification engineers want to describe the soil as closely as practically possible using soil grain size distribution and plasticity characteristics, and the roadway people want to determine a correlation of soil type to subgrade performance as directly as possible.

You may wonder why I haven't reproduced the classification charts used for USCS and AASHTO classifications that are included in every soil engineering or geotechnical text book that you've ever seen. That's the point. There are dozens of books (probably at least two or three of which you own) on the market with classification chart figures. I want you to refer to your other reference books while reading this book. I always read technical books in pairs or sometimes three or more at once. What I'm doing is comparing and contrasting several references at the same time to catch errors and make sure that I'm not misunderstanding their explanations due to their selected phrasing. I told you earlier that I check everything, and this is just another example of that principle.

Do you remember the contrarians and devil's advocates that I described in Chapter 1, Section 1.2? They can turn a monthly geotechnical meeting into a two-hour debate over the second-order terms involved in the water content test. All disagreements and opinions aside, can the standard soil classification tests and the unified soil classification or AASHTO classification of soil samples be misleading? The answer to this question is most definitely, yes! Below, I'll go over some of the complicating factors and issues encountered in apparently simple classification tests.

2.1.3 Examples of Soil Classification Problems

Recall that I mentioned above that Karl Terzaghi (alias Charles Terzaghi) described soils in his early papers as having either the properties of sand or clay. To this very day many engineers think of soils as being either sand or clay materials. This line of thinking suggests that sands are materials having friction and clays are materials having cohesion. If you have a material that has both friction and cohesion, you or the computer program you are using will likely refer to it as silt.

The idea behind classifying soils is not to put them into one of three groups, but rather you want to characterize or identify the essential nature of your soil by performing a few standard laboratory tests, compare the test results with published values of properties, then assign a reasonable, bounded soil design parameter to your soil. Standard laboratory tests most often used for correlation purposes are the water content, sieve analysis, and Atterberg limits. Sometimes additional tests such as specific gravity, swell tests, percent organic content, and peak shear strength tests for cohesion and internal friction angle are included in soil characterization work. Let's start at the beginning with the water content test.

2.1.3.1 Water Content Issues

Now what could be simpler than the water content test? How could this little test mislead you? Just check the box labeled ASTM D 2216 on your lab's test request sheet and you're done. Oh, you want to do your own lab testing. No problem, just weigh the sample wet, put it in the oven at 110 °C overnight, weight the dry sample in the morning, subtract the dry weight from the wet weight to get the weight of water, divide the weight of water by the dry weight of the sample and multiply the result by 100%. Not so fast my engineering friend! The latest 2010 ASTM version of D 2216 is seven pages long, and it references 13 other ASTM standards. Let's just stop here and mention that laboratory testing of soils these days is a specialty that requires trained qualified personnel working with calibrated equipment in a facility that is periodically checked and audited.

Let's move on to what you as an engineer need to know about soil water content. What does the water content of a soil mean? You could repeat the description I gave above and tell me that it is equal to the weight of water divided by the weight of solids in a given soil sample. The resulting decimal value is then multiplied by 100% and the water content expressed as a percentage. For those of you who are civil or structural engineers, I have to admit that geotechnical engineers are famous for flip flopping between decimal and percentage values of water content. I'm not sure why, but I am sure that you have to be on the lookout for this decimal versus percentage water content issue. One easy way to spot when water content should be expressed as a decimal (or divided by 100%) is when a value of $1 + w\%$ is required, such as when the dry unit weight is calculated from the moist unit weight: $\gamma_{\text{dry}} = (\gamma_{\text{moist}})/(1 + w\%)$. This equation should be expressed as: $\gamma_{\text{dry}} = (\gamma_{\text{moist}})/[1 + (w\%/100\%)]$.

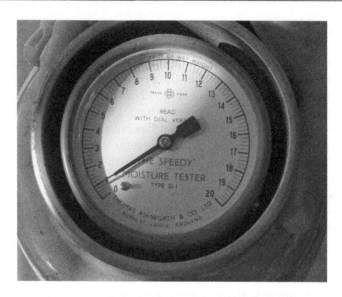

Figure 2.1.9 Speedy moisture meter dial, measures percentage of wet weight

Why do geotechnical engineers leave off the 100%? Are they lazy? I don't think so, but when asked about this equation most engineers I know just say that "you know what I mean."

Another somewhat confusing "water content issue" is the fact that there are three values of "water content" in the literature. The water content defined above is the geotechnical water content, which is called the gravimetric water content in soil science and hydrology references. The second "water content" is calculated using the volume of water in a soil sample divided by the volume of the sample. This second "water content" is called the volumetric water content, and it is typically used by groundwater hydrologists. A third type of "water content" is defined as the weight of water in a sample divided by the total weight of the sample. I call this third definition of water content the agricultural water content. I don't often see this third definition of water content used, but it still exists in the literature. Check closely in Figure 2.1.9, the dial of our retired Speedy Moisture Meter is calibrated to measure this third moisture content value, and I was told that farmers used it to measure the moisture content of grains.

We still need to answer the question, what does the water content of a soil mean? The geotechnical water content is a measure of water that exists in the pores of a soil. When most of us visualize water in soil pores, we think of soil like that shown in Figure 2.1.1. Primarily quartz grains of sand with open spaces between the sand grains. Problems arise when real-world soils don't match the ideal soil illustrated in Figure 2.1.1.

The standard laboratory water content test is assumed to burn off water existing in the soil's pores. Soil grains are assumed to be solid mineral particles that are unaffected by heating to 110 °C. What if your soil includes a hydrated compound like gypsum, which has the chemical composition of $CaSO_4 + 2H_2O$. When gypsum minerals are heated above 60 °C, they release their water of hydration, that is, the $2H_2Os$ in their chemical composition. The water released from gypsum is *not* geotechnical water content, and so if you don't know that gypsum was present in your soil samples, you are fooled by the unusually high water content laboratory value determined from a soil that you classified as dry! The same thing can happen when organic compounds are heated and driven off in the water content test.

Clay minerals such as smectite clays can have the opposite effect of hydrated compounds. Water added to smectite clays can be absorbed into loosely held layers and is called interlayer water. Interlayer water is driven off by heating to temperatures between approximately 125 and 200 °C. You thought that water had been added to the clayey fine sandy soil, but when the laboratory water content tests come back, values were less than 5%. Where did the water go? If you re-run the water content tests at increasing temperatures up to 200 °C, you will see that the samples continue to lose weight. This method of progressive sample heating was used in early clay mineralogy studies to identify clay minerals prior to advanced development of X-ray diffraction technology. As you might expect, clays that absorb water and form interlayers swell significantly. This swelling is not often apparent when small amounts of the clay coat granular particles in your samples.

I'll give you one additional example of water content mischaracterization from our friends up north in Canada. The problem in this case is water in the form of microcrystalline ice. Samples of apparently dry granular soil from cold regions can have small crystals or thin particle coatings of ice that are not obvious to the engineer when the soils are maintained at temperatures below freezing. If frozen samples are not maintained in cold storage and are transported over great distances, the ice particles can melt and drain out of the sample before water content testing is performed in the laboratory. The result is a reported laboratory water content test that significantly underestimates the field moisture content. I understand that some projects in northern regions were built with backfill soils containing ice crystals prior to installing heating systems in the buildings. Sometime after the heating system warmed up the building space, water began to pour out of the building slab subgrade as the melting backfill soil drained. Legal issues between building owners, engineers, and contractors were reported to have followed.

Water content is an important indicator of soil properties, including strength and compressibility. Care must be taken to assure that the proper value of this parameter is used in engineering analyses.

2.1.3.2 Plasticity Index Issues

Liquid and plastic limit tests are performed on remolded soils. Natural clay soils generally have natural shear strength either equal to or somewhat greater than laboratory remolded soils. Some natural clay soils called sensitive clays have much higher natural shear strengths, 8 to more than 15 times higher, in their natural state than when remolded. This sensitivity is usually attributed to a flocculated natural clay structure (i.e., like cards stacked carefully end to side in a pyramid) being broken down into a dispersed soil structure in the laboratory. If laboratory plasticity index values are used to classify properties of sensitive clays without identifying their structure-collapse sensitivity issues, engineers may grossly misunderstand the nature of the soils. Extreme cases of sensitive clays are quick clays that can flow like a liquid when disturbed on site.

Clay samples tested for liquid and plastic limits require moist sample preparation. If your laboratory technician dry prepared your clay samples for Atterberg limits testing, he or she might not have waited long enough for mixing water to soak into the dry clayey sample. The result will be small sand-sized pieces of dry clay acting like sand particles rather than as clay material. Such improper moisture aging of your clayey sample prior to liquid and plastic limit testing will give erroneously low plasticity or non-plastic test results for a clay soil sample that could be a high-plasticity clay. Similar preparation problems can occur with soils containing soft shale pieces. In your compacted fill containing soft shale pieces, they have decades to soak in water, decompose, and turn into mud. If your lab technician does not properly wet prepare samples with soft shale pieces long enough to allow them to break down into clay, you may have thought that your soil was a silty sand, but alas it was actually an unconsolidated sandy clay after shale pieces decompose.

2.1.3.3 Sieve Analyzes Issues

Let's say you have a silty fine-sand sample, or at least that's what it looks like to you. Your lab does a sieve analysis on the sample and it is classified as a silty fine-sand material, so you are happy that the lab confirmed your field observations. Here we go again, the sample had small hard pieces of high plasticity clay that were not detected by the laboratory because they dry sieved your sample, and you didn't think it necessary to specify a hydrometer test on a sand.

To avoid such problems your lab needs to wet sieve your samples, and a hydrometer test would be nice because the sample has to be soaked and deflocculated which should break down and detect hard clay and soft shale pieces in your "sand."

Another common problem with sieve analysis testing results from testing aggregates to check specification compliance for construction. If the material supplier mixed two or three sizes of aggregate, he created a step-graded aggregate sample. If you only check the percentage passing on a few sieves to check the specification numbers you may not realize that the sample is step or gap grading, and is highly prone to washing of fine particle sizes through the pores between larger sample particle sizes. It is better practice to sieve samples on enough sieve sizes to allow plotting of a grain size distribution curve on all of your sieve tests. One look at a well-developed grain size distribution curve will clearly show vertical or near vertical steps in the grading, indicating step-graded samples.

2.1.3.4 Relative Density Issues

Every day geotechnical engineers use relative density when they describe or classify a granular soil as very loose, loose, medium dense, dense, or very dense. Did you

Table 2.1.2 Relative density descriptions and correlations to blow count

Soil Density Descriptor	Relative Density	Standard Penetration Blow Count (N)
Very Loose	Less than 20%	0 to 4
Loose	20 to 40%	4 to 10
Medium Dense	40 to 70%	10 to 30
Dense	70 to 90%	30 to 50
Very Dense	90 to 100%	Greater than 50

know, or do you recall, that a very loose descriptor implies a relative density of less than 20%, loose is a relative density of 20 to 40%, medium dense is a relative density of 40 to 70%, dense is 70 to 90%, and very dense sand implies a granular material with a relative density of 90 to 100%. When I first learned this granular soil relative consistency classification system, it did not occur to me that it had anything to do with actual numerical values of relative density. I thought it was a packing description of granular soil. For example, I thought loose sand was loosely packed and dense sand was densely packed. I knew that these descriptors were tied to standard penetration blow count, but not to relative density values, as shown below. Table 2.1.2 gives implied soil density descriptors with correlations to relative density and standard penetration blow count, N.

My first issue with this common soil classification application of relative density is that it doesn't actually relate to numerical values of maximum and minimum soil densities, like 100 pounds per cubic foot minimum and 125 pounds per cubic foot maximum density. Using this method I only have blow count measurements to use for determining these soil density descriptors. The best we can say is that these descriptors give us an idea of what a loose cohesionless soil is like versus a dense cohesionless soil. But what about selecting soil properties for use in determining soil settlements or bearing capacity, are these descriptors useful in selecting design soil parameters? I say, "Not so much ..."

My second issue with the relative density test is that technicians commonly read relative density specifications, such as 80% relative density, and incorrectly calculate the test value in the same manner as relative compaction. Relative density *is not calculated* by dividing tested dry density by maximum dry density times 100%. Absolutely not! When using measured dry density values, relative density is calculated by the equation:

$$\text{Relative Density } \% = D_R = \frac{\left[\gamma_{\text{dry field}} - \gamma_{\text{dry min.}}\right]}{\left[\gamma_{\text{dry max.}} - \gamma_{\text{dry min.}}\right]} \left(\frac{\gamma_{\text{dry max.}}}{\gamma_{\text{dry field}}}\right) \times 100\%$$

Where

$\gamma_{\text{dry field}}$ is measured dry unit weight in field test,
$\gamma_{\text{dry min}}$ is laboratory tested minimum dry unit weight,
$\gamma_{\text{dry max}}$ is laboratory tested maximum dry unit weight.

If you use the original void ratio definition, relative density is calculated by the equation:

$$\text{Relative Density } \% = D_R = \frac{(e_{\max} - e_{\text{field}})}{(e_{\max} - e_{\min})} \times 100\%$$

Where

$e_{\max}$ is the maximum void ratio, soil at loosest state
e_{field} is the measured void ratio in the field (or laboratory)
$e_{\min}$ is the minimum void ratio, soil at the densest state

Now that we've cleared up how to calculate relative density, how do you actually measure relative density of *in situ* (field) sandy soils? Like I said above, you can get the blow count correlated relative density of undisturbed *in situ* soils as given in Table 2.1.2, but there is often no practical way to measure maximum and minimum values of dry density or void ratio in the field.

There is some modern controversy concerning the application and significance of the relative density test, which has been renamed in international publications as the density index, I_D. Early on in geotechnical engineering practice (Burmister, 1948), it was a common belief that relative density of a natural sandy soil deposit strongly correlated to its engineering properties and behavior. More recently Fellenius (Fellenius and Altaee, 1994) and Briaud (Briaud, 2007) have suggested that soil density or relative density is not sufficient to determine foundation load settlement relationships. I almost hate to go into this topic, because I really love the simplicity of the old Terzaghi and Peck foundation design charts, but I promised to mentor you on modern geotechnical techniques and that is what I intend to do! I'll take you back to 1948 and 1953 and explain what the common understanding of relative density versus foundation-load–settlement performance was at that time (it is still repeated in many modern texts), then I'll explain the foundation-load–settlement concepts which contradict many of our past assumptions.

2.1.3.5 Relative Density, Foundation Loads, and Settlements 1948–1953

In 1948 two representative papers on foundation-load–settlement relationships and the importance of cohesionless soils' relative density were published. One paper by L.S. Chen (Chen, 1948) in the Second International Conference on Soil Mechanics and Foundation Engineering was titled, *An investigation of stress-strength characteristics of cohesionless soils.* Chen's paper emphasized the importance of soil modulus in calculation of foundation settlement, and that studies of cohesionless soils closely correlate relative density to soil modulus. He then went on to describe the close correlation of standard penetration blow count, N, to relative density and soil modulus, and how it could be used to calculate footing settlements.

A second paper published in 1948 by Don Burmister (Burmister, 1948) was presented at the D-18 committee meeting during the Fifty-first Annual Meeting of the American Society for Testing and Materials in Detroit, Michigan. Burmister's paper was titled, *The importance and practical use of relative density in soil mechanics*. Burmister emphasized that relative density was a fundamental concept in the interpretation and correlation of soil behavior. He said that relative density was more useful as a correlation parameter for determining compressibility, angle of friction, permeability, and bearing capacity than rigorous mathematical methods.

In 1953, the first edition of Peck, Thornburn, and Hanson was published (Peck, Hanson, and Thornburn, 1953) including a foundation design chart. This chart was based on measured settlements of moderately sized footings on sands of varying relative densities. Terzaghi and Peck developed this empirical foundation design chart, which conveniently correlated footing width, soil relative density, SPT blow count, and allowable foundation bearing pressure for a settlement of 1 inch. Engineers used this chart to estimate settlement of individual footings, and they used it to estimate differential settlements between footings of different sizes loaded to the same bearing pressure by using a formula correlating the settlement of a one foot square footing to larger sized footings. This formula is given below, but please *don't* use it for anything other than FYI (for your information).

$$\frac{s_{\text{full sized}}}{s_{\text{one foot}}} = \left(\frac{2B}{B+1}\right)^2$$

Where

$s_{\text{full sized}}$ is settlement of the full sized footing with width B
and $s_{\text{one foot}}$ is the settlement of a one foot square footing

2.1.3.6 Relative Density, Foundation Loads, and Settlements 1994–2007

In Fellenius' paper (Fellenius and Altaee, 1994), he says in the third and fourth lines of the abstract, "... analysis of settlement for footings of three sizes placed in two different sand types show that the settlement in sand is a direct function of neither footing size nor soil density. Instead, the settlement should be related to the steady state line of the sand and to the upsilon distance of the sand, that is, the initial void ratio distance to the steady state line at equal mean stress and at homologous points." In his paper's conclusion, Fellenius says, "When comparing the behavior of footings of different size in the same sand, the settlement is not a function of the density per se, nor is it a function of the . . . relative density." What is he talking about? Have you heard of the steady state line or the critical void ratio? We are going to talk about these concepts in Section 2.5, so consider this an introduction.

When a sandy soil is subjected to a state of stresses that causes shearing strains, the sand may increase in volume (i.e., dilation) or it may decrease in volume (i.e., compression), or in a very special case it may deform with no change in volume (i.e., critical void ratio is the void ratio of a soil that is in a steady state of shearing deformation and experiences no change in volume). If the sand acts as a dense sand and dilates or as a loose sand and compresses, you might jump to the conclusion that the first sand had a high relative density, say greater than 90% and the second sand had a low relative density, say 25%. The problem is that you could be wrong. Why? The volume change of a sandy soil depends on variables other than density or relative density. If you have a copy of the first or second edition of Holtz and Kovacs book (Holtz and Kovacs, 1981 page 503 or Holtz, Kovacs, and Sheahan, 2011 page 551), *Introduction to Geotechnical Engineering,* take a look at the Peacock diagram. The Peacock diagram illustrates three primary variables, volume change, void ratio (directly related to sample density), and confining stress. What the Peacock Diagram illustrates is that loose sandy soils can act like dense sandy soils (in triaxial shear) when their confining stress is low. That is, they dilate and have a peak in their stress–strain diagram at low strains. Conversely, dense sandy soils can act like loose sandy soils when their confining stress is high. That is, they do not have a peak in their stress–strain diagram at low strains. The Peacock diagram also illustrates that you cannot predetermine whether a sandy soil will dilate or compress without knowledge of the confining stress and the distance of the sample's void ratio from the critical void ratio at the confining stress you are studying. The relationship between sample void ratio, sample confining stress, and sample volume change is what Fellenius is talking about in the quote given above. We will discuss the steady state line and the impacts of critical state soil mechanics concepts on our understanding and visual model of soils in Section 2.5. Prior to that, I recommend that you read Section 11.4 or 12.4 on the "Effect of Void Ratio and Confining Pressure on Volume Change" in Holtz and Kovacs first or second edition text.

2.1.4 A Word about Units and Normalization

When engineers were trained after World War II in the late 1940s and 1950s, it was understood by all that English or American units were to be used. Feet, inches, pounds, gallons, acre-feet, kips, and tons were the rule. It was common in those days to find engineering equations with odd constants and odd exponents in the terms of the equations. These odd constants were included in equations because the equation terms pertained to a single set of units that were made consistent by the constants. Students were warned constantly that if they didn't use the correct units in these equations they wouldn't get the correct answers. Homework problems often came back with the dreaded "wrong units" marked boldly on the paper with a large X through the incorrect answer.

Today things are different in that we are often presented with equations that do not have units specified. When using these equations you have the choice to use English units or metric units, so long as you are consistent, that is, if you use pounds per square inch, you have to use inches and pounds in all terms requiring length (L) and force (F) units. Please check your equations carefully, because a few of the old kind of unit specific equations still exist, and you need to confirm whether specific units are required for their use.

What is normalization? Have you noticed that many geotechnical plots and equations include terms that have ratios such as (σ_1'/p_a) or (τ/σ_3')? These ratio terms are normalized values that become terms without units because their numerators and denominators have the same units. By normalizing terms, the use of English or metric units is eliminated as an issue. By the way, do you know what p_a is? It is the atmospheric pressure in force (F) per area (L^2). I remember from elementary school that the atmospheric pressure is 14.7 psi and then make unit conversions from there to find the required value.

2.1.5 Soil Classification – Concluding Remarks

I recommend that you consider soil classification of samples as a guide or as a general indicator to help you start to focus in on the nature of the soils that are involved in your particular problem. You can use soil classifications and classification parameters such as moisture content, Atterberg limits, sieve analyses, and relative density to select "estimating" values of soil shear strength and compressibility for your preliminary project analyses and to check the reasonableness of your answers.

With shrinking budgets for engineering design services there is considerable pressure on all types of engineers to work cheaper, faster, and mindlessly. Don't do it! If you have a problem that requires extensive geotechnical investigation, testing and analyses, fight for what is right. If you have strong evidence and feel strongly that a minimal field investigation with little sample testing and textbook correlated parameters is sufficient work effort, then you should go for it, but be ready to defend your position at a later date. Please record your assumptions and reasons for conducting a minimal investigation. It's likely that the client and your managers will not fight you if a minimal investigation is what you believe to be right. Please have a peer review of your analyses and geotechnical report in all cases before sending the final product to your client.

2.1.6 Additional Reading Material

In Jim Mitchell's book (Mitchell, 1976, 1993), *Fundamentals of Soil Behavior*, *Chapter 6 – Soil Water*, Dr Mitchell discusses the often overlooked, yet important interactions of water with soils. I suggest that you get a copy of *Soil Behavior* and study sections on the dipolar nature of water and the interactions of water molecules and clay particles.

I have to admit that Jim Mitchell is my hero. I greatly admire his scientific approach to the study of soils. I love his books and papers on soil behavior. I purchased both the first and second editions of Dr Mitchell's book immediately after they came out in 1976 and 1993. Now there is a third edition. I wish my children or grandchildren had gotten me a copy of Dr Mitchell's third edition of *Fundamentals of Soil Behavior* for Christmas. It was on my wish list, but alas I got a biography of Ben Franklin (still a great present for a history buff!). Guess I'll have to buy my own copy.

References

Burmister, D.M. (1948) The importance and practical use of relative density in soil mechanics. Presented at the Meeting of Committee D-18 on Soils for Engineering Purposes held during the Fifty-first annual meeting of the ASTM, June 22, 1948, 18 pages.

Briaud, J.L. (2007) Spread footings in sand: load settlement curves approach. *ASCE Journal of Geotechnical and Geoenvironmental Engineering*, **133**(8), 905–920.

Cheng, L.-S. (1948) An investigation of stress-strain and strength characteristics of cohesionless soils by triaxial compression tests. Proceedings of the 2nd International Conference on Soil Mechanics and Foundation Engineering, Vol. V, Rotterdam, pp. 35–48.

Fellenius, B.H. and Altaee, A. (1994) Stress and settlement of footings in sand. Proceedings of the American Society of Civil Engineers, ASCE, Conference on Vertical and Horizontal Deformations for Foundations and Embankments, Geotechnical Special Publication, GSP, No. 40, College Station, TX, June 16–18, 1994, vol. 2, pp. 1760–1773.

Hogentogler, C.A. and Terzaghi, C. (1929) Interrelationship of load, road, and subgrade. *Public Roads*, **10**(3), 37–64.

Holtz, R.D. and Kovacs W.D. (1981) *An Introduction to Geotechnical Engineering*, 1st edn, Prentice-Hall, Inc., Englewood Cliffs, New Jersey, 733 pages.

Holtz, R.D., Kovacs, W.D., and Sheahan, T.C. (2011) *An Introduction to Geotechnical Engineering*, 2nd edn, Pearson, Prentice-Hall, 853 pages.

Mitchell, J.K. (1976 and 1993) *Fundamentals of Soil Behavior*, 1st and 2nd edns, John Wiley & Sons, Inc., 422 and 437 pages.

Peck, R.B., Hanson, W.E., and Thornburn, T.H. (1953) *Foundation Engineering*, 1st edn, John Wiley & Sons, 410 pages.

Sienko, M. J. and Plane, R. A. (1961) *Chemistry*, 2nd edn, McGraw-Hill Book Company, Inc., New York, Toronto, London, 623 pages.

2.2

Soil Stresses and Strains

2.2.1 Introduction to Soil Stresses

I have heard the following lament from fellow engineers so often that I cannot attribute it to one person or group of engineers. It goes something like this:

> When I was in school, the professor told us that soils are not linearly elastic like steel. Soil doesn't have a straight line stress–strain curve; in fact, it doesn't even follow the same stress–strain curve during loading and unloading. Soils are not homogeneous and they aren't isotropic like the engineering materials we studied in sophomore year strength of materials class. Just about the time I wrapped my mind around the anisotropy of soils thing, my damn professor turned to the next chapter in our soils textbook and started calculating soils stresses using the Boussinesq theory. Almost matter-of-factly while calculating soil stresses caused by surface loadings my professor mentions that the Boussinesq theory assumes that soils are elastic, homogeneous, isotropic materials. Now wait one darn minute, first he says soils are anisotropic, then with the wave of a hand he changes his mind and says that it is OK to use the elastic material Boussinesq theory. What's up with that? He never answered my question. Why not use an analysis that matches the material?
>
> I'm still confused 10 years after graduation from college. First, I'm told that soils are not linearly elastic; they are not homogeneous or isotropic. Then I'm told to go ahead and use equations that are based on the assumptions of linear elasticity! Come on now, which is it? Does it matter? What are the consequences of using the Boussinesq equation?

These are good questions, and they deserve answers. I'm your mentor in these pages, so here goes.

Geotechnical Problem Solving, First Edition. John C. Lommler.
© 2012 John Wiley & Sons, Ltd. Published 2012 by John Wiley & Sons, Ltd.

2.2.2 Isotropic and Linearly Elastic versus Anisotropic and Non-Linearly Elastic

2.2.2.1 Inelastic – Does it Matter?

The simple answer is that it does matter. Using an equation that is based on assumptions that are at variance with known soil properties will produce differences between calculated and observed results. The big question is how much difference is there between calculated and actual changes in stress in the soil mass given the differences between an ideal material and a soil material.

I'm really eager to jump into a discussion about stress and material anisotropy, but I'm going to restrain myself, and take it slowly, one step at a time.

Since you probably don't want to perform a detailed finite element analysis to find the soil stresses for all of your geotechnical problems, it might be nice if you knew how to do the old-fashioned manual calculations. If you know how to do manual calculation of soil stresses and the limits of the methods, you have a powerful tool for geotechnical analysis and for checking the results of computer solutions.

There are two commonly used methods for manual calculation of soil stresses generated by surface surcharge loadings. These two published simplified cases are: (1) Linearly elastic, homogeneous, isotropic, weightless material, the Boussinesq equations (1883); and (2) Linearly elastic, layered anisotropic material with a Poisson's ratio normally taken as zero (i.e., zero lateral strain), that does not allow soil tension stresses to occur, the Westergaard equations (1938). In general, the Boussinesq equations give higher stresses close to the footing, and the Westergaard equations give higher stresses away from the centerline of the footing because they "spread out" the stresses. At an r/z value of 1.8 (r is the radial distance from the centerline of the footing, and z is depth below the bottom center of the footing), both the Westergaard and Boussinesq equations give the same answer.

If the soils on your site have well-defined horizontal layers of sand and clay, the Westergaard equations are recommended. If the soils on your site are relatively uniform, it is recommended that the Boussinesq equations be used to calculate increases in soil stresses from application of surface loadings. If the soils are not layered and not uniform, which is just about every other type of site, most engineers use the Boussinesq equation. I frequently calculate changes in stresses using both methods and compare the results.

The Boussinesq equations give higher stresses directly beneath the foundation and their use is considered conservative if soft/loose compressible soils lie directly beneath the footing. Boussinesq equations and charts are available in nearly all geotechnical texts for calculating both vertical and horizontal stress increases for point loads, line loads, circular loaded areas, and rectangular loaded areas (Holtz and Kovacs, 1981; Holtz, Kovacs and Sheahan, 2011). You can calculate the change in soil stresses caused by a point load (see Figure 2.2.1) or a small circular load which approximates a point load by used Equation 2.2.1. The change in vertical stress in a soil mass beneath

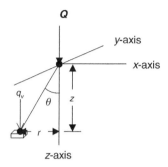

Figure 2.2.1   Terms defining the Boussinesq and Westergaard equations

a point load applied to a semi-infinite half space is:

$$q_v = \frac{3Q\left(\cos^5\theta\right)}{2\pi z^2}$$ (2.2.1)

Where

q_v is the change in stress at point (r,z) due to application of Q
Q is the point loading applied to the soil surface
z is a vertical distance into the soil directly beneath the load Q
r is a horizontal radial distance from the vertical z axis
r and z define a point in the soil where the change in vertical stress is required.
θ is the angle from the vertical z-axis to the point, equal to $\tan^{-1}(r/z)$

The change in vertical stress in a soil mass beneath a point load applied to a semi-infinite half space is given by the Westergaard equation 2.2.2.

$$q_v = \frac{Q\left[(1-2\mu)/(2-2\mu)\right]^{0.5}}{(2\pi z^2)\left\{\left[(1-2\mu)/(2-2\mu)\right]+\left(\frac{r}{z}\right)^2\right\}^{1.5}}$$ (2.2.2)

Where

q_v is the change in stress at point (r,z) due to application of Q
Q is the point loading applied to the soil surface
z is a vertical distance into the soil directly beneath the load Q
r is a horizontal radial distance from vertical z-axis
r and z define a point in the soil where the change in vertical stress is required.
μ is Poisson's ratio, which normally ranges from 0.0 to 0.4 (for triaxial stresses)

Table 2.2.1 Soil stresses generated by Boussinesq and Westergaard directly below applied loading

Depth, z (feet)	Boussinesq, q_v (psf)	Westergaard, $\mu = 0, q_v$ (psf)	Westergaard, $\mu = 0.4, q_v$ (psf)
1	477.5	318.3	954.9
5	19.1	12.7	38.2
10	4.77	3.2	9.6
20	1.19	0.8	2.4

Comparing the soil stresses generated in a soil mass by a surface loading ($Q = 1000$ pounds), we can check and verify in Table 2.2.1 that the Boussinesq equation does indeed generate higher stresses directly beneath the loading than the Westergaard equation, so long as the common assumption of Poisson's ratio equal to zero is used in the Westergaard equation calculations. If a reasonable upper bound Poisson's Ratio of 0.4 is used in the Westergaard equation, then the Westergaard equation yields higher stresses directly beneath the surface loading.

The Westergaard equation is reported in all texts reviewed by this author to spread out the loading laterally into soil further from the vertical axis of loading than the Boussinesq equation. You should know me by now. Why take someone's word for something when we can check it. Checking this by using the equations given above, values for points located five and ten feet horizontally from the axis of loading give values in Table 2.2.2 below.

From a brief study of the data points in the table above, you can see that the simplification often given, "Westergaard spreads out the loading," is not exactly

Table 2.2.2 Soil stresses generated by Boussinesq and Westergaard at various distances beyond the applied loading

Depth, z (feet)	Radial Distance, r (feet)	Boussines q, q_v (psf)	Westergaard, $\mu = 0, q_v$ (psf)	Westergaard, $\mu = 0.4, q_v$ (psf)
1	5	0.14	0.87	0.51
1	10	0.0047	0.11	0.06
5	5	3.38	2.45	2.06
5	10	0.34	0.47	0.31
10	5	2.73	1.73	2.42
10	10	0.84	0.61	0.52
20	5	1.03	0.67	1.48
20	10	0.68	0.43	0.60

correct for all radial distances. It appears that for r/z values greater than 2, the Westergaard values are indeed greater than the Boussinesq values. When r/z equals 1, the Boussinesq values are greater than the Westergaard values, and when r/z is less than 1 the greater values seems to depend of Poisson's ratio used in the Westergaard equation.

I found a figure in Bowles 3rd edition text (Bowles, 1982, Figure 5.1, page 174) that clarifies this situation between Boussinesq and Westergaard equations. Please check your copy of Bowles book, or purchase a used copy to find his analysis of Boussinesq and Westergaard equations. Bowles defined terms $A_{Boussinesq}$ and $A_{Westergaard}$ that are functions of r/z values. What Bowles' table of values shows is that for Poisson's ratio of 0.0, Boussinesq stresses are greater than Westergaard stresses for r/z values less than or equal to 1.5, and for r/z values slightly greater than 1.5 Westergaard stresses are greater. Things are much different if you assume a Poisson's ratio of 0.4. For r/z values less than about 0.4 Westergaard stresses are greater than Boussinesq stresses, up to r/z slightly greater than 2.0 Boussinesq stresses are greater than Westergaard stresses, and for r/z greater than 2.5 Westergaard stresses are again greater than Boussinesq stresses. This may seem a bit confusing, but it illustrates that the simple statements that "Boussinesq stresses are greater directly beneath the surface loading" and "Westergaard stresses are greater at distances beyond the loading axis" are not quite true. Now you understand why I calculate both cases. I have an Excel table that I use for Boussinesq and Westergaard stress calculations. You can easily make one for yourself, it is a good exercise.

Check references included at the end of this section for recommended books where Boussinesq and Westergaard equations, tables and graphs may be found. If you don't have any of these, look in your college texts and there should be some mention of Boussinesq and Westergaard equation. Caution, if you intend to use any of the Boussinesq or Westergaard equations given in your college text please check the terms of the printed equations in at least two or three other references to confirm that there are no misprinted terms. You can always go on the internet and find a copy of Poulos and Davis' out of print book of elastic solutions (Poulos and Davis, 1974).

If you do look up Poulos and Davis book , I suggest that you take a look at Burmister's equation. Burmister's equation was originally developed in the 1940s for paving systems where stresses in a paving over a rough base course were calculated.

2.2.2.1.1 Burmister's Equation

Burmister (1943 and 1945) published his work in the *Proceedings of the Highway Research Board* (1943) and in the *Journal of Applied Physics* (1945), which you might be able to find through a major metropolitan or university library. I suggest that an easier source of Burmister's material can be found in T. H. Wu's soil mechanics textbook (Wu, 1966, page 132–133), or Yoder and Witczak's paving textbook (Yoder and Witczak, 1975, pp. 40–44).

Unlike Boussinesq's equation, which calculates stresses in a homogeneous elastic soil, the equations and charts developed by Burmister were for a two-layered system. In Burmister's analysis, both layers are homogeneous elastic materials; the only difference is that the layers have different moduli of elasticity, E_1 for the upper layer and E_2 for the lower layer. When E_1 is 20 times greater than E_2, the stress at the interface between the upper and the lower layer is reduced from 68% of the surface pressure to 22% of the surface pressure. So Burmister's analysis indicates that a stiff surface layer spreads the surface stresses out laterally into the softer lower layer thus reducing stresses in the lower layer directly beneath the loading. Field tests (Sowers and Vesić, 1962) were conducted to check Burmister's predicted reduction in stresses beneath a flexible paving. The tests showed that stresses at the interface between the paving and the subgrade soil were better predicted by the Boussinesq equation than the Burmister equation. Why the discrepancy between theory and measured results? The answer is not as complicated as you might expect. The Burmister equation required that a tension stress develop between the paving and the soil to maintain an unmoving contact between the layers. Since the silty sand subgrade in the Sowers and Vesić test case had a very low tensile strength, there was movement between the layers and the vertical stresses were not spread out and reduced, as predicted by Burmister.

As it turns out, the major discrepancy between stresses predicted by Burmister's equation and field-measured stresses is the frequent requirement of Burmister's equation for the soil to mobilize tensile stresses. Today Burmister's equation has fallen into disuse by geotechnical engineers because of the problems reported and emphasized by Sowers. But wait, progress may have caught up with Burmister. Today we have a broad selection of soil-reinforcing systems that provide tensile stresses to a reinforced soil mass. The Burmister method could be used to calculate tensile stresses required to be resisted by reinforcements, and published two- and three-layer methods developed by extension from the Burmister method could be used to predict reinforced soil displacements. All you would need is the modulus of the reinforced soil mass. Any graduate students interested?

2.2.3 Anisotropic Materials and Anisotropic Stresses

Isotropic materials have the same properties in all of the three principle stress directions, x, y and z. The word anisotropic means "not isotropic", so anisotropic materials have different properties in two or in all three of the principle stress directions.

The topic of anisotropic soils can be somewhat confusing, so I like to simplify it a bit. I classify anisotropic soils into three categories: (1) Soils with anisotropic fabric that I can see with the unaided eye. An example of soil with visible anisotropic fabric would be a soil with thin alternating layers of clay and fine sand. (2) Soils with anisotropic fabric that I can't see with the naked eye. An example of a soil with invisible anisotropic fabric would be clay that was deposited with a vertical stress much higher than its lateral stress resulting in flat clay plates aligning in a horizontal plane perpendicular to the major principle vertical stress. (3) Soils that may or may not initially have anisotropic fabric that in the future will be subjected to changes in stresses that result in an anisotropic stress state. You couldn't possibly see or measure this third type of anisotropy because the stress changes that cause this anisotropy have not yet occurred.

These three basic types of soil anisotropy have several different names. When the soil material itself has different properties in the principle stress directions, many engineers refer to this type of anisotropy as material anisotropy, anisotropic soil fabric, or inherent anisotropy.

When a soil material is deposited under isotropic conditions (implying that it is nearly a fluid), its principle stresses align with the x, y and z axes. If during loading of this isotropic soil material, the principle stresses change direction and/or the ratio of the principle stresses changes, the change in principle stresses causes my third class of soil anisotropy. Engineers often call this third type of anisotropy, stress anisotropy, induced anisotropy or evolving anisotropy.

Before I go on, please let me digress and tell you about one of the biggest frustrations in geotechnical engineering. Almost every geotechnical expert or geotechnical group seems to come up with their own names for geotechnical concepts. Anisotropy of soils is a great example of this naming problem. I personally prefer to call the three main types of soil anisotropy: (1) soil fabric anisotropy (you can see it); (2) material anisotropy (you can't see it); and (3) stress anisotropy (it hasn't even happened yet). I am a great admirer of Chuck Ladd's work at MIT and I especially treasure his 22nd Terzaghi Lecture titled *Stability Evaluation During Staged Construction*, which you can find in the *ASCE Journal of Geotechnical Engineering* (its name has changed since 1991) April 1991, Volume 117 No. 4, pages 537 to 615. The reason I mention Dr. Ladd's paper is that Sections 4.7, 4.8, and 4.9 on pages 572 to 576 deal with soil anisotropy. Dr. Ladd uses different terms to describe anisotropy, and I often slip into the use of his names for soil anisotropy interchangeably with mine in the following discussion. See if you can pick out which anisotropy is which.

When I visualize material anisotropy in my mind I most often picture a piece of wood, see Figure 2.2.2. Notice that the grain in the wood indicates different material properties in the horizontal versus vertical directions.

When I visualize the anisotropic fabric of wood, I normally imagine generally straight grain, as shown in Figure 2.2.2. The problem is that when I get into a test pit to check the soil deposition, the actual soil fabric exhibits a material anisotropy more like that shown below in Figure 2.2.3

As you might expect, the actual soil fabric is always more like Figure 2.2.3 than Figure 2.2.2 because the actual soil fabric is more complicated than you imagine. The horizontal soil fabric can be wavy rather than horizontally layered and it has pockets of dissimilar materials, all causing highly complex material anisotropy in reality that is not captured by your soil model.

Figure 2.2.2   Anisotropic fabric – wood with straight grain

A common soil with fabric anisotropy is varved clay. Varved clays' horizontal layers of clay and fine silty sand are deposited during winters (clay) and summers (fine silty sand), so you can see the material anisotropy of varved clays are much like the wood grain in Figures 2.2.2 and 2.2.3 above.

Now that you have an idea of what soil fabric anisotropy looks like, let's consider material anisotropy of the second type (you can't see it) that is caused by stress anisotropy during deposition. Natural clay soils are frequently deposited such that they build up vertically and experience one-dimensional consolidation. The ratio of the horizontal stresses to vertical stresses in natural soils is called K_o, the *at-rest* earth pressure coefficient. If the soil were isotropic, K_o would be equal to one. That's not what we find in most natural soils. In most natural soil deposits at any given point in the soil mass, the vertical stress in the z direction is greater than the horizontal stresses in the x and y directions. The horizontal stresses in the x and y directions are often equal or near equal (or they are assumed to be nearly equal). Another way to indicate that this soil type is anisotropic is $K_o \neq 1$. The anisotropy in this case is called

Figure 2.2.3   Anisotropic fabric – wavy wood grain with knot

cross-anisotropic since there are only two different strength directions. If you investigate a site with this type of cross-anisotropy, the soil has inherent material anisotropy that comes from the soil structure developed during deposition on a microstructure level (which is in response to the differing vertical and horizontal stresses during deposition). If the soil has inherent material anisotropy on a microstructure scale it wouldn't have visible horizontally layered clay and fine sand seams, such as varved clays, but I imagine that it has extremely fine layers similar to varved clay.

If you want to visualize material anisotropy you can think of my wood chips in Figures 2.2.2 or 2.2.3 or just imagine a varved, horizontally layered clay which has different properties in the vertical direction than it has in the horizontal directions.

Stress anisotropy is a bit more difficult to visualize. Let's consider an example of horizontally deposited soil. In its horizontally deposited state, vertical stresses are the major principle stress, that is, the horizontal stresses are less and $K_0 < 1$. Imagine you excavate a 100 foot long by 100 foot wide hole, 20 feet deep with 2 to 1 (horizontal to vertical) sloped slides into this horizontally deposited soil deposit. The vertical soil stresses in the bottom of the hole at the toe of the slope would be significantly decreased after the excavation was dug, but the horizontal stresses would not be significantly decreased. The result of reducing the vertical stress while not significantly decreasing the horizontal stress at the toe of slope is that the horizontal stress becomes the major principle stress. Soils beneath the slope from top to bottom experience a rotation of the major principle stress from vertical beneath the top of slope to horizontal at the bottom of slope. Rotation of the major principle stress causes the failure plane where maximum shearing stresses occur to also rotate. As the direction of the shearing plane rotates its angle to the original bedding planes changes and its apparent anisotropy significantly changes. These changes in stress-induced anisotropy change the soil's measured shearing strength, modulus, pore pressure, and compressibility changes.

When considering combined material and stress anisotropy, I like to think again of my piece of wood. Wood has longitudinal grain that makes it stronger in compression in the longitudinal direction (i.e., parallel to the grain) than it is in compression in the lateral direction (i.e., perpendicular to the grain). Most wood can be split easier in the longitudinal direction than in the cross-grain direction due to its parallel fibers in the longitudinal direction. But don't try to spit a piece of American Elm; it has twisted fibers that make it very difficult to split. You're better off using elm as fence posts, not as fire wood.

Before I get off soil anisotropy, I would like to mention that it can be much more complicated than I have described above. There are three principal stress directions and two shearing stresses on each principle stress plane. That makes nine stresses that can interact on each soil element. If all nine of the stresses and the material constants that relate these stresses to soil strains (i.e., Es and Gs) are different, it is quite difficult to predict what a changes in shearing stress on the x–y, x–z, and y–z planes will do to the soil's change in dimension in the x, y, and z directions. In addition, we don't have commercially available laboratory equipment to test for all of the complex combinations of applied stresses resulting in rotations of principle stresses that can occur in the field. The best we can do in modern laboratories is to attempt to replicate initial stress and strain conditions then follow the stress path that is expected to occur in the field in our tests.

2.2.4 Soil Strains

Most engineers think of soil stresses and changes in stresses caused by various loading conditions, but they generally ignore soil strains. Although thinking about soil stresses may be convenient and comfortable because you are used to thinking about and analyzing stresses, it is actually soil strains and resulting deflections that cause changes in soil materials that you can physically observe. If you have consolidated a direct shear sample and started to shear it, you know that the sample has a void ratio less than (i.e., density greater than) the critical state value if the vertical dial gage starts to move upward, indicating dilation of the sample. You can't see the critical stress condition, but you can see that a sample's volume is not changing during shearing.

T. William (Bill) Lambe introduced the stress-path method in 1967 and updated it in 1979 (Lambe and Marr, 1979) to allow engineers to follow the same or similar state of stresses in the laboratory that the soil experiences in the field. I can't recall the name of the fellow who introduced the strain-path method (Baligh, 1985), but it is great idea that hasn't been popular because it involves strains rather than good old familiar stress (Baligh's paper won ASCE's J. James R. Croes Medal in 1987). I'll look up the strain-path method and include it as a reference at the end of this section. If you really want to visualize how a soil is acting, and then interpret observed soil displacements in the field, you have to think about soil strains as well as stresses.

2.2.5 Additional General Information on Soil Stresses and Strains

Soil stresses and strains measured during laboratory tests such as the consolidation tests, triaxial shear tests, and plain simple shear tests generate parameters that are used in equations to evaluate settlements and the shear strength of soil. These relationships between laboratory tests and geotechnical equations are known as constitutive relationships. These relationships are not equivalent to laws of physics or even equivalent to theories of physics, because they are not developed from basic physics principles. These soil relationships are called constitutive because they are constituted or made up from data generated by specific laboratory tests.

Engineers have been aware of this "constitutive relationship" limitation of geotechnical testing for decades, and to resolve this problem, they have tried to simulate real world stresses and strains in the laboratory setting. As I mentioned above, these laboratory tests often don't replicate the same state of stresses and strains as soil in the field.

As I mentioned above, Bill Lambe helped us visualize this problem by developing the "stress-path" method. If you can find a used copy, I recommend that you study the stress-path method presentation in Dr. Lambe's textbook on soil mechanics (Lambe and Whitman, 1969). With this method, the stress path or stress trajectory of soils in a real world loading case could be plotted and studied for later simulation in the laboratory. Please don't forget that real-world conditions are never completely captured by laboratory simulations or computer modeling. Whether factors that are not considered are important or not is the job of the engineer to determine.

2.2.6 Additional Specific Information on Stresses and Strains

Reducing the problem of characterizing a soil's stress–strain relationship to a few simple terms is quite difficult because in reality a soil's stress–strain properties depend on many variables. Most texts describe classical soil stress–strain properties for sands and clays such as modulus of elasticity (E), shear modulus (G), and Poisson's ratio (μ), and then they continue for page after page to describe the exceptions to these simple relationships. These exceptions are not really exceptions, rather they represent the real world's "second-order" terms that influence stress–strain properties. What are these variables that cause so much debate relative to a soil material's stress–strain properties? These variables include:

1. Brittle (sensitive) versus ductile (not sensitive) materials.
2. Effect of the soil's initial density and structure on stiffness and volume changes that occur during material shearing, that is the sample dilates (increases in volume), compresses (decreases in volume), or maintains a constant volume (material at the critical void ratio) depending on its structure or fabric at initiation of shearing.
3. Effect of the soil's confining stresses (i.e., σ_2 and σ_3) during consolidation and shearing.
4. State of stress changes such as compression, tension and "pure shearing."
5. The soil's degree of saturation at initiation of loading.
6. Presence of free water to come into or to flow out of a sample during shearing.
7. Environmental effects such as pH, temperature gradients, ion gradients, electrical gradients, and hydraulic gradients of fluids and gases. This could also include effects of bacteria and vegetation on the soil structure.

In following sections, we will discuss many of these second-order terms that affect soil stress–strain and shear strength properties. If I haven't covered all of the material that you're interest in studying, please start your own search and see what you can find out there in the big world of geotechnical engineering publications. I suggest you start with the *American Society of Civil Engineering's Journal of the Soil Mechanics and Foundations Division* whose name changed in January 1974 to the *Journal of Geotechnical Engineering*. Later they changed the name again to the *Journal of Geotechnical and*

Geoenvironmental Engineering. Don't forget to try and find at least two sources for all of your study materials. If you are studying a particular journal article, check to see if a commentary or discussion of that journal paper exists, you should find it and study diverging opinions on your topic of study. In the end you have to decide for yourself if the author or the commenter is correct.

References

Selected References for Boussinesq and Westergaard Equations:

Bowles, J.E. (1982) *Foundation Analysis and Design*, All five editions, McGraw Hill.
Holtz, R.D., Kovacs, W.D., and Sheahan, T.C. (1981 and 2011) *An Introduction to Geotechnical Engineering*, Prentice Hall.
Poulos, H.G. and Davis, E.H. (1974) *Elastic Solutions for Soil and Rock Mechanics*, John Wiley & Sons, New York, 411 pages (out of print, but available on internet).

Selected References for Burmister's Equation:

Burmister, D.M. (1943) Theory of stresses and displacements in layered systems and applications to the design of airport runways. Proceedings of the Highway Research Board, No. 23, p. 126.
Burmister, D.M. (1945) The general theory of stresses and displacements in layered soil systems. *Journal of Applied Physics*, **16**, 296–302.
Sowers, G.F. and Vesić, A.B. (1962) Vertical Stresses in Subgrades Beneath Statically Loaded Flexible Pavements. Highway Research Board, Bulletin 342.
Yoder, E.J. and Witczak, M.W. (1975) *Principles of Pavement Design*, 2nd edn, John Wiley & Sons, Inc., 711 pages.
Wu, T.H. (1966) *Soil Mechanics*, Allyn and Bacon, Inc., Boston, 431 pages.

General References

Baligh, M.M. (1985) Strain path method. The American Society of Civil Engineers. *Journal of Geotechnical Engineering*, **111**(9), 1108–1136.
Lambe, T.W. and Marr, W.A. (1979) Stress path method: second edition. American Society of Civil Engineers. *Journal of the Geotechnical Engineering Division*, **105**(6), 727–738.
Lambe, T.W. and Whitman, R.V. (1969) *Soil Mechanics*, John Wiley & Sons, Inc., 553 pages.

2.3

Soil Shear Strength

2.3.1 Introduction to Soil Shear Strength

"When I was in school, the professor said clays have cohesion and sands have friction ... sounded simple enough to me."

"Now I am confused, I'm told that clay has cohesion and friction and that sand has cohesion and friction, or that sand has cohesion and clay has friction."

These are comments I heard from practicing geotechnical engineers at a recent conference after a young professor's talk on soil shear strength and dilation. He really riled up the crowd when he concluded that there is no such thing as cohesion. What does clay and sand look like (see Figure 2.3.1)? What does cohesion look like? Please read on.

2.3.2 Soil Cohesion and Friction

The problem of characterizing soil shear strength is a perennial issue that confuses engineers at all levels of training from BS to PhD The initial problem with describing soil strength is a plain and simple mix up of terms. The words "cohesion" and "friction" describe physical properties. When a substance is cohesive or sticky like glue or chewing gum, you know it because the material sticks to your fingers or to the bottom of your shoe. You also know when a material has friction such as fine sandpaper or coarse sandpaper by the abrasion of your fingers while running your hand over its rough surface.

The terms "c" and "ϕ" used in soil shear strength equations are *not* these physical properties of stickiness or roughness. Geotechnical engineers do not care if a soil is sticky or if it has friction. Engineers want to know what the shear strength of a soil is for use in calculations and designs. The term "c" is called cohesion, and the term "ϕ" is called angle of internal friction. These terms are used to calculate the strength

Geotechnical Problem Solving, First Edition. John C. Lommler.
© 2012 John Wiley & Sons, Ltd. Published 2012 by John Wiley & Sons, Ltd.

(a)   (b)

Figure 2.3.1 Topsoil and gray-red clay over bedrock and a jar of "running" sand

of a soil. They are not necessarily intended to describe a soil's physical appearance. In the following material, I will describe how c and ϕ are determined and how they are used to calculate soil shear strength.

2.3.3 Soil Shear Strength

The first thing you need to know is that there are several alternate methods of defining and calculating a soil's shear strength. Let's start first with the Mohr–Coulomb equation.

The basic, standard soil shear strength equation used in geotechnical engineering is the Mohr–Coulomb equation, which is illustrated below in Equation 2.3.1.

The shear strength of a soil, S, is:

$$S = c + \sigma_n \tan \phi \qquad (2.3.1)$$

Where

    S is soil shear strength with units of force divided by area or F/L^2
    c is the cohesive component of strength with units of F/L^2
    σ_n is the normal stress acting perpendicular to the failure plane with units of F/L^2
    ϕ is the angle of internal friction with units of degrees (that is angular 0 to 360 degrees, not temperature degrees)

When evaluating the shear strength of a soil for design purposes, you need to simulate the stress–strain conditions generated in your design problem in laboratory testing used to determine values of S, c, and ϕ. On the most basic level, this generally involves answering two primary questions about the testing used to determine S, c, and ϕ. The

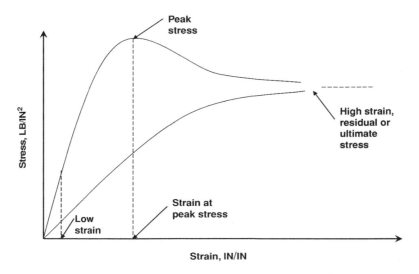

Figure 2.3.2 Stress–strain curves – low, peak and residual strength from triaxial compression tests

first question is: what strain or deflection is associated with the soil strength values in the problem? Are those strains associated with peak stress values, with ultimate strength values (sometimes called residual strength values), or some limited strain controlled strength values, see Figure 2.3.2.

The second question is: what saturation state and drainage condition will the soil sample have during testing; does it replicate the soil in the field in your problem? You need to know if the soil is required to be saturated or unsaturated during testing, or if it starts unsaturated during the first stages of testing and becomes saturated during later stages of testing. Soil saturation and drainage conditions are very important because they directly related to the soil's shear strength. Remember the effective stress equation developed by Terzaghi ($\sigma_{\text{effective}} = \sigma' = \sigma_{\text{total}} - u_{\text{pore water pressure}}$). Soil shear strength depends on effective stresses, and effective stresses depend on pore water pressures in the soil during shearing. If the soil is unsaturated and/or it is drained during loading, then excess pore water pressures (i.e., pore pressure changes) don't develop. If excess pore water pressures do not develop (don't forget that pore pressure changes could be positive or negative), total stresses are equal to effective stresses. If excess pore pressures do not develop during testing, the soil's shear strength is considered to be "drained shear." A soil's strength determined by a drained shear test can be expressed as follows:

$$S = c' + \sigma_n \tan \phi' \tag{2.3.2}$$

If the soil sheared during a drained shear test is normally consolidated, that is when the maximum past consolidation stress is less than the consolidation stress used in the shear strength test, the cohesion term c' in Equation 2.3.2 should be equal to zero, and the equation reduces to: $S = \sigma_n \tan \phi'$.

If the soil tested in the drained test is saturated and loaded rapidly enough to prevent changes in pore water pressure during shearing (even though the drainage line is open), the soil strength results can be considered to be from an undrained shear test. With a series of rapid undrained tests, the soil's strength can be expressed in terms of c only because ϕ equals zero, as follows:

$$S = c_u + \sigma_n \tan 0 = c_{\text{undrained}} = S_{\text{undrained}} \tag{2.3.3}$$

I know what you are thinking, "It would be easy if I could just use c only or ϕ only when calculating soil shear strength. When do I have to consider soil strength to be c only, ϕ only or with both c and ϕ?" I expect that you have figured it out by now that geotechnical research engineers can't help but complicate your life!

There are several reasons why soil tests give results indicating both a c and a ϕ component of soil strength. Some of these reasons are functions of primary variables and are quite reasonable to consider, others are second-order terms and are quite esoteric. If you are curious or interested in advanced topics related to soil shear strength, please refer to the Sections 2.3.4, 2.3.5, and 2.3.6, "Additional Information." If you would like to skip the advanced material, the primary thing you need to know is that some soils have been exposed to stresses in the past that are greater than those presently existing. These are called over-consolidated soils. If the current equilibrium soil stresses are equal to the greatest value of stresses experienced by the soil, it is referred to as a normally consolidated soil. Both normally consolidated soils and over-consolidated soils that are going to be re-consolidated and sheared at test stresses greater than their maximum historical stress state (called the preconsolidation stress) and are loaded in drained shear tests, their strength can be characterized by ϕ only as described above, please refer to Figure 2.3.3.

If over-consolidated soils are loaded to stress levels below their preconsolidation stress, their drained shear strength should be characterized by both c and ϕ, please refer to Figure 2.3.3.

If a saturated sample is tested unconsolidated and undrained, no matter how much you increase the confining stress the pore water takes the stress because it is incompressible. If there is no sample drainage, and you don't allow dissipation of excess pore water pressures, the shear strength of the sample will remain the same. In this case of unconsolidated, undrained shear called the UU triaxial test, the sample's shear strength can be expressed as c-only, as shown in Figure 2.3.3.

2.3.4 Additional Soil Shear Strength Information – General

Soil strength tests such as direct shear tests, triaxial tests, and plain simple shear tests generate parameters c and ϕ that are used in equations to determine the strength

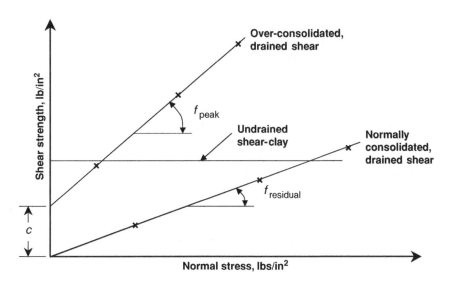

Figure 2.3.3　Drained and undrained shear strengths

of soil. The use of parameters c and ϕ to characterize a soil's strength derives from plotting the Mohr–Coulomb failure envelope, as illustrated in Figure 2.3.4, which assumes that the failure envelope is a straight line. There is no physical reason that soils should be so kind as to line up for us. In fact, they don't. Most soils have a curved failure surface, which is illustrated by data points that do not line up in Figure 2.3.4.

As mentioned above, the Mohr–Coulomb failure envelope is not a straight line, but is a curved line over most of the ordinary range of normal stresses. The fact that the Mohr–Coulomb line is not a straight line is not new news. The classic geotechnical book, *Soil Mechanics in Engineering Practice*, second edition, 1967 by Terzaghi and Peck

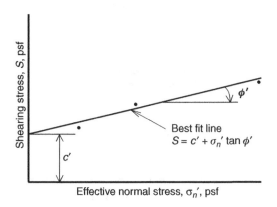

Figure 2.3.4　Mohr–Coulomb failure plot

(Terzaghi and Peck, 1967) shows a Mohr–Coulomb failure envelope on page 102, and clearly describes on page 103 that the rupture line for soils is curved. Many engineers born since 1967 have not gotten the word. Ralph Peck told me that if younger engineers are lacking in anything, it is my generation's fault because we failed to mentor them. Ralph, I'm doing my best to remedy the situation! See Section 2.3.6 below for my answer to the question, why is the Mohr–Coulomb failure envelope curved?

Looking back at Figure 2.3.4, the individual shear strength tests are plotted as points. What do these points represent? Unless the testing laboratory or reporting engineer says differently, these points represent the peak shear strength measured in each test. In standard geotechnical practice the friction angle ϕ reported from sample testing of peak values is used as a correlation parameter. When you see ϕ values reported in charts of typical soil properties or correlated to Atterberg limits, they represent peak values, unless otherwise noted.

Do you think that the peak friction angle value is the only shear strength parameter that you need? Take a moment and look at Figure 2.3.5, where peak and residual shear strength test results from the same samples are plotted.

My protractor placed on Figure 2.3.5 reads 22 and 43 degrees for residual and peak values of friction angle respectively. If the stress–strain analyses for your project indicate that the soil may experience high strains resulting in residual strength values, what would be the error in using the peak value? OK, given that the soil shear strength is a function of the tangent of the friction angle, let's compare the tangent 22 degrees to tangent 43 degrees. Tangent 22 degrees is 0.404 and the tangent 43 degrees is 0.933. If you used the peak friction angle when the residual value was appropriate to calculate

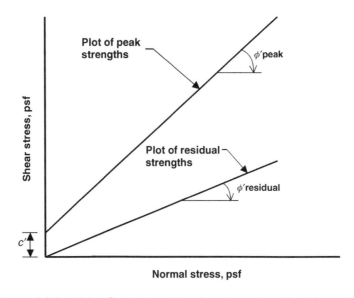

Figure 2.3.5 Mohr–Coulomb plots of peak and residual strengths

the soil shear strength, the actual strength developed would be 0.404/0.933 or 43% of the anticipated value. If you anticipated having a factor of safety of 2.5 when you incorrectly used the peak value, you actually have a factor of safety of 1.08 when high strains generate the residual strength. Ouch, that is too close to failure for me!

2.3.5 Additional Information and Second-Order Terms

Reducing the problem of characterizing a soil's shear strength to a few simple terms is quite difficult because in reality a soil's strength depends on many primary variables and several second-order variables. Most soil texts describe classical soil strength for sands and clays which depend on void ratio, grain sizes, moisture content, and OCR (over-consolidation ratio), and then they continue for page after page to describe the exceptions. These exceptions are not really exceptions, rather they represent the real world's second-order terms that influence shear strength. What are these second-order considerations that affect a soil material's strength? The second-order variables impacting soil shear strength include:

1. **Brittle (sensitive) versus ductile (not sensitive) materials.**
   Brittle soils have a significant reduction in shear strength after reaching their peak strength, while ductile soils do not have a significant loss of shear strength at higher strains. Brittle soils are often over-consolidated soils and ductile soils are often normally consolidated soils, but there are exceptions.
   The term sensitivity is used to define the loss of shear strength as a soil's initial *in situ* fabric is remolded by shearing stresses. Best I can tell this phenomenon of loss of soil strength during shearing was initially considered by engineers studying landslides in clays. They defined sensitivity (S) of a clay sample as the ratio of the undrained shear strength of an undisturbed sample ($s_{u\ undisturbed}$) divided by the undrained shear strength of a remolded sample, that is $S = s_{u\ undisturbed} / s_{u\ remolded}$. Of course, the undisturbed and the remolded samples had to be tested at the same water content. Using this definition of sensitivity of a clayey soil, engineers could communicate the potential of experiencing a significant loss of strength during shearing of clays on their projects. Of course, one problem with defining sensitivity was engineers around the world didn't get together and decide on one set of terms to classify soil sensitivity. In the United States we normally call clays with $S = 2$ to 4 low-sensitivity clays, and clays with $S = 8$ to 16 high sensitivity clays. In between are medium sensitivity clays and over 16 are extremely sensitive clays. In Canada and Scandinavia where they have extremely sensitive clays they use different classifications.
   Although use of the sensitivity concept considers only clays, this idea of loss of strength as larger and larger shearing strains remold soil fabric was extended to sands during development of critical-state soil mechanics theories.

2. **Critical void ratio and the effect of the soil's initial density and structure on stiffness and volume changes.**

   Dense soils dilate during material shearing, that is the sample increases in volume, and loose soils compress during shearing, that is the sample decreases in volume. A sample that maintains a constant volume during shearing is a special case where the soil is at the critical void ratio. Density of a soil has a significant effect on its shearing strength. Granular soils have been reported to have effective friction angles that vary as much as 15 degrees from their loosest to their densest state. Additional discussion of the critical void ratio and the steady state line are included in Section 2.5.

3. **Effect of the soil's confining stresses (i.e., σ_2 and σ_3) during consolidation and shearing.**

   Triaxial shear test samples in the laboratory have equal confining stresses all around (i.e., $\sigma_2 = \sigma_3$) due to the hydraulics used to apply confining stresses. The consolidation test has equal horizontal stresses all around its soil sample because the consolidation device is a rigid ring. In the field, it is unlikely that confining stresses in the perpendicular lateral directions are equal like the confining stresses used in soils laboratory testing. Does it matter? Yes it does.

   Assume that the vertical stress is the major principle stress, σ_1, and that the horizontal stresses are the minor principle stresses, σ_2 and σ_3. See the discussion on plain strain shearing in item 4 below, but for now assume that the horizontal stress σ_2 is along the length of an infinite slope and that the horizontal stress perpendicular to the length of the slope is stress σ_3. If the slope is in a state of active stress, the horizontal stress σ_3 is equal to $(K_a)(\sigma_1)$. Along the length of the slope the soil is in a plain strain stress state, and so σ_2 is equal to $(K_o)(\sigma_1)$. If the internal friction angle of the soil is equal to 30 degrees, then K_a is equal to 0.33 and K_o is equal to 0.50, which can be interpreted as indicating that the ratio of σ_2 to σ_3 is 0.50/0.33 or the intermediate principle stress σ_2 is 1.5 times greater than the minor principle stress σ_3. Soils with a friction component to their shear strength have greater strength if they have greater confining stresses. Since triaxial samples have $\sigma_2 = \sigma_3$ and our slope's plain strain case has $\sigma_2 > \sigma_3$ I would expect that the plain strain stresses existing in our field slope example would have greater strength than the laboratory triaxial tests. See the discussion of plain strain stresses discussed below in item 4, reported examples indicate that most typical soils sheared in plane strain (when $\sigma_2 > \sigma_3$) do have shear strengths greater than the same soils tested in triaxial shear (Lade and Duncan, 1973).

   What if you have estimated the confining stress σ_3 and assumed that the intermediate stress $\sigma_2 > \sigma_3$, and it turns out that $\sigma_2 < \sigma_3$, or in other words that σ_2 is actually the minor principle stress? Your triaxial test results may over-estimate the soil strength in this case. It is good practice to consider the intermediate principle stress if at all possible.

4. **State of stress changes such as compression, tension and "pure shearing," or triaxial stresses versus plain strain shearing.**

Given three identical clay samples tested in triaxial compression, triaxial tension, and direct simple shear (DSS) "pure shear" states of stress, the shear strength of the three samples will be different. Ladd's plot of data illustrates in Figure 15 in the 22nd Terzaghi lecture on page 573 (Ladd, 1991), that a low-plasticity clay with a plasticity index of 10 has a DSS strength that is 72% of the triaxial compression value, and a triaxial extension strength that is 53% of the triaxial compression value.

Most if not all commercial and research soil shear strength tests are performed on relatively small cylindrical samples tested in triaxial, direct shear, or direct simple shear devices. In the field, where footings can be tens of feet long, and landslides are very wide, the stress conditions are closer to plain strain (i.e., primary motion in one lateral direction with zero or near-zero strain in the perpendicular lateral direction). Published data comparing triaxial shear friction angles ϕ_{ts} with plain strain shear friction angles ϕ_{ps} for sandy soils (Lee, 1970) show that plain strain friction angles can be up to 8 degrees greater than triaxial test values. The greatest difference between plain strain and triaxial friction angles is found in dense granular soils tested at low confining pressures. I like to visualize plain strain as the long direction supporting the soil that is shearing in the short direction. Since the soil is being supported in one lateral direction resulting in no strain, it is stronger and has less volume change than a soil in the triaxial state of stress. A good summary of plain strain behavior of granular soils is included in the second edition of Holtz and Kovacs book in Section 13.11 (Holtz, Kovacs, and Sheahan, 2011).

5. **The soil's degree of saturation at initiation of loading.**
   If a soil is unsaturated at the initiation of loading, it has what I call dry strength. Dry strength is really the soil strength derived from soil suction in unsaturated soils. If unsaturated soils are wetted and lose their dry strength, the soil will experience self-weight settlement, that is, settlement without adding surcharge loading to the soil. Many cases of house settlement distress in dry regions of the southwestern United States are due to this type of collapse upon wetting settlement. We will discuss unsaturated soils further in Section 2.7.

6. **Presence of free water to come into or to flow out of a sample during shearing.**
   If a sample dilates during shearing, it is increasing in volume. If the sample is saturated, but is not in contact with free water, an increase in its volume could result in air being drawn into the sample, making it unsaturated. In a granular soil this could result in increased shear strength due to increased soil suction. Presence of free water that is available to maintain soil saturation is also a very important factor related to potential frost heave of freezing soils. If freezing soil has limited access to free water, its potential to heave during freezing is also limited.

7. **Environmental effects such as pH, temperature gradients, ion gradients, electrical gradients, and hydraulic gradients of fluids and gases. This could also include effects of bacteria and vegetation on the soil structure.**
   Environmental effects on soil behavior are a special class of soil issues. I recommend that you get a copy of Jim Mitchell's soil behavior textbook to study these affects (Mitchell, 1976, 1993, and Mitchell and Soga 2005).

2.3.6 Additional Information – Non-Linear Failure Envelopes

We started into this topic earlier, but now is a good time to get into more detail. Why is the Mohr–Coulomb failure envelope curved? The simple answer is that the shear strength of soil used to construct the Mohr–Coulomb failure envelope is a non-linear function of many variables.

The first shear strength variable considered in Mohr–Coulomb equation is the normal stress. Vertical normal stresses are not constant in the ground. With depth the weight of soil becomes greater and the vertical normal stresses become greater. As the vertical stresses become greater with depth, they exert increasingly greater confining stresses on the soil ($\sigma_{horiz} = K \sigma_{vertical}$). A homogeneous uniformly graded soil becomes stronger with depth as the confining stresses increase. Stronger soil with depth, does that mean that just the confining stress is increasing or is the internal friction angle ϕ changing? Is the friction angle constant, is it increasing, or is it decreasing with depth? This effect of increasing soil strength with depth can be included in the shear strength equation by using a function of ϕ that changes with depth.

To include the curved nature of the soil failure envelope, several engineers have proposed equations that modify the friction angle as a function of the confining stress, $(\sigma_2 + \sigma_3)/2$, or the average normal stress, $(\sigma_1 + \sigma_2 + \sigma_3)/3$. There is a body of evidence that indicates that the effective stress friction angle ϕ' in granular soils does not increase as you might expect, but it actually decreases as the confining stress increases, (Duncan and Wright, 2005, p. 36). This reduction of ϕ' with increasing confining stress results in a curved strength envelope. The equation often used for representing this reduction in ϕ' is given below as Equation 2.3.4.

$$\phi' = \phi'_o - \Delta\phi' \log_{10} \frac{\sigma'_3}{p_a} \tag{2.3.4}$$

In Equation 2.3.4, ϕ' is the secant effective stress friction angle, ϕ'_o is the reference effective stress friction angle for a confining stress (σ_3) equal to one atmosphere (2116 pounds per square foot), and $\Delta\phi'$ is the reduction of effective stress friction angle for a 10-fold increase in confining stress (i.e., from 2000 to 20 000 pounds per square foot). The term p_a is equal to standard atmospheric pressure or 2116 pounds per square foot. The term p_a is used to normalize the confining stress σ_3' by making the term inside the logarithmic function dimensionless.

References

Duncan, J.M. and Wright, S.G. (2005) *Soil Strength and Slope Stability*, John Wiley & Sons, Inc., Chapter 5, pp. 35–55.

Holtz, R.D., Kovacs, W.D., and Sheahan, T.C. (2011) *An Introduction to Geotechnical Engineering, Chapter 13 – Advanced Topics in Shear Strength of Soils and Rocks*, 2nd edn, Prentice Hall Pearson, pp. 614–764.

Ladd, C.C. (1991) Stability evaluation during staged construction. The 22nd Karl Terzaghi Lecture, American Society of Civil Engineers. *Journal of Geotechnical Engineering*, **117**(4), 537–615.

Lade, P.V. and Duncan, J.M. (1973) Cubic triaxial tests on cohesionless soil. American Society of Civil Engineers, *Journal of the Soil Mechanics and Foundations Division*. **99**(SM10), 793–812.

Lee, K.L. (1970) Comparison of plane strain and triaxial tests on sand. American Society of Civil Engineers, *Journal of the Soil Mechanics and Foundations Division*. **96**(SM3), 901–923.

Mitchell, J.K. (1993), *Fundamentals of Soil Behavior*, 2nd edn, John Wiley & Sons, Inc., 437 pages.

Mitchell, J.K. and Soga, K. (2005), *Fundamentals of Soil Behavior*, 3rd edn, John Wiley & Sons, Inc., 577 pages.

2.4

Shear Strength Testing – What is Wrong with the Direct Shear Test?

2.4.1 Introduction to Direct Shear Testing

Geotechnical texts often state that you should use some form of the triaxial test to determine the drained or undrained shear strength of soil. The direct shear test is dismissed as having an uncontrolled stress state due to rotation of principle stresses.

The direct shear test is simple, inexpensive, and doesn't require an 18-inch thick uniform, high-quality sample (Lambe and Whitman, 1969, see page 120). If you are testing remolded samples, such as sands for compliance with Federal Highway guidance on MSE (mechanically stabilized earth) reinforced backfill, it's easier to compact three similar samples of sand for a direct shear test than prepare three comparable samples for triaxial testing. Besides, FHWA publication FHWA-NHI-10-024, November 2009, recommends use of the direct shear or triaxial tests to evaluate the friction angle of reinforced fill material (specifying that it be at least 34 degrees, as determined by the direct shear test). Earlier versions of the FHWA MSE design guide required use of the direct shear test only, and didn't even mention use of the triaxial shear test.

So the question is: What is wrong with the direct shear test and why worry about rotation of principle stresses? You might not recall from college days what rotation of principle stresses is, so I'll refresh your memory.

2.4.2 Direct Shear Rotation of Principle Stresses

Principle stresses on a soil element are tension or compression stresses acting on planes that have zero shear stresses. In Figure 2.4.1, stresses σ_{xx}, σ_{yy}, σ_{zz} are normal

Geotechnical Problem Solving, First Edition. John C. Lommler.
© 2012 John Wiley & Sons, Ltd. Published 2012 by John Wiley & Sons, Ltd.

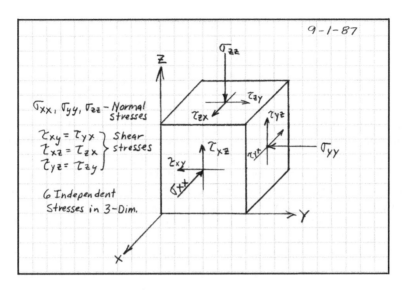

Figure 2.4.1 Normal and shear stresses on soil element – 1987 sketch

stresses, meaning they act perpendicular to the x, y, and z planes. These normal stresses are principle stresses by definition when $\tau_{xy} = \tau_{yx} = \tau_{xz} = \tau_{zx} = \tau_{yz} = \tau_{zy} = 0$, that is when all of the shearing stresses on the x, y, and z planes are equal to zero. If you are like me, you have probably forgotten the stress subscript convention, so I'll give it to you for clarity: σ_{ij} is a stress where i is the direction of the normal to the plane where the stress acts, and j is the direction that the stress is acting. For example σ_{xx} is a stress acting on the plane that is perpendicular to the x-axis, and this stress is acting in the x-direction. The stress τ_{xy} (which we could call σ_{xy}, but I prefer using the tau symbol because it is a shearing stress) is acting on the plane that is perpendicular to the x-axis, and it is acting in the y-direction. When the subscripts i and j are the same, like xx, yy, or zz, the stress is a normal stress, and when the subscripts are different, the stress is a shearing stress.

You might notice that Figure 2.4.1 is a hand-drawn and hand-lettered figure. This figure is from my 1 September 1987 class notes for advanced soil mechanics. This was a few years after the Cleveland Museum of Art installed a sculpture by Isamu Noguchi, and I went to see Noguchi and the accompanying exhibition of his work. Noguchi said that his work always included the H-factor. Afraid to ask the master himself, I asked my sculpture professor friend Carl Floyd, "What is the H-factor?" Carl told me that Noguchi's sculptures were so perfect that he had been accused of making them by machine. So to prove that a human hand was involved, he always included a defect or chop mark on the sculpture's surface to confirm that a human had made it. The H-factor meant that it was done by a human. Consider Figure 2.4.1 my inclusion of the H-factor to this work.

OK, let's get back to the direct shear test. When you start a direct shear test, the vertical compression stress (i.e., σ_{zz} in the z-direction) is a principle stress because the horizontal plane perpendicular to the z-axis has no shear stress. Lateral bulging in response to the vertical sample loading causes lateral stresses on the sample (i.e., in the x-and y-directions). Lateral stresses in the x-and y-directions are the other two principle stresses at the beginning of the test because there are no shearing stresses on vertical planes through the sample. I might mention here that we can calculate the magnitude of the vertical normal stress in the z-direction at the beginning of the test, but we don't know the magnitude of normal stresses in the x- and y-directions in a direct shear device. Some people assume that the lateral stress in the direct shear sample is $(K_o) (\sigma_v)$, but hold on now, how do you know what K_o is before you run the test when you don't know what the friction angle ϕ is prior to testing three samples and drawing the failure envelope. You need to know ϕ to determine $K_o = 1 - \sin \phi$. Then again, how do you know that the direct shear sample is in an at-rest condition in the direct shear box? If the sample is clayey enough to be extruded into the direct shear box, there will probably be a gap between the sample and the sides of the box. Having a gap between the sample and the shear box will surely not replicate field at-rest conditions in the laboratory test. If the sample is a cohesionless sandy sample, it will likely have to be compacted in the laboratory in the shear box, again resulting in a different compacted lateral stress in the shear box than in the field. If you have an *in situ* sandy sample that is suitable for extruding into the shear box, or compact a sandy sample in a mold and then extrude it into the shear box, you will not likely replicate field at-rest conditions. During shearing of a direct shear sample there are serious lateral stress concentrations that are not quantifiable. So in conclusion, I prefer to say that we do not know what the lateral normal stresses are in a direct shear test.

As the direct shear test starts, a lateral shearing force is applied to the soil sample, which generates shearing stresses on a horizontal plane through the sample. By definition, the vertical compression stress on the horizontal plane can no longer be a principle stress because shearing stresses on the horizontal plane are no longer zero. The principle stress direction has rotated from vertical, as shown in Figure 2.4.2. If the

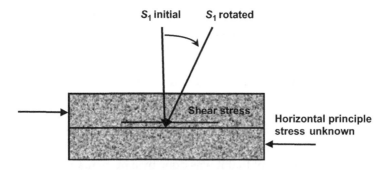

Figure 2.4.2   Direct shear test – rotation of principle stress when shearing

stress state in your foundation problem maintains the orientation of principle stresses in the vertical and horizontal directions, the direct shear test does not replicate the stresses generated during loading of your footing. Since the soil's internal friction angle measured during a laboratory test depends on the state of stress generated during the test, it is important to replicate as closely as possibly your real-world problem's state of stresses in your laboratory tests.

2.4.3 Use of the Direct Shear Test to Determine Internal Friction Angle, ϕ

Many soil material specifications include required values of the internal friction angle ϕ that must be equaled or exceeded. An example is the FHWA mechanically stabilized earth (MSE) wall reinforced backfill, which is specified to have an internal angle of friction equal to or greater than 34 degrees. These types of specifications have generated a problem in the industry and considerable confusion among engineers, contractors, owners, and yes I regret to include, lawyers. What am I talking about? Didn't you think that the internal friction angle of a given soil was a fixed, characteristic soil property?

Where do I start? First I'll answer the question directly, and then I'll explain. The answer is no, the internal friction angle ϕ is not a characteristic soil property. The value of ϕ depends on the shear strength tests used, how the tests are run (saturated versus unsaturated, slowly versus rapidly, drained versus undrained, to name a few), density and fabric of the soil in the lab versus in the field, stress state in the lab versus in the field, and shearing strains generated in the lab versus the actual strains experienced in the field.

Before we go on, I'd like to explain what those tables of soil types with listed friction angles in standard geotechnical references mean. I know that they surely look like prescriptive internal friction angles for listed soil types. For an example, let's go to page 107 of Terzaghi and Peck's 2nd edition of *Soil Mechanics in Engineering Practice* and look at their Table 17.1 (Terzaghi and Peck, 1967). Representative values of internal friction angle ϕ are given for round-grained uniform sands, angular-grained well-graded sands, sandy gravels, silt sand, and inorganic silt. Values given in Terzaghi and Peck's Table 17.1 (Terzaghi and Peck, 1967) for each soil type are listed for both the loose and dense states. These reported drained shear, friction angle values are peak values for effective vertical stresses less than 10 000 pounds per square foot. Although not discussed, it is likely that these internal friction angles are from direct shear tests with vertical effective stresses greater than about 500 to 1000 pounds per square foot (Note: I know that Terzaghi preferred triaxial shear tests, but my experience is that most reported friction angle values in early texts are from direct shear testing). At the time these table values were generated, it was not common to determine high-strain, residual values of internal friction angle, and so many engineers just took the friction angle from the loose state as the residual, high-strain value. Many laboratories doing soil testing in the 1950s and 1960s were using direct shear equipment that did not have the capability of measuring shear strength values beyond the peak value. Also at this time the values of ϕ were reported to be the same for wet and dry samples.

2.4.4 The Direct Shear Test – Details

The direct shear test is still the commercial laboratory test of choice for testing granular soils for most geotechnical applications. The ASTM standard for direct shear testing is ASTM D 3080, and the AASHTO standard for this test is T-236 (AASHTO, 2006). According to AASHTO, T-236 is similar, but not technically identical to ASTM D 3080. The only difference that I'm aware of between ASTM D 3080 and T-236, is sample saturation. The AASHTO tests are all required to be saturated prior to testing, and the ASTM direct shear tests can be dry, moist, or saturated, so long as the engineer specifies test conditions that replicate field conditions.

The direct shear test is intended by specification to be a consolidated drained shear test, although the rate of shearing of fine-grained samples has to be appropriately slowed to prevent generation of excess pore water pressures during testing.

The standard test is conducted at a controlled rate of horizontal shearing on a sample held between a top and bottom portion of a shear box, see Figure 2.4.3. The shear plane is assumed to be a single horizontal failure surface generated between the top and bottom sections of the shear box. Observation of the actual shearing surface after conducting a direct shear test indicates that it is not planar, but more of a "concave up" curved surface. The curved nature of the actual failure surface in

Figure 2.4.3 Direct shear device

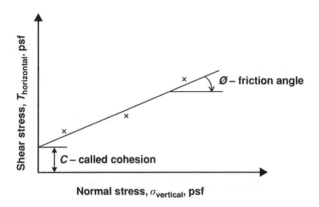

Figure 2.4.4   Direct shear plot – three tests

the direct shear test is explained by rotation of principal stresses that changes the orientation of sample failure stresses as the test progresses, thus causing changes in the orientation of the failure surface.

The direct shear test can be modified to apply gravity-actuated fixed horizontal loads to evaluate creep properties of clayey soils, but such tests are considered to be special, engineering tests and are not covered by standard ASTM or AASHTO specifications.

Results of a single direct shear test are not often used alone, but rather three or more tests at varying vertical consolidating stresses are performed, and the peak shear strength values from these tests are plotted on a graph of vertical stress (on the x-axis) versus shearing stress (on the y-axis) to determine the soil's internal friction angle. The soil's friction angle is defined as the slope of the line through the plotted shear stress versus vertical stress points, see Figure 2.4.4.

Please note that since you don't know what the principle stresses are at sample failure, you cannot plot Mohr's circle of stresses for the direct shear test. Also note that since the direct shear test measures horizontal deflection during loading and not horizontal shearing strain, a plot of stress versus strain cannot be generated from this test. This is a significant drawback, because determination of the shear modulus from a shearing stress versus shearing strain plot would be useful. Attempts to design a square direct shear box to allow determination of shearing strains have not been successful due to non-uniform distribution of strains on the shearing plane, and changes in stress orientations during the test. Yes, it's the old rotation of principal stresses issue again.

Since by both ASTM and AASHTO standards, the design engineer must specify test conditions to replicate field conditions, there can be no such thing as a standard direct shear test. You can't call up the lab and order one like a cheeseburger. You need to specify the vertical stresses for the three test points, when to consider the vertical consolidation to be complete, when to saturate the shear box, and what shearing rate of loading to use. Although not included in some soil laboratories' test reports, you should require plots of vertical dial gauge readings both during consolidation of the sample and during shearing of the sample. You should also request a plot of lateral deformation versus lateral shearing stress for all tested samples. In addition to the vertical stress versus horizontal shearing stress plot of points used to determine the internal friction angle, you should also have laboratory data for the dry unit weight and degree of saturation of each sample both before and after shear testing.

2.4.5 How Can the Direct Shear Test Go Wrong?

There are likely dozens of things that can go wrong in a direct shear test, but I'll focus on our part first, the engineer's issues. I mentioned it before, but I'll say it again, you can't order a direct shear test from a laboratory like you order a ready to eat cheeseburger from McDonald's. You have to specify the details of the test, the consolidation vertical stresses, shearing rate, sample saturation, and so on. I suggest you closely review ASTM D 3080 or AASHTO T-236 before you send your direct shear order to the laboratory.

One example of misapplication of the direct shear test that sticks in my mind is one that resulted in a costly failure for a local engineer (not me). The fill soil proposed for backfill was an on-site clay shale material. A good granular backfill may have been an easy choice, but costs were a significant project factor and the engineer thought, "if the direct shear test gives me a high enough friction angle I can design around it." The engineer had samples of on-site clay shale material delivered to a soils laboratory and he asked for a direct shear test. He didn't specify any of the required details of the test to simulate anticipated on-site conditions. The laboratory technician did his best and guessed that since this was going to be a fill soil application that the test samples should be compacted to 95% standard Proctor at 2% below optimum moisture content. The direct shear test was done on three samples using vertical stresses that seemed good to the laboratory technician. The samples were run "saturated" in that the laboratory technician filled the chamber around the direct shear samples with water immediately before shearing. The result of this test was a reported friction angle of 45 degrees.

Upon review of the direct shear test results and after checking his college textbook, the engineer called the soils laboratory to suggest that the test was in error because a clay shale material has a friction angle less than 45 degrees. He questioned the laboratory technician about saturating the samples, and recommended that he wait a while for water to soak in before shearing the clay shale soils. The engineer then directed the technician to rerun the direct shear tests. The laboratory technician reran the tests exactly the same as before, but he added water and waited two hours before shearing each sample. When asked later why he waited two hours for the water to soak into the clay shale samples, he said that he wanted to finish the tests in a single work day. Results of the second set of direct shear tests was a reported friction angle of 24 degrees. Upon receiving results of the second set of direct shear tests, the engineer was satisfied that now he had a reasonable friction angle value for design. He designed the fill structure using the 24 degree friction angle, and about a year later the structure failed. During forensic analyses of the failed structure, direct shear tests were performed on the clay shale material, and several sets of samples were soaked until the vertical dial gage indicated that swelling had stopped, which took about 36 hours per sample. The reported friction angle of this third set of tests was 11 to 12 degrees.

What can we learn from this application of the direct shear test to design? If the vertical confining stresses are too high, that is they are much higher than will be experienced in the field, the indicated swell during saturation of samples can be reduced or eliminated. What can be a real problem is if two tests do swell and the third one has a high vertical stress and doesn't swell. The result is that the soil will swell in the field but not in the test. Another problem with excessive vertical test stresses is that the first test and second test may dilate during shearing, but the third test does not dilate due to high stresses. The result is three points that don't come close to a straight line. Time is very important in direct shear tests of clays and clay shales. If the samples don't soak long enough to reach equilibrium, they will not fully soften. Also if these highly impermeable samples are sheared too rapidly, they will generate excess pore pressures and will not represent the consolidated drained conditions that you expect from a direct shear test.

2.4.6 Evaluating Results of Direct Shear Tests

When evaluating the results of direct shear tests, your main concern has to be uniformity of the three test specimens. If they are not identical, you cannot draw a line through the three test results and compute shear strength parameters, c and ϕ. Shear strength of the three test specimens used in the direct shear test should be a function of vertical effective stress and nothing more. If other unrecognized variables are present that affect the shear strength of one or more of the test samples, the shear strength parameters developed by the test are affected in undefined ways.

Check that the dry unit weight of all three samples is equal (or very close to equal) prior to applying the vertical loading. Dense samples are stronger than looser samples. If you have to fit a line through three points, weigh your decision based on dry unit weights.

Check that all three samples are handled and processed the same. For remolded soils, all three samples should be processed and compacted by the same process. For *in situ* samples, they should be as similar as possible, and they should all be handled in exactly the same way for each test. Changes in compacting samples in the shear box or in handling and inserting samples into the shear box will cause disturbance and affect soil fabric, which changes the soil shear strength.

Check to be sure that all three samples have reached equilibrium after consolidating and after flooding the shear box. After flooding, the vertical displacement dial gage readings should be taken until the sample stops moving. Make sure the laboratory technician does not flood the samples and test immediately, or flood the samples and wait 15 minutes. It is not the lab technician who should define the correct sample soaking time, it should be the sample! Guidance for consolidation and soaking times are given in ASTM and AASHTO standards.

The lab should report the vertical dial gage readings during sample soaking and during shearing. The lab should also plot horizontal shearing stress versus lateral deflection for your study and inclusion in the geotechnical report. In a constant rate of shearing test, the test should be run until a constant residual shearing stress is achieved. If the lateral displacement reaches 10% of the sample diameter, the test should be stopped to prevent excessive reduction of the sample's shearing cross-sectional area. For a standard test, pick the peak shearing stress for construction of the shearing stress versus normal stress plot, see Figure 2.4.4. If you need a residual shear strength value, you can plot the shearing stress at maximum deflection, but I doubt you would have a true high strain residual value (Duncan and Wright, 2005, page 50), or you can do a stress reversal loading where the sample is sheared back and forth until a high strain residual shearing stress value is reached. I know what you're thinking, "What about rotation of principle stresses when you reverse shearing directions back and forth several cycles?" I know it's not the best large strain test, because it is highly unlikely to represent field conditions, *but* it is better than nothing and it does provide residual values that approximate high strain shear strength.

2.4.7 Concluding Remarks about the Direct Shear Test

The direct shear test is cost effective and useful when you don't have enough identical soil samples to perform triaxial testing. The direct shear test has several pitfalls and can be misleading if you are not careful when specifying test conditions and interpreting test results. Like everything else in geotechnical engineering, the direct shear test is OK if you use it appropriately. Unlike many other standard ASTM or AASHTO tests, the direct shear test requires several test conditions be specified by the project geotechnical engineer. This is because the direct shear test conditions should simulate or replicate field conditions for the project specific application as closely as possible (within reason). This required interaction between the geotechnical engineer and the laboratory is often misunderstood or ignored by both the engineers and laboratory personnel. Rotation of principle stresses will occur in your direct shear tests, but remember that some data is better than no data. If you are careful and don't believe that direct shear values of c and ϕ are exact models of your soil's shear strength, you will be able to use direct shear data in your analyses and designs.

References

Standard Method of Test for *Direct Shear of Soils Under Consolidated Drained Conditions*, AASHTO Designation T 236-03 and ASTM Designation D 3080-72 (1979).

Duncan, J.M. and Wright, S.G. (2005) *Soil Strength and Slope Stability*, John Wiley & Sons, Inc. Direct Shear Test, pp. 50–53.

Lambe, T.W. and Whitman, R.V. (1969) *Soil Mechanics*, John Wiley & Sons, Inc. , 553 pages.

2.5

What is the Steady State Line?

2.5.1 Introduction to the Steady State Line

It might be a bit of a stretch, but I'm leaning toward the opinion that understanding of the steady state line is the doorway to advanced geotechnical understanding and advanced geotechnical practice. To predict soil response to a change in stresses you have to understand the interaction of several variables, including stress-history-induced soil fabric, changes in normal stresses, changes in shearing stresses, and changes in soil volume. The steady state line helps you start to understand these interactions (Altaee and Fellenius, 1994).

Let me say right now that one line is not going to solve the problem of characterizing soil performance. By definition a function drawn on an x–y plot can only consider two variables. A three dimensional x–y–z plot can be drawn, but it only considers three variables. If four or five variables are involved, you can't use a single simple black and white plot, although I'm aware that computer visualization experts use colors, textures, and vector arrows to enhance three-dimensional plots. I am a fan of the Peacock diagram, which is a three-dimensional state diagram shown in Holtz and Kovacs (1981, 2011), but it has been my experience working with students and practicing engineers that a series of two-dimensional plots works better than three-dimensional plots when developing an understanding of soil performance concepts. Use of two-dimensional plots better fits the graded approach.

Let me introduce you to the steady state line. The steady state line (SSL) is a plot of the critical void ratio of a soil versus the $\log_{10}$ of the confining pressure or the mean normal stress , see Figure 2.5.1. The critical void ratio is the void ratio of a soil that is in a steady state of shearing deformation and experiences no change in volume, no

Geotechnical Problem Solving, First Edition. John C. Lommler.
© 2012 John Wiley & Sons, Ltd. Published 2012 by John Wiley & Sons, Ltd.

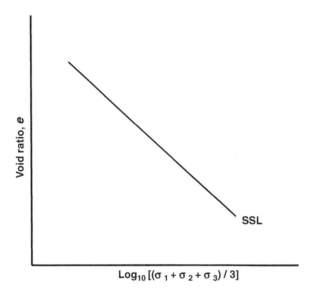

Figure 2.5.1 The steady state line, SSL

change in effective normal stresses, no change in shearing stresses, and deforms at a constant velocity (Poulos, 1981).

Recall from Section 2.3 that in general we may consider loose soils as those that compress during shearing and dense soils are those that dilate during shearing. If shearing continues from either a loose or a dense state, the soil eventually reaches the critical void ratio when shearing strains are large. Looking at Figure 2.5.1, you probably notice that the SSL plot looks similar to the classical consolidation plot.

2.5.2 Hydrostatic Stresses and Volume Changes

Critical state soil mechanics principles (Schofield and Wroth, 1968) were originally developed for clays and then extended to sandy soils. I currently live in a sandy soil region, so I'll explain the concept using sand first and then we will turn to clay. I think it will be easier for you to visualize the critical state concepts for sand, because clay has more issues to consider than sand. What do we mean by a soil state and how can it help you understand soil behavior? Let's start from the beginning.

Imagine a cubic foot of sandy soil that is confined by an all-around hydrostatic pressure, that is, where all normal stresses are equal and there are no shearing stresses. In this case, the three normal principle stresses on the x-, y-, and z-axes are equal compressive stresses, $\sigma_z = \sigma_y = \sigma_x = p_{hydrostatic}$ (a constant), see Figure 2.5.2(a). You might notice that my imaginary one cubic foot of sandy soil with equal all-around

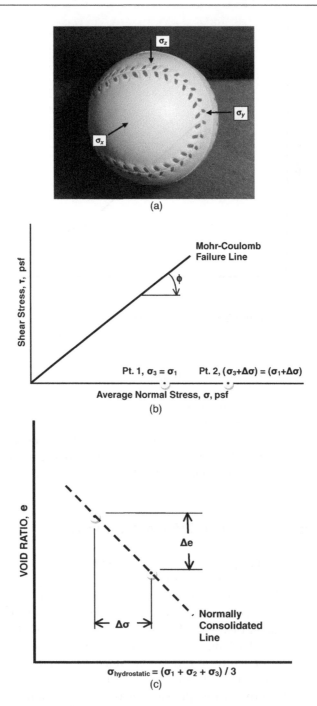

Figure 2.5.2 (a) Hydrostatic stresses equal all around the soil element. (b) Mohr's circle plot of hydrostatic compressive stresses. (c) Plot of increased hydrostatic stress on void ratio versus confining stress curve

compressive stresses looks like a baseball. It's late winter, almost spring, and I'm ready for baseball.

If you plot this initial state of hydrostatic stress on a Mohr–Coulomb plot, see Point No. 1 to the left on Figure 2.5.2(b), you will notice that the plot is a single point, since vertical and horizontal normal stresses are equal. This stress state of hydrostatic compression generates no shearing stresses, because the Mohr's circle has a radius of zero.

If you allow drainage from the soil and start at a low hydrostatic state of stress and then increase the hydrostatic state of stress, you cannot fail the soil sample because you haven't generated any shearing stresses. The result of increasing hydrostatic stresses in compression while allowing drainage of pore fluids is a reduction in soil volume, reduction in soil void ratio, reduction in water content, and an increase in soil unit weight and density. If the soil void volume is reduced and the soil solids volume remains the same, the soil will experience an increase in modulus and strength. On the Mohr–Coulomb plot the point moves to the right, as shown by Point No. 2 on Figure 2.5.2 (b), but the increases in soil strength and reduction in void ratio are not apparent. This is a result of the problem I mentioned earlier, that multiple numbers of variables cannot be represented on a single two-dimensional plot. What we need is a second plot of void volume change versus confining stresses. Rather than use void volume, we use void ratio on the vertical axis because the solid volume is constant and the void ratio, $e = V_{voids}/V_{solids}$ is dimensionless. On the horizontal axis, we use the average normal stress [$\sigma_{ave} = \sigma_{hydrostatic} = (\sigma_x + \sigma_y + \sigma_z)/3$] which is equal to the hydrostatic stress. Checking Figure 2.5.2(c) we note that the void ratio decreases when the average confining stress increases.

If we were to plot Figure 2.5.2(c) as void ratio versus the logarithm of the vertical effective stress, we would have what looks like a standard consolidation curve. If we plotted the void ratio versus the logarithm of the consolidating pressure in a triaxial test, we could use $\log_{10}(\sigma_3')$ or $\log_{10}[(\sigma_2' + \sigma_3')/2]$ or $\log_{10}[(\sigma_1' + \sigma_2' + \sigma_3')/3]$ because they are all equal when a hydrostatic pressure is used to consolidate our soil sample. You will see all three of these values used on the horizontal axes of e versus log p plots in text books if you look at several of them. Not to worry, they are all used to explain the same concepts.

Why do we use the vertical consolidating effective stress, σ_1' in the standard consolidation test plot rather than the average consolidation stress? We use the vertical effective stress when plotting consolidation test data because we don't know what the lateral stresses are in a consolidation test. All we do know is that the sample is confined in a rigid ring in a state of confined compression and the lateral strains during loading are approximately equal to zero. I said approximately equal to zero, because undisturbed samples extruded into the consolidation ring may have a slight gap between the soil and the ring at the start of the test that requires some initial seating strains during initial sample loading. During loading of the sample in the consolidation test, we do not know the lateral stresses, but we do know that they are not equal to the vertical consolidating stress. When $\sigma_1 \neq \sigma_2 = \sigma_3$ the state of consolidating stresses is call the anisotropic state of stress, that is they are no equal. The hydrostatic state of stress that we described above is called the isotropic state of stress, that is, they are equal. If no shearing stresses are generated in the soil sample during isotropic consolidation, then an anisotropic state of consolidation does generate shearing stresses in the sample. If the state of stresses generated during consolidation of a sample is different, the soil fabric developed during sample consolidation is different. We will discuss the topic of anisotropic affects on soil shear strength and its affect on slope stability in Chapter 5.

2.5.3 Shearing Stresses and Volume Changes

Consolidating soils under hydrostatic compression increases their density and strength, but you have no idea of the resulting soil strength gains based on plots of void ratio versus consolidating stress. Soil strength can only be evaluated by increasing shearing stresses in a shear strength test. Shear strength of the two sample points plotted on Figure 2.5.2(b) can be tested in triaxial compression by maintaining the horizontal stresses σ_2 and σ_3 equal to the hydrostatic confining stress, while increasing the vertical stress σ_1. To evaluate when a soil sample has failed in triaxial compression, we use a third plot of vertical compression stress versus vertical compression strain, as shown in Figure 2.5.3 below.

Notice in Figure 2.5.3 that the two samples have different strengths. Sample No. 1 with the lower consolidation pressure has a lower compressive strength, and Sample No. 2 with the higher consolidation pressure has a higher compressive strength. Both

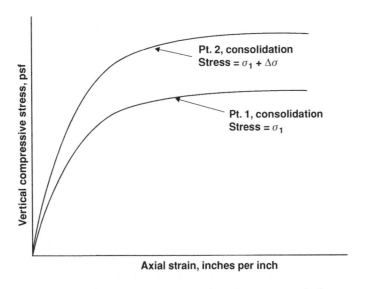

Figure 2.5.3 Vertical compression stress versus strain

samples have similar stress versus strain curves because they were both normally consolidated. By normally consolidated I mean that both samples were never subjected to consolidating stresses greater than that shown in Figure 2.5.2(b). Note that both curves in Figure 2.5.3 initially have a steeper stress–strain curve that reaches a maximum value and then flattens out. Samples that reach a maximum compressive strength without a pronounced peak value and maintain that maximum value with increasing strains are called ductile materials.

During the initial steeper portion of their stress–strain curves these samples have a soil fabric determined during consolidation, and they act as elastic or near elastic materials. Near maximum compressive strength on their stress–strain curves, these samples start to yield and generate plastic strains. Plastic strains remold or reorder the soil fabric from the initially consolidated fabric to a new yielded soil fabric . The point on the stress–strain curves where the soil fabric transforms from the initial fabric to the plastically remolded fabric is the critical stress point. If the soil fabric remains a yielded fabric with continued high strains, the critical stress point is equal to the residual stress. If very high strains align soil particles in a highly ordered structure, the residual stress reduces from the critical stress to a somewhat lower limit value of residual stress.

To illustrate peak/maximum strength, critical stress and residual stress consider the house of cards in Figure 2.5.4 (a)–(c).

In Figure 2.5.4(a) we have a house of cards with a "soil fabric" that is determined by the way we built it. As we apply a lateral force to the card house structure, it starts to deform horizontally and we can plot lateral force versus horizontal movement. At a relatively low lateral strain (implying it is a brittle structure), the card house structure collapses into a much smaller volume. The card house "soil structure" shown in Figure 2.5.4(b) is a soil structure that has reduced in volume during shearing such that the critical state soil fabric has a lower void ratio than the initial loose structure shown in Figure 2.5.4(a).

Now consider Figure 2.5.4(c) where the card house structure has been subjected to very high horizontal movements which aligned all of the cards reducing the volume of the cards "soil fabric" to a minimal value. The void ratio of the initial soil fabric was very high, shearing stresses were increased to a critical value which remolded the soil fabric and reduced the void ratio to its critical state value. Very high strains aligned the soil particles (cards) in a fabric that reduced shear strength and void ratio to a minimal residual value. Notice in this case of a normally consolidated soil, the initial void ratio e_o was greater than the critical state void ratio e_{cr}, and that the critical state void ratio is very slightly higher than or even equal to the residual state void ratio e_{ss}. If you plot the critical state void ratio for each consolidating stress you will have the critical state line. If you plot the residual state void ratio for each consolidating stress you will have the steady state line. In this case of a normally consolidated granular soil the critical state line is very nearly equal to the steady state line, see Figure 2.5.5. (Note: As we have seen before, these lines may be curved rather than straight over a large range of stresses.)

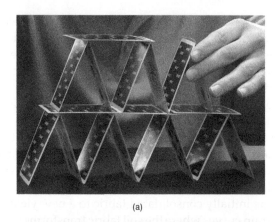

(a)

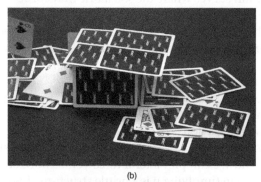

(b)

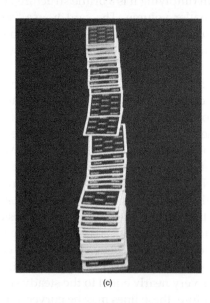

(c)

Figure 2.5.4 (a) House of cards with a "soil-like" fabric. (b) Shearing stresses remolded house of cards structure. (c) High shearing strains aligned cards in residual structure

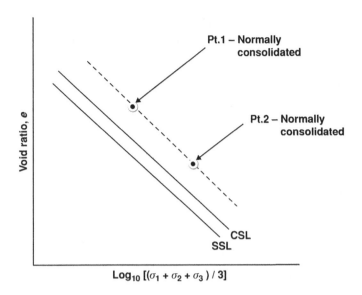

Figure 2.5.5 Steady state line and critical state line with normally consolidated points 1 and 2

Both hydrostatically normally consolidated samples Points 1 and 2 have void ratios that plot above the steady state line. This is a fundamental soil property concept. Any soil sample with a void ratio greater (i.e., plots above) than its steady state and critical state lines will compress or reduce in volume when sheared. Obviously if this soil is saturated and sheared rapidly, the incompressible water cannot drain fast enough to get out of the way of soil fabric compression, the result is generation of excess (i.e., positive) pore water pressures.

Now let's consider shearing of an over-consolidated soil sample. If we take the second soil sample in Figure 2.5.2(b) that has be consolidated by an isotropic stress equal to $\sigma_1 + \Delta\sigma$, and reduce its confining stress to $\sigma_1 + 0$ before starting tri-axial compression loading, the stress versus strain curve will look something like Figure 2.5.6. Notice that the sample's stress–strain curve now exhibits a peak stress value, then reduces to a critical stress value and then at higher strains it reduces further to the residual stress value. Triaxial compression tests of over-consolidated samples exhibit brittle material behavior and have clearly defined peak strength values as shown in Figure 2.5.6.

What do you think happens if a very dense, over-consolidated soil with a void ratio plotting below the steady state line is sheared? Answer: It increases in volume or dilates until it reaches the critical void ratio where plastic shearing takes place and its initial consolidated fabric is remolded. With additional shearing strains the soil fabric becomes aligned and residual soil strength is reached. When a soil is highly over-consolidated, that is when the past consolidation stress is much higher than the

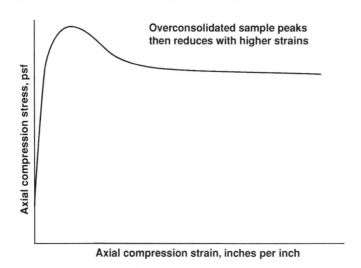

Figure 2.5.6 Stress versus strain curve for over-consolidated sample

current confining stress, the stress strain curve has a pronounced peak value, see Figure 2.5.6.

Let's consider the two points shown in Figure 2.5.2(b), and plot them on the steady state line plot, as shown in Figure 2.5.7 below. If we shear both samples while maintaining their initial consolidating stress, they will both move vertically down (compress) on the stress versus void ratio plot until they reach the steady state line. Once they reach the steady state line they can shear to very high strains at a constant void ratio equal to e_{ss}. Now let's take the second soil sample that was consolidated to higher stress ($\sigma_1 + \delta\sigma$) and reduce its confining stress to that of the first sample, see Figure 2.5.7. Since the stress changes have been a reduction in compression stresses only, shearing has not occurred and point 2 on the steady state line plot moves horizontally until it comes directly below point 1. But now look at Figure 2.5.7 again and notice that point 2 has crossed the SSL and is now located below the SSL. When you shear point 2 with a confining stress equal to the confining stress used to consolidate point 1, the sample dilates or increases in volume and void ratio until it reaches the steady state line. Additional work is required for shearing stresses to remold sample 2's fabric, given a lower confining stress, resulting in a pronounced peak on the stress strain curve (Figure 2.5.6) and the critical state void ratio is somewhat higher than the steady state, residual void ratio, see Figure 2.5.7.

During shearing dilation of the soil, a soil suction develops as the soil volume increases. If unlimited water is available to the sample during dilation it will suck water into its increasing voids to fill the increasing volume until the sample comes into equilibrium. If the soil is partially saturated and water is not available, the soil suction will remain until air or water can be sucked in to neutralize the soil suction. As long as soil suction is active on the soil sample, it acts as an added confining stress

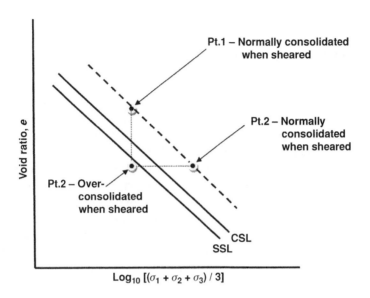

Figure 2.5.7 Shearing of an over-consolidated sample with a void ratio below the SSL

and it gives the soil added strength to the frictional component of strength (shear strength $S = \sigma'$ tan Φ and $\sigma' = \sigma - u$, if u is negative, then $\sigma' = \sigma - (-u) = \sigma + u$). This increasing soil strength due to soil suction is effective until the soil is nearly completely dry or until it is nearly saturated.

2.5.4 Clays – SSL versus CSL

Is the steady state line (SSL) the same thing as the critical state line (CSL)? Although they look alike, they are defined differently as we discussed above for sandy soils. A normally consolidated, sheared sandy soil will likely have a CSL that is either equal to or slightly above (higher void ratio) than the SSL. For normally consolidated clayey samples, the plots of CSL and SSL are also about the same. The major difference between the CSL and the SSL comes into effect when we consider highly over-consolidated, stiff fissured clays and clay shales.

What are stiff fissured clays? Basically they are stiff, hard clays that have a macro system of cracks that allow water to freely flow into and through the clay material. The permeability of stiff fissured clays is not controlled by the low permeability of the clay material, but rather by the cracks in the soil mass. To the best of my experience, stiff fissured clays are always medium- to high-plasticity clay materials that soften when water soaks into the clay material. Can we use our understanding of the SSL to predict the performance of stiff fissured clays? Let's try.

A stiff fissured clay is a high over-consolidated clay so it is likely much denser than a soft normally consolidated clay. Being dense hard clay, the void ratio of stiff fissured clays is likely quite low, and should plot below the SSL (I won't let you guess, this observation is true, the void ratio of stiff fissured clays does plot below the SSL). Since stiff fissured clays' void ratio plots below the SSL, they should experience dilation or increases in volume during shearing. Soil suction generated by dilation during shearing will suck water into the clay material if the soil mass is wetted, because water flows easily through fissures in the clay mass. Water will suck into the clay until soil suction generated during shearing is neutralized. When water is sucked into the clay mass, it soaks into the fissured clay causing it to swell and soften as moisture is drawn into the clay. This bit of logic based on the prediction of shear dilation and suction drawing water into the mass of stiff fissured clay predicts that this soil will soften and lose shear strength with time until the soil becomes fully softened when all suction is neutralized by intrusion of water into the soil mass. This is exactly what happens when excavations are dug into deposits of stiff fissured clays. It often takes many months or even decades for water to fully soak into the soil after an excavation is made, often it takes an especially rainy season or rainy year, but eventually soil softening occurs and many times excavations designed for peak soil strength in stiff fissured clays fail (Skempton, 1964).

References

Altaee, A. and Fellenius, B.H. (1994) Physical modeling in sand. The Canadian Geotechnical Society. *Geotechnical Journal*, **31**, 420–431.

Holtz, R. D. and Kovacs W. D. (1981) *An Introduction to Geotechnical Engineering*, 1st edn, Prentice-Hall, Inc., Englewood Cliffs, New Jersey, 733 pages.

Poulos, S.J. (1981) The steady state of deformation. The American Society of Civil Engineers. *Journal of the Geotechnical Engineering Division*, **107**(GT5), 553–562.

Schofield, A.N. and Wroth, C.P. (1968) *Critical State Soil Mechanics*, McGraw-Hill Publishing Company Limited, London, England, pp. 1–310.

Skempton, A.W. (1964) Long-term stability of clay slopes. Geotechnique, **14**(2), 77–101.

2.6

Static Equilibrium and Limit States

2.6.1 Introduction to Static Equilibrium and Limit States

Static equilibrium is a basic principle of physics that much of civil engineering is based upon. If an object is not moving relative to the earth, it is in a state of static equilibrium. To be in a state of static equilibrium an object must not have any unbalanced forces or moments acting on it. The principle of static equilibrium is the basic principle of the study of statics, which is often the first engineering course that most engineers study. I used to tell my statics students that non-engineers do not carry their arithmetic books around with them for reference because they need to know how to add and subtract wherever they go, and engineers don't carry their statics books with them because they need to know it wherever *they* go!

Analysis of retaining walls, spread footings, piles, and slope stability, to name a few geotechnical topics, all require the summation of forces in the x-, y- and z- directions equal to zero and the summation of moments about the x-, y- and z-axes also be equal to zero. The problem that we have to keep in mind is that we must not forget any of the applied forces or resistances acting on our geotechnical structure.

2.6.2 What Are Limit States?

If we were talking mathematics, limit states would be upper and lower bound values of a function. Knowing maximum and minimum values in any situation is extremely valuable information. If you are trying to lift a large package, and the package is marked with "weight equals 146 pounds," and you know that you cannot safely lift more than 100 pounds without risking injury, then you automatically know that you

Geotechnical Problem Solving, First Edition. John C. Lommler.
© 2012 John Wiley & Sons, Ltd. Published 2012 by John Wiley & Sons, Ltd.

need to find a second healthy person to help you lift the package. In geotechnical engineering we use the limit state concept to define lateral earth pressures. If we know what the lateral earth pressure is when the soil is in a state of tension failure and we also know what the lateral earth pressure is when the soil is in a state of compression failure, we have the upper and lower bound earth pressures defined.

We call the tension failure state or minimum lateral earth pressure the active earth pressure, and we call the compression failure state or maximum lateral earth pressure the passive earth pressure.

In Figure 2.6.1(a), we have a point in the ground 10 feet below grade, pt. 1, that has a vertical effective stress of 1200 pounds per square foot and a horizontal effective

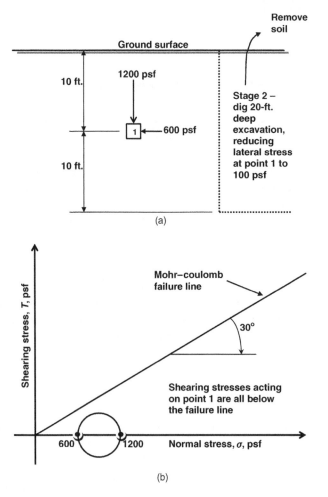

Figure 2.6.1 (a) Vertical and lateral stresses at point 1; (b) Mohr's circle of stresses for point 1; (c) Mohr's circle of stresses for point 1 after excavation lateral stress relief causes shear failure

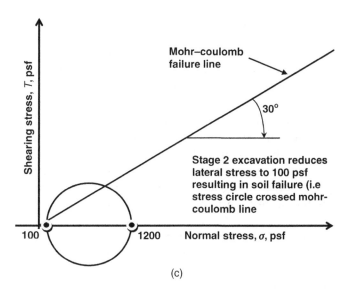

Figure 2.6.1 (Continued)

stress of 600 pounds per square foot. In Figure 2.6.1(b), a Mohr's circle for pt. 1 is constructed. Assuming that the soil around pt.1 is uniform laterally and vertically, and has an internal angle of friction of 30 degrees, the Mohr–Coulomb failure envelope is drawn on Figure 2.6.1(b). Notice that the initial state of stresses at pt. 1 does not generate shearing stresses that equal or exceed the failure envelope line. This means that the soil at pt. 1 is not in a state of failure. Now let's dig an excavation in front of pt. 1 to a depth of 20 feet, and take lateral stress measurements at pt. 1. Assume that the lateral stress at pt. 1 has reduced to 100 pounds per square foot due to the unloading resulting from digging the 20 foot deep excavation. Plotting the new Mohr's circle of stress on Figure 2.6.1(c), we notice that the resulting circle intersects the Mohr–Coulomb failure line. This result means that the soil has or is currently in a state of failure due to the lateral unloading or state of tension stresses induced in the soil adjacent to the excavation. By definition, no point on the Mohr's plot can cross the failure line, because it would fail as soon as the stress point touched the failure plane, see Figure 2.6.1(c). (The measured lateral stress of 100 pounds per square foot is less than the lower bound earth pressure and is theoretically too low.) If you plot the Mohr's circles that are tangent to the failure line, you have an equation that gives the minimum lower bound state of stress defined by the Mohr–Coulomb failure criterion. This equation is given as the active earth pressure coefficient K_a times the vertical soil stress, where $K_a = \tan^2(45° - \phi /2)$. This minimum state of stress is the active earth pressure stress. Please don't forget that the soil must move outward laterally a sufficient amount to generate the active earth pressure. The amount of lateral movement required to generate a minimum lateral earth pressure depends on the consistency of the soil and the height of the retaining wall used to resist lateral earth stresses. Assuming that the height of the retaining wall is H, the amount of lateral movement required to generate lateral earth pressure is approximately 0.002 to 0.004 H for loose sandy soils and approximately 0.001 to 0.002 H for dense sandy soils. What about the required lateral deflection for clayey soil? I'm old fashioned when it comes to retaining structures, so I don't recommend using clayey soil behind retaining walls. When clayey fill becomes wetted, bad things often happen.

There is a limit state that I have not yet mentioned. It is the "at-rest" earth pressure state. The "at-rest" earth pressure state is in effect when the soil is experiencing zero lateral movement, that is, zero lateral strains. This is the state of soil stresses in the ground prior to excavating a hole or installing any structures. For normally consolidated soils, the "at-rest" earth pressure K_o is given by the Jaky equation, where $K_o = 1 - \sin\phi$.

What is the lateral earth pressure in between the active and the "at-rest" states of stress? The answer is we don't know. We only know the lower bound, active state of stress, and the zero lateral strain, "at-rest" state of stress values of lateral earth pressure. You could try to interpolate between K_a and K_o values of lateral earth pressure coefficients, but I wouldn't recommend it. I suggest you use the active value if the wall is anticipated to be appropriately flexible, and if the wall has limited lateral movement, I suggest you use the "at-rest" earth pressure coefficient. These values are relatively close, 0.33 and 0.50 for a friction angle of 30 degrees, and should be suitable for standard design practice. If an advanced practice solution is required for determining a lateral earth pressure that is more than active and less than "at-rest," I suggest you use a computer program such as SIGMA first, and if it is really an expert level practice problem, you should find someone who is expert in a computer program such as FLAC. Please do not attempt to learn FLAC or Plaxis on a limited project budget with a short schedule. Use your experts to solve the problem and show you how they did it.

2.6.3 Competition between Rankine and Coulomb Equations

During the eighteenth and nineteenth centuries there was considerable competition between scientific communities in France and England. For example, both French and English scientists claim that they invented photography. There is evidence to support the claims on both sides of the photography argument (as there always is in a good argument), so the competing claims from the French and English are still debated into the twenty-first century.

A similar situation exists with the development of lateral earth pressure equations. In 1776, the famous French scientist C.A. Coulomb published his earth pressure theory, which included equations for computing active and passive earth pressures (Coulomb, 1776). In 1857 William Rankine, the famous English scientist, published his earth pressure theory (Rankine, 1857) which looks different than Coulomb's work, but upon close examination, it may be found that Rankine's equations can be derived from Coulomb's equations by making some simplifying assumptions. No matter that Coulomb's equations apply to more general cases of earth pressure calculation, to this very day English-speaking nations seem to prefer Rankine's equations and French-speaking nations seem to prefer Coulomb's equations. Both equations have their pluses and minuses, so it important for you to understand their proper application to solving the earth pressure problem. One warning I must mention for your own good. Please use at least two references when looking up the Rankine and Coulomb equations. I'm not sure why, but I know from experience that typographical errors are common in texts that replicate these equations, especially the Coulomb equations. So please check to make sure you have the correct equations before starting to input the required parameters of your problem. This warning also covers use of tables giving values of earth pressure coefficients K_a and K_p. For a detailed discussion of the use and limitations of the Rankine and Coulomb equations please turn to Section 4.1.

References

Coulomb, C.A. (1776) *Essai sur une application des règles de Maximus et Minimis à Quelques Problèmes de Statique, Relatifs à l' Architecture*. Mémoires de Mathématique et de Physique, Présentés a l' Académie Royale des Sciences, par Divers Savans, et lûs dans ses Assemblées, Paris, Vol. 7, pages 143–167.

Rankine, W.J.M. (1857) On the stability of loose earth. *Proceedings of the Royal Society, London*, **VIII**, 185–187.

2.7

Unsaturated Soils

2.7.1 Introduction to Unsaturated Soils

Historians and librarians have the job of classifying what the rest of us do for a living and when we did it. Art historians write texts defining when famous artists were working in their pink period, their blue period or their charcoal period. Relative to my career, they will probably classify it by the five-year structural period, the 20-year saturated clay period, the 20-year unsaturated sand period, and the 20-year teaching/mentoring period. I started as a structural engineer and went into geotechnical engineering in Ohio where we had saturated clays. I moved to New Mexico and had to learn unsaturated soil mechanics in silty sandy soils. During most of this time, I taught classes in mechanics, structural engineering and geotechnical engineering to students and practicing engineers. Currently I have transitioned into the last 20-year period of mentoring practicing engineers and geotechnical faculty. But no matter what the classification categories are called, I have always thought that I was solving problems and having fun!

When I was an undergraduate engineering student in the 1960s I never heard of unsaturated soil mechanics (to the best of my recollection). When I was a graduate student I heard about capillarity and strength of unsaturated soils due to suction (like a ball of moist sand on the beach), but we didn't do anything with this information in design classes. During the first 20 years of my professional geotechnical engineering practice, I generally ignored unsaturated soil mechanics. Thanks to a move to New Mexico, since about 1992, I have worked almost exclusively with unsaturated soils, and have had to climb the steep curve of knowledge required to understand unsaturated soils (with considerable help from my colleagues on and consultants to the UMTRA project).

Even though my current geotechnical practice is in the arid southwestern region of the United States, I still often hear from practicing engineers: "When I was in school,

Geotechnical Problem Solving, First Edition. John C. Lommler.
© 2012 John Wiley & Sons, Ltd. Published 2012 by John Wiley & Sons, Ltd.

the professor never discussed unsaturated soils. He either talked about saturated clays or dry sands. He mentioned capillary action, but never explained how capillarity affected soils that we would see in the field."

An article by Sandra Houston (Houston, 2011) in January/February 2011 *Geo-Strata Magazine* makes the case that we have achieved limited adoption of unsaturated soil mechanics principles in geotechnical engineering practice and that we do not include adequate coverage of unsaturated soil mechanics in undergraduate soils and foundations classes. Why?

A bit of historical perspective is in order to answer this question. In 1925, *Engineering News-Record* magazine published by McGraw-Hill, printed eight articles on the "Principles of Soil Mechanics" (Terzaghi, 1925) written by Charles Terzaghi. The author's name was actually Karl Terzaghi, but times were different just after World War 1 (WW1) and it was common to Americanize ones name to not sound too German. These eight articles represent the beginning of modern geotechnical engineering in the United States. A reprint of these articles is available for your review in the ASCE Geotechnical Special Publication No. 118, Volume One, *A History of Progress – Selected US Papers in Geotechnical Engineering* (ASCE, 2003). The first four of Terzaghi's articles were about clay. In the fifth article, he introduced the topic of the differences between sand and clay. Terzaghi described grain size, uniformity curves, particle shapes, void ratio differences, cohesion, and plasticity of sands and clays in this article. None of these topics included consideration of unsaturated properties of sands and clays, and Terzaghi's conclusions relative to these topics became the basis of soil mechanics and geotechnical engineering taught in the United States for decades (even up to the present day). But wait, Terzaghi did mention capillarity while explaining shrinkage of a clay soil. He said that volume change shrinkage of clay is produced by a pressure generated by capillary forces, and that this capillary compressive stress (or pressure) depends on the grain size of the soil. Terzaghi said that the maximum capillary compressive stress that causes the reduction in volume observed during drying of a clayey soil, what he called shrinkage, is a function of the maximum capillary rise observed in soil. He said that the maximum capillary pressure in kg cm^{-2} is equal to 10% of the maximum capillary rise (h) expressed in meters. I know the units seem weird, but the idea is interesting. Terzaghi said that the capillary rise in a very fine sand is approximately 0.05 meters (about 2 inches), giving a maximum capillary pressure of 0.005 kg cm^{-2} (about 10 pounds per square foot), and that the capillary pressure in a CH clay is 200 kg cm^{-2} or greater (400 000 pounds per square foot or greater). Based on the small compressive pressure generated by capillarity of the sand and the huge compressive pressure generated by capillarity of the clay, Terzaghi concluded that this explains why high plasticity clays shrink and silty sands don't shrink. I don't know if clays generate quite that much compressive stress, but I do know from Fredlund (Fredlund, Xing, and Huang, 1994) that soil suction can approach 1 000 000 kPa (more than 20 million pounds per square foot). Limiting suctions and pressures notwithstanding, it is interesting that Terzaghi introduced the topic of capillary suction and resulting tension stresses in unsaturated soils in 1925 and this topic didn't significantly impact the geotechnical literature until the 1990s (my dating, others may disagree), and is still missing from most university curricula if Sandra Houston (Houston, 2011) is correct. I think she is.

2.7.2 Unsaturated Soil Mechanics – Soil Suction and Soil Tension

Do we really have to consider unsaturated soils differently than the total and effective stressed general soils that we studied in school? Does it really matter? The simple answer is that it does matter. Let's take a silty, fine sandy soil as an example. When our silty, fine sandy soil sample is completely dry, that is a saturation (S_r%) of 0%, it has a soil suction of approximately 20 million pounds per square foot and a soil tension stress (or compressive stress on the soil skeleton) due to the soil suction of zero. Does that seem reasonable to you? Consider that our silty sand is completely dry. It pours from your hand like table salt, see Figure 2.7.1. In this condition it has no cohesion strength. When we add water to the same dry sand it develops apparent cohesion and can be formed into a ball as illustrated in Figure 2.7.2. What difference does the water make?

When the sand is dry as shown in Figure 2.7.1, it has an extremely high suction potential. Suction is like your potential to hit a home run. You have the necessary

Figure 2.7.1   Dry flowing sand – no apparent cohesion

Figure 2.7.2 Added moisture to sand – now has apparent cohesion

strength, hand-eye coordination and desire, but you lack a bat. You need something to help you hit the ball to score a home run. You can't think a home run, you can't hope a home run, what you need is a physical means to hit a home run. It's the same with soil suction and water. Water is the physical means for soil suction to apply tension stresses onto the soil skeleton, resulting in increased effective confining stress and increased soil strength. When a soil is completely dry with zero water content and zero degree of saturation, the soil suction is just sucking wind. No matter how high the suction's potential, it needs a little water to coat soil particles, form a membrane that acts like a balloon to develop tension stresses as the suction pulls in the water membrane "balloon." As suction pulls in the water membrane, the membrane's tension stresses apply confining stress to the soil grains which increases strength of the soil and acts as apparent cohesion. When water is present to coat soil particles, soil suction can act to increase confining stresses in the soil structure, as shown in Figure 2.7.2. When water is absent, there are no increased confining stresses in the soil structure as shown in Figure 2.7.1. The formal name of this air–water interface membrane is the contractile skin (Fredlund and Morgenstern, 1977).

As more and more water enters the voids of an unsaturated soil, air is eventually driven out of the sample. With no air voids, there is no soil suction or contractile skin, and the soil loses its apparent cohesion. You can see that effect, if you take a moist ball of sand like that shown in Figure 2.7.2 and drop it into a bucket of water. Once the ball of sand hits the water surface in the bucket, it loses all apparent cohesion and crumbles into a collection of sand grains.

Between a soil saturation of zero and a saturation of 100%, the soil increases in apparent cohesive strength until it is about 80 to 90% saturated, and then it rapidly loses strength until it is back to zero cohesive strength at approximately 100% saturation.

In their 1977 paper, Fredlund and Morgenstern (Fredlund and Morgenstern, 1977) formally describe unsaturated soil as a four-phase mixture of solid soil particles, water, air, and contractile skin. In that 1977 paper, the contractile skin is defined as an independent phase having properties different from air or water. Given that unsaturated soil is a four-phase mixture, how do we analyze unsaturated soils for stress changes, strength changes and volume changes?

2.7.3 Analysis of Unsaturated Soils

How do we analyze unsaturated soils and still stay true to the classical soil mechanics effective stress principle? For that matter, how do we stay true to laws of physics as we incorporate unsaturated soils into our soil mechanics theories? Remember, mechanics is a part of basic physics. Answers to these questions and many others have been developed by Del Fredlund during his studies under Professor Norbert R. Morgenstern, and during Dr. Fredlund's subsequent studies of unsaturated soils. I consider Dr. Fredlund's book (Fredlund and Rahardjo, 1993) to be the classic text on unsaturated soil mechanics and I recommend it for your library.

A very important concept in determining the shearing strength or compressibility of an unsaturated soil is finding the independent stress state variables. Do you know what independent stress state variables are? Ok, let's break it down. First, we need to figure out which variables are independent and which are dependent variables. Then we need to separate stress state variables from material properties and develop an equation of the form:

$$\{Stress\ State\ Change\}\ \{Material\ Properties\} = \{Strength\ Change\}$$
$$\{Stress\ State\ Change\}\ \{Material\ Properties\} = \{Volume\ Change\}$$

Notice that our idealized equations use stress state changes and appropriate material properties to calculate both unsaturated soil shear strength and volume changes. If we mix up our stress state variables with material properties, we will have difficulties solving different types of geotechnical problems. I'll try to explain how this works.

What if I decided to develop my own theory of effective stresses for unsaturated soils and use it to define shear strength? I might start out by stating that the shear strength of an unsaturated soil depends on its water content, porosity, soil suction, friction angle, cohesion, and confining stresses. My proposed theory is immediately in trouble. The water content is related to the porosity and to the soil suction and as such they are dependent on each other. If I try to use my proposed "theory," I need to separate soil stresses and calculate stress changes, how do I do that when my theory has interdependence of soil properties with soil stresses? If I manage to come up with a way to calculate changes in unsaturated soil stresses, then I have to have appropriate soil material properties that define the shear strength changes in response to stress changes. If I mix parameters that define the soil stress state with soil material properties, I'll end up with a soil shear strength that is not independent of the tests I use to define the stresses and the strengths. Another way to say this is that the soil stresses will not be independent of the stress path used in testing soil properties. If you need further clarification of the requirement to separate soil stresses from soil material properties when characterizing unsaturated soils, please read Fredlund's text (Fredlund and Rahardjo, 1993) Chapter 3 carefully.

To calculate effective stresses in unsaturated soils, the soil suction needs to be defined in basic independent terms. We need to separate stress terms and define them as basic values of total stresses, water stresses, air stresses and contractile skin stresses. Any dependent terms in our equations of stress need to be eliminated to develop useful stress state parameters. That separation into independent and dependent stress and material variables helps us express soil shear strength and compressibility of unsaturated soils in their most basic terms. That is exactly what Dr. Fredlund did in his work. First he defined the three independent stress state variables of an unsaturated soil: (1) $(\sigma - u_a)$, (2) $(u_a - u_w)$, and (3) (u_a), where σ is the normal stress on a soil element in the x-, y- or z- directions, u_a is the pore air pressure (when gauge pressure is used the pore air pressure is zero), and u_w is the pore water pressure (in unsaturated soils the pore water pressure is negative, i.e., a suction). Then he proved that only two of these three independent stress state variables are required to solve the stress state for an unsaturated soil, and he selected the net normal stress $(\sigma - u_a)$ and the matric suction $(u_a - u_w)$ for use in analyzing unsaturated soils.

I can hear some of you saying, "Alright, that's interesting. Let's get on with it. Give me the equation for the effective stress so I can start working."

I'm sorry, but unsaturated soils are not a single equation type of problem. I told you there would be complications. Stop and think about this for a moment. The equations using net normal stress *are* effective stresses because the unsaturated soil has air in some voids and water in other voids. Only saturated soils require the subtraction of pore water pressure from total stress to calculate the effective stress. The complication of unsaturated soils is the independent requirement to consider the matric suction, $(u_a - u_w)$. Matric suction is an additional confining stress that can increase the shear strength and stiffness of the soil if it increases (increased suction means it is becoming a larger negative number, i.e., its absolute value is larger) and reduce the shear strength and stiffness of the unsaturated soil if it decreases. As a result you have to consider two separate steps, first you have to determine how changes in matric suction affect the soil's confining pressure, and then you can consider how changes in normal and shearing stresses affect the soil's shear strength and compressibility. Let's consider shear strength of an unsaturated soil first.

Using the first two stress state variables, net normal stress and matric suc-tion, as suggested by Fredlund, the shear strength of an unsaturated soil may be expressed as:

$$s_{ff} = c' + (\sigma_f - u_a)_f \tan \phi' + (u_a - u_w)_f \tan \phi^b \tag{2.7.1}$$

where,

s_{ff} = the shear strength on the failure plane at failure
c' = the intercept of the Mohr–Coulomb failure envelope at the vertical shear stress axis. This intercept is where both the net normal stress and the matric suction are equal to zero.
$(\sigma_f - u_a)_f$ = net normal stress on the failure plane at failure
ϕ' = angle of internal friction that generates shearing strength from changes in the net normal stress
$(u_a - u_w)_f$ = matric suction on the failure plane at failure
ϕ^b = angle of internal friction that generates shearing strength from changes in the matric suction on the failure plane at failure

The unsaturated soil shear strength Equation 2.7.1 transitions directly into the saturated soil shear strength equation as the soil's saturation increases. As soil saturation increases, the negative pore water pressure approaches the pore air pressure, at saturation the matric suction in the third term of the equation becomes zero and the pore air pressure in the second term of the equation becomes equal to the pore water pressure, thus converting the equation into that of a saturated soil.

How do you determine the terms included in the unsaturated soil's shear strength equation? You have to understand the concept of Mohr–Coulomb failure envelope extended to include both net normal stress and matric suction. In Fredlund and Rahardjo's text (Fredlund and Rahardjo, 1993), Chapter 9, pp. 228–235 there is a very good presentation of this material. Dr. Fredlund shows a three-dimensional plot that extend the standard Mohr–Coulomb failure envelope from two variables to three variables and defines c', ϕ', and ϕ^b. I was tempted to repeat the Fredlund material here, but I just couldn't convince myself to do it. Why not? I have a confession to make to you. Every time I see a three-dimensional plot in a geotechnical text, I get a case of brain freeze. I just stop and stare at the figure. What does it all mean? I was afraid that you might have the same reaction looking at three-dimensional, three-variable plots. So I decided to use a different approach; an approach to dealing with three-dimensional plots that I have been using for years. Please allow me to digress for a moment.

Working example

I was talking with my friend Bob Meyers the other day and he told me and some other friends a story about back in the early 1990s he remembers coming into my office and seeing a Peacock diagram on the white board next to my desk. It was labeled "Peacock Diagram," it was drawn as a three-dimensional plot, and there were three separate two-dimensional plots drawn next to it, each showing a section view of the Peacock diagram with a pair of variables on each two-dimensional plot. The Peacock diagram was first constructed in 1967 by William Peacock (Holtz and Kovacs, 1981), and it shows volume change, initial void ratio and effective confining stress of a sandy soil under drained triaxial shear testing. It illustrates the concept of dilation, compression, critical void ratio, and critical confining stress, and I kept it on my white board to remind me that even loose sands can dilate if their confining stresses are low enough to allow it. Anyhow, the point of the story is that I have always had to dissect three-dimensional plots into separate component two-dimensional plots to grasp a full understanding of the physical principles involved. Given that background, let me show you how the three independent variables of matric suction, net normal stress, and soil shear stress are related.

Consider Figure 2.7.3 which is a plot of effective normal stress versus effective shearing stress. Mohr–Coulomb Line No. 1 is a typical plot of a series of tests with the matric suction equal to zero. Zero matric suction implies that these tests are a series of saturated tests.

Point 0 is the point where the effective normal stress, the effective shearing stress and the matric suction are all equal to zero. The intercept of Mohr–Coulomb Line No. 1 with the vertical axis at point No. 1 is called the effective cohesion, c'. The effective cohesion is the minimum value of this soil's cohesion, where both the net normal stress and the matric suction are equal to zero.

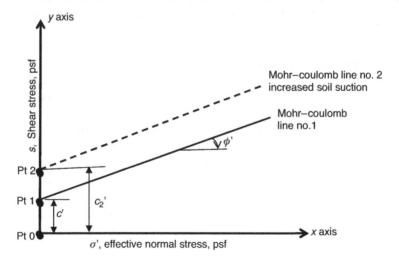

Figure 2.7.3 Mohr–Coulomb plot, net effective normal stress versus shearing stress

Figure 2.7.4 Frontal view of roof left, side view of roof right, you need to see the side view to measure the slope of the roof. With permission from John Wiley & Sons

Now let's dry the soil a bit to a saturation S_r value less than 100%. Some air has entered the soil pores and the pore water pressure has started to become negative, which means that the soil's matric suction has started to develop. When soil dries, its matric suction increases as the pore water pressure goes negative because matric suction equals $(u_a - u_w)$ and when u_w becomes negative $((u_a - (-u_w)) = (u_a + u_w)$. As a result of increased matric suction, the apparent confining normal stress on the soil mineral structure increases, resulting in increased soil shear strength ($\Delta S = \Delta \sigma$ tan ϕ). Upon an increase in the soil's matric suction, the apparent cohesion of the soil moves from c' to c_2' on the vertical axis because the effective normal stress, which is equal to the net normal stress, has not changed and is still equal to zero at the origin. The result of this increased cohesion intercept value is a shift in the Mohr–Coulomb line from Line No. 1 to Line No. 2 as shown in Figure 2.7.3. I have just illustrated three variables on a two-dimensional plot. Everything is done to illustrate the increase in shear strength due to soil matric suction, right? No not really. What we have done in Figure 2.7.3 is like looking at the front of a roof, see Figure 2.7.4. You can tell from looking at the front of a pitched roof that it is rising, but you need to look at the end of the roof, again in Figure 2.7.4, to see how steeply the roof is pitched, that is how flat or how steeply it is inclined.

When you are plotting independent parameters on the x- and z-axes, that is net normal stress on the x-axis and matric suction on the z-axis, you can plot them separately and add their affects on soil shear stress separately, as we did in Equation 2.7.1. So let's go ahead and look at the plot of matric suction on the z-axis versus soil shear stress on the y-axis, see Figure 2.7.5.

By looking at Figure 2.7.5 you can see the slope of the line of matric suction versus shear stress, like the slope on the roof in Figure 2.7.4. You can see from Figure 2.7.5 that the shear strength of the soil increases as the soil's matric suction increases. This is not a big discovery, because most of us know that drying a soil increases its shear strength, but now we have a way of illustrating and calculating how much the shear

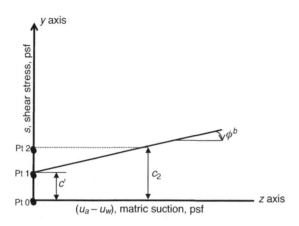

Figure 2.7.5 Matric suction versus shear stress

strength increases with a given decrease in water content (because the soil water characteristic curve gives you soil suction versus water content, see below). The slope of the matric suction versus shear strength line in Figure 2.7.5 is ϕ^b which is the term we need in Equation 2.7.1. Published data (Table 9.1 in Fredlund and Rahardjo, 1993) suggests that for a variety of soils the value of ϕ^b is less than the friction angle ϕ'.

I recommend that you study Fredlund and Rahardjo's text (Fredlund and Rahardjo, 1993) Chapter 9 for additional material on determination of shear strength of unsaturated soils.

Now when the laboratory technician calls you and asks if you want to run the direct shear test in an "as compacted" unsaturated state with a dry direct shear box, or do you want to run the direct shear test with the box saturated, what do you say? If you want lower bound shear strength values of c' and ϕ', and you cannot provide assurance that the compacted soil will *never* become wetted, I'll bet you'll think of Figures 2.7.3 and 2.7.5 and ask the laboratory technician to make sure that the samples are thoroughly saturated before shearing each sample.

2.7.4 Expansive and Collapsible Soils

Two special cases of volume change of unsaturated soils are of particular interest to geotechnical engineers: (1) expansive soils and (2) collapsible soils. I'm going to discuss collapsible soils in depth in Section 3.2, so let me do an introduction to expansive soils here. First let me say that I don't particularly like the topic of expansive soils, and I'm not quite sure why. So what have I got against an inanimate object like a chunk of expansive soil? It might not be the soil at all, but rather the current rather heated debate in the geotechnical community about expansive soils that has me a bit down on the topic. I'm old fashioned, in case you didn't notice, and I really don't like to argue. I'm OK with discussing differing opinions, but I really don't like getting into an argument. You and I can disagree about any number of technical issues and still be friends. I have never been afraid to change my mind. If something new comes along to shed additional light on a geotechnical topic, I will gladly change my position if I'm proven to be wrong. It's not personal! It's professional. Several of my heroes, Einstein and Terzaghi among them, are famous for changing their minds. There is a famous case where Dr. Terzaghi is on the witness stand, and the opposing lawyer can't wait to point out that what Terzaghi just testified to was in conflict with his 1943 book, *Theoretical Soil Mechanics*. When confronted by the conflict between his testimony and his written opinion, Terzaghi said confidently, "Do you think that I haven't learned anything since 1943!"

The best way to visualize an expansive clay soil is to realize that this clay has very high soil suction when relatively dry. This soil suction generates high confining stresses that maintain the expansive clay in a compact state. As the expansive clay is wetted, the high confining stresses generated by this clay's soil suction reduces as the soil sucks in water. As the expansive clay's moisture content increases and its soil suction decreases, the expansive clay increases in volume, that is, it expands. If the same soil were dried, its soil suction would increase again, resulting in increasing confining stresses and decreasing volume, that is, it would shrink until it reached the shrinkage limit. Expansive clays are almost like living beings, they expand and contract as they take in and expel water in response to the environment. There are a couple of things to remember about the limitations of expansive clays. When expansive clays are wetted and prevented from increasing in volume they generate a swell pressure that is a function of the activity of the clay minerals present. At some depth in the ground the vertical effective stress equals the swell pressure of the soil and swelling does not occur at or below that depth. Free swell is when the clay is allowed to suck in as much water as it can and swell as much as it can without restraint. If you allow an expansive soil to freely swell, it does not generate measureable swell pressure. If after free swelling, you press the clay back down to its original volume you generate a vertical pressure that is usually less than the confined swell pressure. If you remold and highly compact an expansive clay on the dry side of optimum moisture content (i.e., modified Proctor compaction at minus 4% below optimum moisture content), you have just built a nice new expansive fill. If you moderately compact expansive clay at or above optimum moisture content (i.e., 90 to 95% of Standard Proctor compaction at 0 to 4% over optimum moisture content), you will initially have a weaker, more compressible fill, but if wetted it will not be a highly expansive fill. You may be able to lime treat an expansive clay to reduce its expansion potential, but be careful to make sure that it does not have significant amounts of sulfates (I personally try to keep the sulfates below 2000 ppm, but others quote the National Lime Association recommendation of 3000 ppm). Sulfates and lime grow a mineral called ettringite that causes chemical swell of the mixture during curing. There are construction techniques described in books published by the lime industry that describe how to temper lime-treated sulfate soils to help reduce chemical swelling. I personally prefer to avoid mixing lime with sulfates, but if you must, I suggest looking up the writings of Dr.

Dallas N. Little for the National Lime Association. My advice is when in doubt, do your own testing of lime treatment on your project's soils, and maintain close quality control of the lime treatment process in the field.

2.7.5 Depth of Wetting

To estimate potential soil collapse or swell due to wetting, we need to test representative samples to determine the percentage collapse or percentage swell due to wetting, and we need to estimate or measure the depth of wetting. Determination of the depth of seasonal wetting is also important in evaluating the performance of landfill cover systems.

If you are not familiar with determining the depth of wetting, you might expect that it is no problem, no big deal, a slam dunk, and so on. I can hear you explaining how to solve the soil wetting problem:

> Water flows onto the ground surface at a known rate of flow in inches per hour, gravity pulls the water down into the soil until the inflow stops, and then evaporation from the surface occurs until the liquid water is drawn out of the ground and the liquid phase changes into water vapor. If rainfall events are weeks or months apart like in the arid Southwest, and the evaporation rate is high, the ground will dry out completely. The depth of water penetration prior to reversal of inflow by evaporation is the depth of wetting . . . right? Then all you have to do is calculate the velocity of inflow by using Darcy's law, which says velocity of flow $v = (k)\,(i)$. Where (k) is the soil permeability, and (i) is the inflow flux rate equal to $\Delta H\,/\,L$. The hydraulic head loss ΔH in feet is the head loss while the water is flowing along the flow path of length L. I can get a lab to do a permeability test on the soil, and I'll use an upper bound limit value of gradient, $i = 1$. Using the inflow velocity, I can calculate how far the water penetrated . . . right?

Au contraire, not so fast, I told you geotechnical engineering could get complicated, and the depth of wetting issue is a prime example. Dr. John Nelson in a recent discussion (Nelson *et al.*, 2011) in the ASCE *Journal of Geotechnical and Geoenvironmental Engineering* gave the following definitions of the depth of wetting:

1. *Depth of Seasonal Moisture Fluctuation* – This depth is that zone of soil within which water contents can change owing to climatic changes.
2. *Depth of Wetting* – This is the depth to which water contents can increase owing to external factors. Such factors could include capillary rise after elimination of evapotranspiration from the surface, infiltration from irrigation or precipitation, or water from off-site.

If I have interpreted Dr. Nelson's terms correctly, these definitions separate climate effects of wetting and drying from external factors which I interpret are somehow

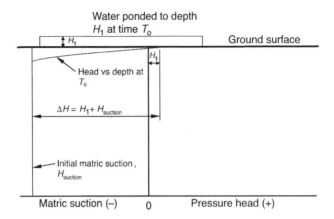

Figure 2.7.6 Hydraulic head at ground surface of dry soil

generated by the actions of man. I have noticed that the term depth of wetting is used in the literature for all kinds of soil wetting, whether it be by nature or affected by man. I agree with Dr. Nelson that it is prudent to pay attention to the sources of wetting, no matter what you call it. I wish we had consistent terms for geotechnical processes. It would surely make life easier. But alas we don't.

I'll just use the term depth of wetting and let you capture all of the potential sources of water involved in your problem. Ok, so what happens when water comes in contact with a relatively dry ground surface? Upon initial contact, the hydraulic gradient is very high. The hydraulic head $\Delta H = H_1 - (-H_{suction}) = H_1 + H_s$, that is change in hydraulic head is equal to the height of water above the ground surface (H_1) plus the amount of soil suction acting at the soil surface, see Figure 2.7.6

Let's walk you through a rainfall event where the rain falls on an unsaturated soil surface. What is the infiltration rate of water into an unsaturated soil surface? As you might expect, the infiltration rate depends on the rainfall intensity and on soil properties. First consider a very light rain, where the rainfall intensity is less than the saturated hydraulic conductivity (a lower bound value of influx) of the soil. In this case all of the water that falls each second is sucked into the soil, and the infiltration rate equals the rainfall intensity. This is intuitively true. Using terminology of unsaturated soil mechanics, the infiltration rate is equal to the rainfall rate, since the minimum infiltration rate of the soil is equal to the saturated hydraulic conductivity (often called permeability) under a hydraulic gradient of one (Wilson, 1997). Whenever the rainfall intensity is less than the saturated hydraulic conductivity under a gradient of one ($v = k * 1$), all of the rainfall enters the ground surface immediately upon falling on the ground surface.

Next take the case that I discussed above, where there is a heavy rainfall and water ponds on the ground surface. Water ponds on the ground surface (assuming it doesn't run off to a drainage ditch) because the rainfall intensity is much greater than the soil's saturated hydraulic conductivity. This results in water piling up at the surface like a crowd trying to rush into an auditorium through too few doors. At the start of rainfall the hydraulic head at the ground surface is zero or a small positive number as water piles up while the suction just below the ground surface is very high. The soil suction for a very dry soil expressed as negative head may be as high as −33 500 feet (i.e., 100 000 kPa). Due to the extremely high hydraulic gradient between zero feet at the surface and −33 500 feet just below the ground surface, there is very rapid inflow of water into the ground. As time goes on and rainfall water penetrates deeper and deeper into the ground surface, the rate of inflow at the ground surface decreases because the hydraulic gradient decreases as the suction decreases due to wetting and as the distance between ground surface and wetting front increases.

After the rainfall stops, water starts to evaporate from the ground surface and flux in the saturated surface soil reverses from flow downward toward the soil suction to flow upward toward the atmosphere. During a given year rainfall water and snow melt soaks into the ground surface and evaporates back to the atmosphere. The depth of change of moisture content during annual seasonal precipitation and evaporation events is often referred to and used as the depth of wetting. This wetting is the "depth of seasonal moisture fluctuation" that we were discussing above.

When the surface of the ground is artificially wetted by irrigation or ponding of water, saturated/unsaturated soil infiltration models are required to estimate the artificially induced additional depth of wetting.

During my work on the landfill projects, I learned about another method of increasing depth of wetting. We used riprap rock cover over our infiltration barriers on landfills to resist rainfall runoff erosion from severe storm events. The riprap rock acted like a mulch and retarded evaporation from the surface of the landfill cover. Retarding evaporation from the ground surface has the same effect as watering the ground surface: the depth of wetting into the ground increases.

2.7.6 The Soil Water Characteristic Curve

The soil water characteristic curve (SWCC) is the basic tool used by engineers to understand an unsaturated soil. The basic soil water characteristic curve is a plot of volumetric water content θ_w ($V_{\text{water}}/V_{\text{total}}$) versus matric suction. Why do most groundwater engineers use volumetric water content rather than our normal geotechnical water content ($W_{\text{water}}/W_{\text{solids}}$)? I expect it is because they use porosity n ($V_{\text{voids}}/V_{\text{total}}$) in many of their equations, and $\theta_w = n$ when the soil is 100% saturated.

Besides being a tool used to determine soil suction for a given water content value, the SWCC is used to model response of unsaturated soils to moisture changes. For example, permeability of unsaturated soils can be predicted by use of the soil water characteristic curve (Fredlund, Xing, and Huang, 1994).

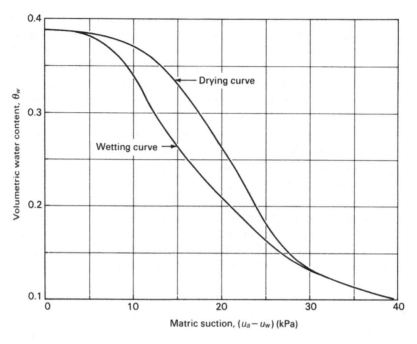

Figure 2.7.7 Soil water characteristic curves for wetting and drying, (Fredlund and Rahardjo, 1993)

In similar fashion to permeability, the SWCC is used in several computer models to estimate unsaturated soil material properties. The soil water characteristic curve for soil wetting is different than the curve for soil drying, see Figure 2.7.7 for typical wetting and drying SWCC curves for a sandy soil (from Fredlund and Rahardjo, 1993).

2.7.7 Additional Study of Unsaturated Soils

The American Society of Civil Engineers has held several regional and national conferences on unsaturated soils and expansive soils. As a start, I recommend reading Geotechnical Special Publications numbers 39 and 68. Once you start to get some ideas about the topic, you will want to get a copy of Del Fredlund's book (Fredlund and Rahardjo, 1993) *Soil Mechanics for Unsaturated Soils*. If you want to look up papers in geotechnical publications about unsaturated soils and expansive clays, I suggest you start by searching for papers authored by Del Fredlund, Sandra Houston, Gordon McKeen, John Nelson, Warren Wray, Ward Wilson, and Robert Thompson. This is far from a complete list, but it will get you started.

References

ASCE (2003), *A History of Progress – Selected US Papers in Geotechnical Engineering*, American Society of Civil Engineers, Geotechnical Special Publication No. 118, vol.1, 1092 pages.

Fredlund, D.G. and Morgenstern, N.R. (1977) Stress state variables for unsaturated soils. American Society of Civil Engineers. *Journal of the Geotechnical Engineering Division*, GT5, **103**, 447–466.

Fredlund, D.G. and Rahardjo, H. (1993) *Soil Mechanics for Unsaturated Soils*, John Wiley and Sons, 544 pages.

Fredlund, D.G., Xing, A., and Huang, S. (1994) Predicting the permeability function for unsaturated soils using the soil-water characteristic curve. *Canadian Geotechnical Journal*, **31**(3), 533–546.

Holtz, R.D. and Kovacs, W.D. (1981) *An Introduction to Geotechnical Engineering*, 1st edn, Prentice-Hall, Inc., Englewood Cliffs, New Jersey, 733 pages.

Houston, S.L. (2011) Commentary: advancing the practice of unsaturated soil mechanics. Geo-Institute of the American Society of Civil Engineers. Geo-Strata Magazine, Issue No. 1, Vol. 15, January/February 2011, pp. 16–17.

Nelson, J.D., Chao, K.-C. and Overton, D.D. (2011) Discussion of "Method for evaluation of depth of wetting in residential areas" by Walsh, Colby, Houston and Houston, *ASCE, Journal of Geotechnical and Geoenvironmental Engineering*, **137**, 293–296.

Terzaghi, C. (1925) Principles of soil mechanics, I–VIII. *A History of Progress – Selected US Papers in Geotechnical Engineering*, American Society of Civil Engineers, Geotechnical Special Publication No. 118, vol.1, edited by W. Allen Marr, pp. 2–38. Original articles in Engineering News Record 1925 Volumes 19, 20, 21, 22, 23, 25, 26 and 27, reprinted by the American Society of Civil Engineers.

Wilson, G.W. (1997) Surface Flux Boundary Modeling for Unsaturated Soils. American Society of Civil Engineers, Geotechnical Special Publication No. 68, pp. 38–67.

3

Foundations

3.1

Settlements of Clays

Introduction

As the saying goes, we must pay tribute to tradition, and discuss settlement of clays first. Personally I would prefer to discuss sands and granular materials first, but Karl Terzaghi introduced modern soil mechanics to the United States in 1925 with eight papers (Terzaghi, 1925), the first six discussing clays. He didn't introduce a discussion about the properties of sandy soils until the seventh paper, and even then sand didn't get top billing in the seventh paper because Terzaghi discussed the differences between sands and clays. Who am I to break with this tradition?

3.1.1 A Brief Geotechnical History and Overview of Clay Settlement

Some stories are handed down from father to son to grandson and become family oral history. These stories are not intended to be closely researched and analyzed; rather they are stories that are intended to make an impact on the family. My first geotechnical mentor Neil Mason was a student of Terzaghi's at Harvard University, and he told me this story. If Terzaghi is the father of soil mechanics, and Neil Mason is one of his student sons, does that make me a grandson of soil mechanics?

Working example

Terzaghi was working on a concrete dam in Europe that failed. Initially the dam was fine, but with time it settled and cracked. Part of the problem was piping and part of the problem was consolidation and shear strength of clayey soils. Terzaghi was very upset with the late nineteenth century technology used to design this dam, because it did not consider time-dependent settlement of

Geotechnical Problem Solving, First Edition. John C. Lommler.
© 2012 John Wiley & Sons, Ltd. Published 2012 by John Wiley & Sons, Ltd.

structures. In fact no one at the time seemed to acknowledge time-dependent settlement of geomaterials.

Terzaghi moved from central Europe to Istanbul to work and teach at the American University. Driven by the settlement failure of his concrete dam, Terzaghi studied settlement of clayey soils, and developed the theories of effective stress and consolidation. Having a good mathematics background, he developed the equations for time rate of dissipation of excess pore water pressures from application of the equation of heat flow from thermodynamics.

During his work on consolidation Terzaghi developed or invented the direct shear machine and the consolidometer. The consolidometer was intended to evaluate a function that related settlement to time.

While Terzaghi was concerned with time rate of settlement of clays, other engineers in Europe and the United States were concerned with the stress–strain properties of soils. These engineers treated soils much like other engineering materials, and developed stress–strain curves, modulus of elasticity and Poisson's ratio values, and so on.

The problem with characterizing soils by their elastic moduli and Poisson's ratio values is the fact that soils are non-linear materials. Dr. Nilmar Janbu (1920–present), a professor at the Norwegian Technical University, Trondheim, Norway attacked this problem straight on by studying the stress–strain properties of soils. He considered soils as having three types of stress–strain curve: (1) over-consolidated, nearly elastic soils as those having an approximately linear stress–strain curve in the range of ordinary foundation loadings; (2) plastic, soft, normally consolidated soils as those that have approximately a plastic stress–strain curve in the range of ordinary foundation loadings; and (3) soils with curved, non-linear stress–strain curves, such as granular soils or unsaturated soils. Most real soils have non-linear, curved stress versus strain functions, but any model that closely approximates the true stress–strain curve of a soil can be used to calculate the strains that the soil will experience during loading.

For clayey soils, the most important piece of information that you can determine is the preconsolidation pressure. The preconsolidation pressure is the vertical effective stress that separates the virgin compression portion of the consolidation curve from the previously loaded portion of the curve. If the clay's current vertical stress is equal to the preconsolidation pressure it is normally consolidated, and if it is less than the preconsolidation pressure the clay is over-consolidated. The compression modulus or stiffness of the reloading portion of the curve when the sample is over-consolidated clay is high and the modulus or stiffness on the virgin compression portion of the consolidation curve when the clay is normally consolidated is low. To calculate the settlement of your soil on your site you have to know which portion of the loading increment follows the reloading portion of the curve and which portion follows the virgin compression portion of the curve. Using the original e versus log

σ'_v version of the consolidation curve plot you calculate consolidation settlement by using Equation 3.1.1 if the loading goes from below the preconsolidation pressure to above the preconsolidation pressure:

$$S = C_r \frac{H_o}{1 + e_o} \log \frac{\sigma'_{vo} + \Delta\sigma_1}{\sigma'_{vo}} + C_c \frac{H_o}{1 + e_o} \log \frac{\sigma'_p + \Delta\sigma_2}{\sigma'_p} \qquad (3.1.1)$$

where,

$S =$ consolidation settlement in the same units as H_o

$C_r =$ recompression index, the slope of the reloading portion of the consolidation curve

$H_o =$ initial height of consolidating layer prior to change in stress at the center of the layer, $\Delta\sigma$

$e_o =$ the initial void ratio of layer H_o prior to application of loading stress $\Delta\sigma$

$\sigma'_{vo} =$ the initial vertical effective stress on layer H_o prior to application of loading stress $\Delta\sigma$

$\Delta\sigma_1 =$ the change in vertical stress from σ'_{vo} to the preconsolidation pressure σ'_p

$C_c =$ the compression index, slope of the virgin compression portion of the consolidation curve

$\sigma'_p =$ the preconsolidation pressure at the center of layer H_o

$\Delta\sigma_2 =$ The change in vertical stress from the preconsolidation pressure σ'_p to the maximum loading stress at the center of the layer.

I mentioned earlier that it is not my intention to reproduce material here that you can easily assess in your college soil mechanics or foundation engineering text books. I taught soil mechanics at Cleveland State University in the early 1980s and used Holtz and Kovacs' text (Holtz and Kovacs, 1981). I was impressed that they thoroughly covered the consolidation material, including consolidation plots of e versus log σ'_v (void ratio versus the log base 10 vertical effective stress) and ε versus log σ'_v (vertical strain versus the log base 10 vertical effective stress), and I recommend their book if you don't have a good college text in your library.

3.1.2 Time Rate of Consolidation Issues

Accurate calculation of the amount of consolidation settlement is a function of obtaining and testing high-quality consolidation samples. The calculation procedure is straight forward. So you have a good estimate of consolidation settlement, but how long will it take for this settlement to occur. This has always been a difficult problem. Remember that this is the primary problem that Terzaghi set out to solve when he first tackled the consolidation settlement problem.

Working example

From my earliest experiences with calculating time rate of settlement (check out my story about the 100 000 gallon fuel tank in Section 3.2), it was clear that we always calculated a time of consolidation that was too long. While I was at Ohio State University studying under T. H. Wu, we did some time rate of consolidation studies for ODOT (the Ohio Department of Transportation) highway embankments planned to be constructed over soft varved clays. We studied settlement rates both with and without vertical sand drains. While working with samples of the varved clays I noticed that water seeped out of thin seams of sand in the samples. The point that I took away from this observation was that although layers of clay may be logged as 5 to 10 feet thick, the presence of thin sand seams in the clay layers reduces their drainage distance to much less than half the layer thickness. Based on this observation I closely inspected my boring samples for sand seams and used the distance between these seams to calculate the drainage distance. As a result, my time rate of consolidation calculations based on locations of thin sand seams gave much shorter times to 90% consolidation that better matched field measurement of settlement versus time. This was a significant finding, but there was a problem. To identify and locate these sand seams in thick clay layers we needed to obtain continuous samples from the top to the bottom of the clay layers. Although split-spoon samples would work for identifying thin sand seams, it was better to have Shelby tube samples and here is the kicker...we needed to get 100% recovery on all sampling. Apart from research projects where we could drill and sample multiple borings at the same spot to get 100% sample coverage of the compressible clay layers, it was impractical to use this method of predicting time rate of consolidation.

Eventually we had a technological breakthrough that solved the soil sampling problems. This technological advancement was called the piezocone. The piezocone measures pore water pressure at the tip of a CPT cone. How do pore water pressure measurements at the tip of a CPT cone help resolve the time rate of settlement problem? When the CPT cone is pushed through soft saturated clays, the piezo-element measures high excess pore water pressures. When the CPT cone encounters a seam of sand in a thick clay deposit, the pore water pressures quickly drop to near zero because the pore water pressures dissipate. As soon as the cone leaves the sand seam, pore water pressures abruptly jump back up as excess pore water pressures are generated in the clay. I have found that if you carefully measure the distance between sand seams indicated by the CPT pore water pressure curve, and use these distances between sand seams as the distance between drainage layers, the calculated time rates of consolidation are quite accurate, at least they are much more accurate than earlier calculations.

Now what if some of the sand seams indicated by the piezocone CPT pore water dissipation curves are discontinuous and do not actually drain the clay layer. How do you correct for mistakes in interpretation of the CPT pore water dissipation curves? The answer to this question is found in the Asaoka method (Asaoka, 1978).

3.1.3 Time Rate of Consolidation Corrections – The Asaoka Method

Did you ever try to accurately shoot a rifle with one hand? It's not so easy is it? Everyone knows you need a second hand to help support the rifle. But then what if you could walk half way to the target and shoot again? Your shot would be a bit more accurate as you move closer to the target. Then for a third time you walk to within 10 feet of the target. Your aim and accuracy would be much better after you moved very close to the target. This in a nutshell is how the Asaoka method works. You make a prediction of settlement for two months from now, then after two weeks you check the predicted settlement for two weeks and re-adjust your estimate of settlement for the originally predicted two-month settlement. You keep taking settlement measurements and adjusting the prediction as time goes along. At some time before the settlement prediction period is over you can predict very accurately the amount of consolidation that will occur after two months.

Never heard of the Asaoka method you say? Well, I never heard of it either, not until it was brought to my attention two years ago by my friend Kevin Scott. Kevin works on design-build projects such as highway interstates where large roadway embankment fills are required to be built over soft clay deposits. In most of these cases, a preload fill is required to be added to the required embankment fill height to speed up consolidation of soft clay deposits. Often the state DOT requires that the design-build team guarantee that the embankment fills constructed adjacent to bridge abutments not settle more than 2 inches in the two years following completion of the project. To keep on schedule and budget, it is very important for design-build teams to know when they can remove the preload fill and continue on with roadway construction. If you predict the preload settlement will be done in six months and it is done in two months, you just wasted four months of the design-build team's schedule. If you predict the preload settlement will be done in two months and the preload should have been kept on for six months, the team's guarantee to the DOT will be violated, and there will be penalties which translate into loss of profits. As you can see, adjusting predictions of settlement and obtaining more accurate results are worth the additional cost of making field settlement measurements and adjusting predictions to match reality.

3.1.3.1 The Asaoka Method

Since this method of refining settlement predictions was developed by Akira Asaoka (Asaoka, 1978) in the 1970s and published by the Japanese Society of Soil Mechanics and Foundation Engineering, it is likely that you may not have heard of it. Of course, you may have heard of this method if you are Japanese or if you know Kevin Scott. Either way, I will go over the Asaoka method here to help introduce you to this method and give you some references for further study.

The Asaoka method (Asaoka, 1978) is a method that analyzes field settlement observations to predict the time to the end of primary consolidation settlement where only secondary compression or creep settlement continues. I know that theoretically the time to the end of primary consolidation is mathematically infinite, but we're talking here about the practical time to the end of primary consolidation, as determined by the Casagrande fitting procedure (where the tangent and the asymptote to the dial reading versus log time curve intersects thus defining the end of primary consolidation and the start of secondary compression). To perform the Asaoka method on field settlement data, no laboratory data is required. If you plot field-measured settlements (y-axis) versus time (x-axis) on a natural scale, the curve is initially quite steep, as shown in Figure 3.1.1. With extended time after loading the curve of settlement versus time flattens out. If you take settlement measurements at regular time intervals, the difference between settlement readings will be relatively great for the initial readings, but the difference between readings will become much less as you approach the end of primary consolidation. In fact after you reach the practical end of primary consolidation at time t_p, the difference between any two settlement readings will be zero, that is $S_n = S_{n-1} = S_{max \, @ \, t \, primary}$.

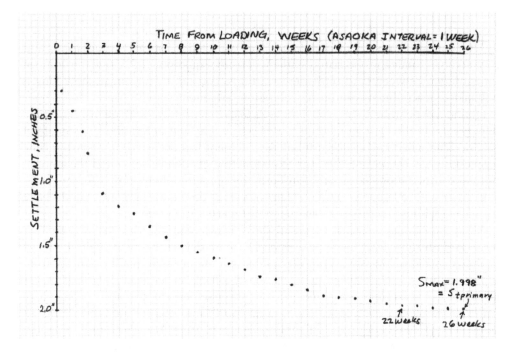

Figure 3.1.1 Field settlements versus time for the Asaoka method

Asaoka's derivation (Asaoka, 1978) shows that consolidation settlements measured at fixed time intervals after loading can be approximately by the equation:

$$S_n = \beta_o + \beta_1 S_{n-1} \tag{3.1.2}$$

where,

$S_1, S_2, S_3, \ldots, S_n =$ are settlement observations

$\qquad S_n =$ settlement measurement taken at time t_n

$\qquad \Delta t =$ a uniform, constant time interval between settlement readings and is equal to $(t_n - t_{n-1})$

β_o and $\beta_1 =$ constants of the Asaoka method. For this method we plot S_n on the y-axis and S_{n-1} on the x-axis, as shown in Figure 3.1.2. β_o is the intercept of this plot with the vertical axis and β_1 is the slope of the S_n versus S_{n-1} plot.

When the plot of S_n versus S_{n-1} data intersects the plot of a 45 degree line (that is when S_n equals S_{n-1}), that means that primary consolidation is done. Since the maximum value of primary consolidation (S_{max}) would be the same value for all readings at and after t_p that means that $S_n = S_{n-1} = S_{max}$ at times t_n and t_{n-1}, and we can substitute these values into Equation 3.1.2, giving:

$$S_{max} = \frac{\beta_o}{1 - \beta_1} \tag{3.1.3}$$

After enough field settlement data has been collected to estimate values for β_o and β_1 you can estimate the maximum settlement using Equation 3.1.3. Using the uniform time interval between field settlement readings, you can then estimate how long it will take to reach the maximum consolidation settlement.

I found a paper (Huat *et al.*, 2004) that suggests that plotting settlement versus time data for use in the Asaoka method works best if you first determine an estimate of t_{50} (the time when half of the estimated settlement has occurred) and start at time t_{50} plotting and estimating the time to completion of consolidation and the amount of total consolidation settlement. I'm normally not sure whether my estimate of t_{50} is good or not, so I start plotting settlement versus time data right away and I have increasing faith in my projected estimates as I reach and exceed t_{50}.

All of the Asaoka material discussed above is nice, but you can take this method a few steps further. First using piezocone CPT data determine the distance between drainage layers. Using laboratory consolidation data on high-quality samples from representative soft clay layers, using standard consolidation analysis procedures to calculate the maximum estimated consolidation settlement and the coefficient of consolidation c_v for each layer. Using a uniform time interval between estimated settlements, calculate a predicted settlement versus time plot. From the settlement versus time plot, take adjacent readings and prepare a predicted Asaoka plot of S_n versus S_{n-1}. Draw the $S_n = S_{n-1}$ (45 degree slope line) and a vertical line from the

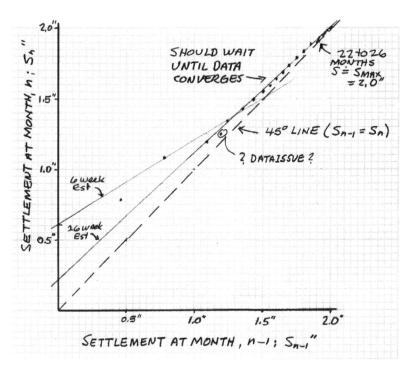

Figure 3.1.2 Asaoka plot of settlement reading

S_{n-1} axis representing your maximum estimated settlement. After fill construction and placement of surcharge fill loading, start your settlement readings at the predetermined time intervals. After you have several points plotted you can start making estimates of β_0, β_1, and S_{max}. Experience indicates that your estimates will start to get better and better after t_{50} (Huat et al., 2004), just look at my example in Figure 3.1.2.

Using field measurements and continuing to adjust estimates of maximum settlement and time to achieve maximum settlement by the Asaoka plot, you can start to make corrections to your original time rate of settlement calculations by making corrected estimates of c_v to replicate measured values of settlements.

3.1.4 A Few Lessons Learned from Field Measurements of Settlement

Working example

One afternoon, I was visiting Ralph Peck and we started a detailed discussion about a dam in Canada (I can't remember which one). Ralph went into his garage and pulled down the file from his investigation of the project (this was before Ralph sent his papers to the Peck Library at the Norwegian Geotechnical Institute). I remember he pulled out a hand-drawn plot of settlements versus time for several settlement plates located in and around the dam's embankment. I asked him if he was familiar with spreadsheets and he said that he was. So then I asked Ralph why he took the time to plot settlements by hand when he could do it easier with the computer-generated spreadsheet. His answer got me thinking. He said, "When I plot profiles and graphs of field data by hand it gives me a better feel for the information. It gives me a chance to think about the information while I'm hand plotting it on paper."

I had given up hand plotting of data 15 years earlier, but I started doing it again (see Figures 3.1.1 and 3.1.2) just to see if I could better visualize the site profile and the trend of field data as Ralph suggested. I don't always do a hand-plotted profile or a hand plot of field data, but when I do a hand plot, it does give me a chance to think. Thinking is good, and I do feel that I see the site and the data more clearly in my mind if I do a hand plot of data. You should try it sometime.

Working example

One problem with settlement measurements that I have observed from my own projects and from participating as a peer reviewer on other engineers' projects is a gross under- or over-estimation of the time to achieve the end of primary consolidation in the field. Two projects that I reviewed (I can't mention the project names) had project requirements that end of primary consolidation be accurately determined.

The first project was a landfill disposal cell that project engineers had estimated end of primary consolidation to be between 3 and 5 years. Thanks to regulatory requirements, the landfill owner was required to provide a detailed description of disposal cell settlement monitoring. One of the requirements was monitoring settlements until primary consolidation had ended and the rate of secondary compression was confirmed. In the landfill license was the requirement that the slope of measured settlement data plotted on a semi-log plot of time had to clearly change in a manner often illustrated in soil mechanics text books (one slope for primary consolidation followed by a flatter slope for secondary compression). Nearly 15 years after the disposal cell was completed, the owners and their engineers were still monitoring settlements, plotting settlements, and submitting results to regulators.

I was asked to review the data because the slope of the settlement curve had not changed and regulators were questioning why the consolidation settlement had taken much longer than the predicted three to five years. After a brief review of the data, I had the answer to their question. They didn't start monitoring settlement of the landfill disposal cell until after the primary consolidation was completed. I estimated that consolidation was completed in about six to nine months and they didn't start taking settlement readings until nearly one year had passed. All of the settlement readings that they had been taking for nearly 15 years were secondary compression settlement readings. The criterion written into their landfill license would never occur! They missed measuring the initial consolidation settlement data so it was impossible to see a change in slope as the regulators required.

My second example is similar to the first, and it also surprised project engineers. A large highway embankment was planned for a project over a deposit of soft clay. Calculations indicated that settlements of over a foot would occur, and that it would take two to three years for the primary consolidation to be completed. To allow the project to proceed on schedule, project engineers designed a surcharge loading to accelerate settlements and hopefully reduce long-term remaining settlements to less than 2 inches. There was a bit of a coordination problem between the contractor building the embankment and the engineers installing the settlement-measuring instrumentation. Without getting into the details, the settlement-measuring instrumentation was not installed until somewhat more than a month after completion of construction of the embankment plus surcharge fill. Settlement readings were taken week after week, and no significant change in settlement was measured. The engineers checked and rechecked their instrumentation and were confused by the failure of the embankment plus surcharge to settle. The contractor started getting nervous, because he didn't know when to remove the surcharge and continue construction. I was asked to see if I could figure out what was wrong with the instrumentation. I started by reviewing the time rate of settlement predictions. I found that embankment settlement analyses ignored CPT piezocone data and only used test boring profiles with laboratory consolidation test results. The CPT data showed that numerous sand seams were present in the thick clay deposit. Reworking the time rate of consolidation analyses using the drainage distances indicated by the CPT data gave a time to 95% consolidation of less than 1 month. The project engineers had installed the instrumentation to read consolidation settlement after the settlement had occurred.

Just about when I thought I had figured out actual time for completion of consolidation settlement was much less than my calculated values, I ran into a case similar to that illustrated in Figure 3.1.2 where the actual time for completion of consolidation settlement was nearly two years. I was hoping that unseen sand seams would reduce the time for consolidation to something like four months. Hoping is one thing, and having data is another, I really needed CPT piezocone data and was forced to get by with standard boring data due to client restraints.

If you are getting the notion that all consolidation settlement occurs in a couple of years or less, I suggest you review the case of the Shin-Ube power station included in the Asaoka paper (Asaoka, 1978). Primary consolidation of the clays beneath the Shin-Ube power plant took between six and seven years to complete settling, and the total settlement was nearly 80 inches.

3.1.5 Closing Remarks on Clay Settlement Calculations

Calculating the amount of consolidation settlement is a function of the quality of your sampling and laboratory consolidation testing. It is important to have a good estimate of the preconsolidation pressure to make accurate settlement estimates.

Time rate of consolidation settlement has been a difficult problem because historically, estimated times to reach 90 to 95% consolidation have been poor. Generally these estimates were much longer than observed times in the field, but not always. I have two suggestions that will help you solve this problem of accurately calculating the time rate of consolidation settlement. First use piezocone CPT tests to refine your drainage distance estimates, and second use regularly spaced (in time) field measurements and the Asaoka method to adjust your settlement estimates as your project progresses. Adjustments to your settlement estimates help reduce uncertainties that always exist on construction projects.

References

Asaoka, A. (1978) Observational procedure of settlement prediction. *Soils and Foundations*, **18**(4). Japanese Society of Soil Mechanics and Foundation Engineering. 87–101.

Holtz, R.D., and Kovacs, W.D. (1981) *An Introduction to Geotechnical Engineering*, 1st edn, Prentice-Hall Inc., *Englewood Cliffs, N.J.*, 733 pages.

Huat, B.K., Hoe, N.C., and Munzir, H.A. (2004) Observational methods for predicting embankment settlement, *Perlanika Journal of Science and Technology*, **12**, 115–128.

Terzaghi, C. (1925) *Principles of Soil Mechanics, I – VIII*, original articles in Engineering News Record 1925 Volumes 19, 20, 21, 22, 23, 25, 26 and 27, reprinted by the American Society of Civil Engineers, Geotechnical Special Publication No. 118, Volume One, *A History of Progress – Selected U.S. Papers in Geotechnical Engineering*, Edited by W. Allen Marr, pp. 2–38.

3.2

Settlement of Sands

3.2.1 Introduction

Calculating the settlement of a building footing, a bridge abutment, a bridge approach fill, a machine foundation, a landfill cover or a concrete dam are a few of the areas where settlement calculations are required. If your client cares about the performance of his structure, he will definitely want to know how much settlement or differential settlement it is anticipated to experience. I know from my experience as a structural engineer that my first request of the geotechnical engineer was an allowable bearing pressure value to use in sizing and designing concrete footings. After I had sized my footings, I needed an estimate of anticipated settlements to use in my structural steel frame model. Foundation settlements were very important in that model because settlements impacted redistribution of loading in the structural members. Calculation of settlements is an important part of geotechnical engineering practice, and it is one of the most difficult problems to tackle. How accurately must the settlement prediction be determined, and do we need to include time rate of settlement estimates? Settlement predictions are another case where the graded approach to problem solution is required.

When I took graduate geotechnical courses in the late 1960s, we calculated total foundation settlement as the sum of immediate settlement plus consolidation settlement plus secondary compression settlement. This total settlement calculation was assumed to apply to clayey soils since sandy soils were considered highly permeable and were not expected to experience consolidation settlement. So for sandy soils we added a calculated immediate or elastic settlement to a time-dependent secondary compression or creep settlement to find the total settlement.

There were problems with our settlement calculations based on this procedure. When we added immediate, consolidation and secondary compression settlements for clays or added immediate and creep settlements for sands to obtain a value of

Geotechnical Problem Solving, First Edition. John C. Lommler.
© 2012 John Wiley & Sons, Ltd. Published 2012 by John Wiley & Sons, Ltd.

total foundation settlement, it was too large, that is the estimated settlement greatly exceeded measured settlements. Every time we compared calculated settlements to measured foundation settlements, the calculated value was greater than the actual measured value.

Working example

For example, I had a great opportunity to check my calculations of immediate and consolidation settlement for a 100 000 gallon ground-supported fuel tank. The soil at this site was an unsaturated clayey fill over a saturated clay deposit. To calculate immediate tank settlement directly after tank filling, I used a layer thickness weighted average E_s value based on modulus values from unconfined compression tests multiplied by 4.0, as suggested by Bowles in his first edition text (Bowles, 1968). My calculated immediate tank settlement after filling was 3/8 of an inch. Since my calculated consolidation settlement was about 4 inches and since fuel was valuable even in those days, the client hired a surveyor to monitor settlements to make sure that the tank was not damaged by excessive settlements. I never got to see what the consolidation settlements were in this case. After initial tank filling, the immediate settlement was reported by the surveyor to be 1/8 of an inch. After immediate settlement, the surveyor detected zero consolidation settlement for the first month and a half. Seeing no significant settlements for 45 days, the surveyor convinced the owner to stop measurements and save money on surveying. I was young and not too convincing, because the client did not listen to my pleas to continue tank settlement monitoring. I told him consolidation settlement takes time to occur, but he chose not to listen. Two geotechnical lessons I learned: (1) my E_s value estimated from unconfined compression testing was too small by a factor of 12, and (2) time rate of consolidation estimates are extremely sensitive to field drainage conditions (not assumed drainage conditions). I also learned some lessons about engineer–client communications.

3.2.2 Settlement of Sands – General

The problem of calculating soil settlements is first an identification problem (what kinds of soils are present on my site?), second a characterization problem (what are my site's soil layer thicknesses, material properties etc.?), and thirdly it is a calculation problem (how do I model the problem and estimate a likely foundation settlement?). In this Section 3.2, we will discuss granular soil or "sandy soil" settlements and in Section 3.3 we will discuss self-weight settlement and metastable collapsible soils.

To avoid confusion, clarification of sand settlement terms is in order. The first type of sand settlement can be caused by a wetting-induced change in the state of the soil's structure under self-weight, which is called soil collapse. Soil collapse settlement does not require loading of the soil and it does not require wetting of the soil in some rare cases. Wetting-induced self-weight settlement is the most common type of collapse settlement. Self-weight settlement of wetted granular soils is discussed in Section 3.3.

Loading of a granular soil causes a second type of settlement that I will call ordinary granular soil settlement. We will discuss ordinary loading settlement in this section.

Granular soils' load-induced settlements are due to increases in normal and shearing stresses in the soil mass. These changes in soil stresses can be caused by large area fill loadings, shallow footing loadings, deep foundation loadings, both vertical and lateral, earthquake loadings, traffic surface loadings, and hydraulic loadings, to name a few. In this section, we will discuss settlement due to stress changes, assuming that the soil moisture content and soil suction-induced stresses remain constant.

3.2.3 The Granular Soil Identification Problem

Considering granular soils, the identification problem is a matter of determining what the nature of the granular soil is: (1) metastable, (2) loose, or (3) dense. A metastable granular soil has an unstable structure that can collapse upon wetting and if it has a very weakly cemented metastable structure, it could collapse upon loading. A loose granular soil has a void ratio that plots above the steady state line (SSL, see Chapter 2, Section 2.5), see Figure 3.2.1. Loose granular soil can also experience self-weight settlement from wetting and settlement from increases in shearing and normal stresses. A dense granular soil has a void ratio that is below the SSL, and can experience settlement upon increases in confining stresses and dilation (i.e., expansion) upon increase in shearing stresses.

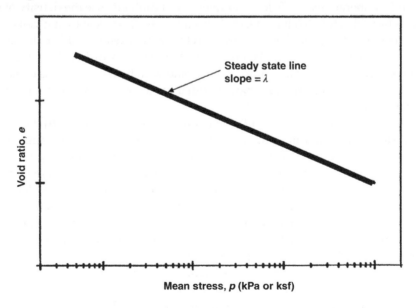

Figure 3.2.1 The steady state line

3.2.4 Identification of Loose Granular Soils

In general, you can't identify loose sandy soil by looking at it. Oh all right, if it is freshly deposited wind-blown sand, or a sand that was washing into a lake last week; it is surely loose sand. Putting those cases aside, you need a measure or an indicator of sand consistency to determine if it is loose sand. There are problems in defining terms when discussing sand consistency by using the terms soil density and "relative density." Sand consistency refers to the packing of the sand particles. If the sand particles are tightly packed, they form densely packed or "dense" sand. If the sand particles are not tightly packed, they form loosely packed or "loose" sand.

If soil materials were a solid like gold and aluminum, and you weighted a cubic foot of the soil, you would always get the same answer. Pure gold always weighs about 1206.1 pounds per cubic foot and pure aluminum always weighs about 168.5 pounds per cubic foot, so why doesn't sandy soil have a fixed unit weight. The simple answer is that sand is not a solid material. Sand is an accumulation of mineral particles with air- and water-filled voids between the sand's solid particles. The structure of one sand is never exactly like another. They may look very similar in color and texture, but they are never exactly alike.

If a cubic foot of sand has a large volume of voids and a small volume of solid particles, then it has a low unit weight and is considered loose sand. If sand has a relatively small volume of voids and a relatively large volume of solid particles, it is considered dense sand and has a greater unit weight.

It's not good enough to say relatively small or relatively large volume of voids when describing the consistency of sand, you need a frame of reference to know if a given sand's mineral mixture is loose or dense. You can't just use the density of sand in pounds per cubic foot to characterize it, because one mixture with a given density could be loose sand and another mixture with the same density could be dense sand. If mixtures of minerals for two layers of sand come from different geologic rock formations yet are found on the same site, the sand with heavier minerals could have a loose structure and weigh more than the other sand composed of lighter minerals with a dense structure.

So how do you identify loose sand? First you have to be familiar with the geology and consistency of sands in your area. If you are experienced and familiar with local sands, you can use SPT blow count, CPT cone bearing, dry density, void ratio, porosity, and relative density to identify your sand deposit as loose. Personally I like to use the old SPT method, if the blow count is between 2 and 4 it is very loose, and if it is between 4 and 10 it is loose. I don't generally consider loose soils as being potentially dilative unless their confining stress is well below 2000 pounds per square foot. Remember the term loose is an indicator that your sand is contractive (i.e., it will settle when sheared) under your anticipated loadings.

Now if you are up to date, you should ask me which N values are these. Given the yellow color of the paper that has my Rules of Thumb, you might guess the answer ... N_{55}. Want to see what a working engineer's "Rules of Thumb" chart looks like, check out Figure 3.2.2. I didn't clean it up. I like it the way it is with the charm of a working document. I've had this piece of paper for nearly 40 years. You can use it as a guide to check the reasonableness of an answer, but please never use it for final design.

3.2.5 Identification of Dense Granular Soils

So how do you identify dense sand? Again you have to be familiar with the geology and consistency of sands in your area. Just like loose sands, I like to use the old SPT method to identify dense sands. If the blow count is greater than 30 I usually consider it to be dense sand unless the confining stresses are estimated to be greater than about 20 000 pounds per square foot. Again you can check out my "Rules of Thumb" chart to evaluate your dense sand, but don't forget to adjust your SPT energy to N_{55}. If you have highly variable sands in your area, sands with differential cementing and weathering, or highly aged sand deposits, you will have to use several indicators to help identify which sands will likely act as loose sands and which ones will act like dense sands. If you have any doubts about sands on your project (I always maintain a certain amount of skepticism) make sure to check your direct shear and triaxial compression testing volume changes during shearing.

3.2.6 Analyzing the Sand Settlement Problem

A building footing or a highway embankment is constructed over a deposit of sandy soil; how do we calculate the settlement that may be expected to occur due to the increased soil stresses? Figure 3.2.3 shows a typical foundation settlement scenario.

Often elementary geotechnical texts suggest using a simple elasticity equation to calculate settlements, using the soil modulus E_s. To use these elasticity equations to calculate foundation settlements you have to divide the soil into layers, calculate the changes in stress in each layer by use of the Boussinesq or Westergaard chart, and then assign an appropriate value of soil modulus to each layer.

What is the soil modulus? The soil modulus E_s is the slope of the soil's stress versus strain curve. Determining the slope of the stress versus strain function is pretty easy if the stress–strain curve is a straight line, but it is not so easy for most soils because their stress–strain curves are curved! If we had a field or laboratory test method of determining the stress––strain curve, we could measure the modulus of elasticity of the sandy subsurface soils, but where do we take the slope of the curve? We have several choices that we will discuss in a moment.

Once you select an appropriate soil modulus, foundation settlement may be calculated from the change in layer stress $\Delta\sigma$ and modulus E_s by determining the change in strain $\Delta\varepsilon$ experienced by each soil layer, where $E_s = \Delta\sigma / \Delta\varepsilon$ or $\Delta\varepsilon = \Delta\sigma / E_s$. Then you determine the settlement simply by multiplying the resulting layer strain by each layer thickness and sum up the settlement of all of the layers. The main problem

Rules of Thumb

$CME \doteq N_{70}$
Safety
Hammer

$N_{90} = $ Auto hammer $\Rightarrow N_{90} \times 1.286 = N_{70}$

$N = N_{70}$

COHESIVE SOILS

Consistency	N_{55}	qu (tsf)
Very Soft	< 2	< 0.25
Soft	2-4	0.25-0.50
Medium Stiff	4-8	0.50-1.00
Stiff	8-15	1.00-2.00
Very Stiff	15-30	2.00-4.00
Hard	> 30	> 4.00

Meyerhoff Equation degrees

$\phi = 27 + \dfrac{10N}{35}$

COHESIONLESS SOILS

Compactness	N_{55}	Peck ϕ	Meyerhoff ϕ
Very Loose	2-4	< 28	30°
Loose	4-10	28-30	30-35°
Medium Dense	10-30	30-36	35-40
Dense	30-50	36-41	40-45
Very Dense	> 50	> 41	>45

$kg/cm^2 \sim tons/ft^2$

Soil	qc/N_{70}		Rock Hardness	N
Clay	1.0		Soft	< 50/6-8"
Silty Clay	1.5	clayey silt 1.5	Medium	50/3-4"
Sandy Silt	2.0		Hard	50/1-2"
Silty Sand	3.0		Very Hard	> 50/0"
Sand	4.0			
Sand & Gravel	6.0			

* qc/N values for mechanical cone, still works OK with piezocone.

TRACE	1-10%
MINOR	11-20%
SOME	21-35%
AND	> 35%

Modulus Subgrade Reaction from N_{70},

$ky = 4.23 \times qcon$

$ky = (4.23)\left(\dfrac{qc}{N}\right)(N_{70}) = lbs/in^3$

Figure 3.2.2   Geotechnical rules of thumb – I didn't clean it up!

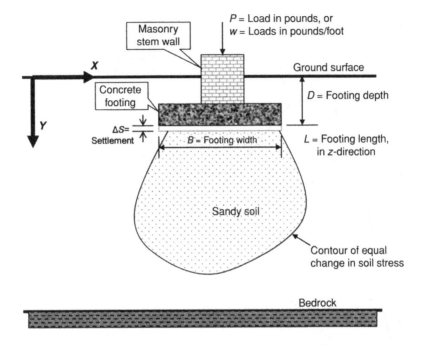

Figure 3.2.3 Footing settling in sandy soil

encountered with this solution strategy is the variety of soil moduli, illustrated in Figures 3.2.4 and 3.2.5. Other problems include calculations of the changes in stress in each layer with depth, and deciding which value of soil modulus should be used for each layer, the initial, the peak or the residual secant modulus.

To solve these problems you need to evaluate the distribution of stress increases generated in the soil by the loaded footing, and you have to model the soil settlement generated by increases in soil stress. There are several competing analysis methods to accomplish calculation of sand settlement; examples include the elastic method, the Schmertmann method, the Janbu method, and many others.

Two simple, straightforward methods of calculating sand settlements are the elastic method and the Schmertmann Method (which also uses E_s values for soil layers). Both of these methods are illustrated in standard geotechnical texts such as Bowles (5th edn, 1996), Das (6th edn 2005), and many others. The Schmertmann Method is also described in FHWA (Samtani and Nowatzki, 2006) and The US Corps of Engineers Manuals.

3.2.6.1 The Elastic Method

To use elasticity to calculate the settlement of a footing, we need to start with the basics. Young's modulus or the modulus of elasticity (E) is defined as: $E = \sigma / \varepsilon$, where

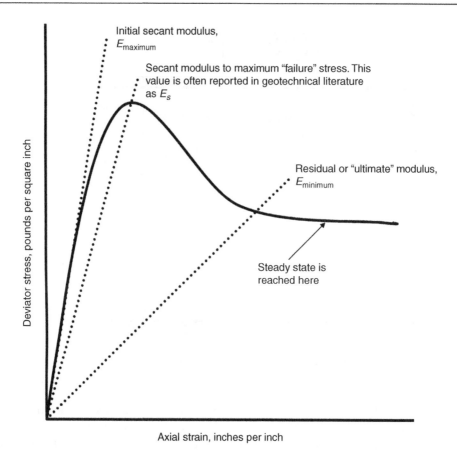

Figure 3.2.4 Secant modulus for dense sand in triaxial compression test

σ is the normal stress (foundation bearing stress) in units such as pounds per square inch, and ε is the soil strain generated by foundation loading stress σ with units of inches per inch. Strain ε is in the same direction as the applied normal stress σ. Stress σ is equal to force P divided by cross-sectional area A, and ε is equal to the change in length ΔS (soil layer settlement) divided by thickness H of the stratum compressed. You can substitute into the equation for modulus and get

$$E_s = \frac{P/A}{\Delta S/H}$$

or reordering terms

$$\Delta S = \frac{PH}{AE_s}$$

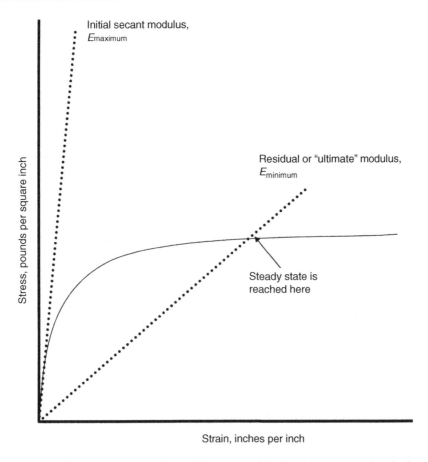

Figure 3.2.5   Secant modulus of loose sand in triaxial compression test

which is a simple form of an equation to calculate the settlement ΔS of a soil layer with thickness H and material modulus E_s, loaded by a force of magnitude P applied to a footing with bearing area A. You can see right off that we need help in using this simplified elastic settlement equation. First we need to know how deep or thick (H) the soil stratum affected by the footing loading extends into the soil. We also need to know how to determine the change in average stress in the compressible stratum. To do this we need to know how deeply the footing is embedded into the soil, and the depth of influence of the footing's applied stresses. Another issue is calculating immediate settlement and separating it from time rate or creep settlements.

The elastic method of calculating soil settlements works fine if you have an appropriate soil modulus E_s that is representative of stress levels of your foundations and the soils present at your site. To get an appropriate value of E_s you need a stress–strain curve for your soils that is representative of your problem. If strains are anticipated to be very low, that is on the initial, approximately linear part of the soil stress–strain curve, you can use the initial secant modulus for E_s, as shown on Figure 3.2.4. If the service loadings are anticipated to be approximately 1/3 of the peak soil strength, you can select a point on the stress–strain curve one-third of the way to the peak value, and draw a secant line to that point for determination of E_s. If you anticipate maximum or even residual stresses will be generated in your problem, then you need to select points on the stress–strain curve representing those conditions to determine your E_s value. This process may not be practical for simple projects, because you should use high-quality, relatively undisturbed, triaxial compression tests to determine your stress–strain curves. High-quality undisturbed samples of sandy soils are very difficult to obtain in the field and are difficult to maintain in an undisturbed state throughout the complete laboratory testing process. You can see why many labs use recompacted samples for triaxial testing, or engineers use correlated values of E_s obtained from SPT or CPT data. If you are making rough estimates of settlements in sandy soils you could use textbook values of E_s that approximate the soil types and conditions at your site.

Obviously you can't use unconfined compression tests on sandy soils to obtain stress–strain data. Even if your sand has enough clay to hold up in an unconfined compression test, it is my experience (see Section 3.2.1 above) that your measured E_s values will be too small by a factor of about 4 to 13, resulting in gross over-estimation of settlements.

The elastic settlement equation can be written as

$$S = C_1 C_2 q B \frac{1 - \mu^2}{E_s}$$
(3.2.1)

or as

$$S = q B \frac{1 - \mu^2}{E_s} I_w$$
(3.2.2)

where,

S = settlement of footing in units consistent with q, B, and E_s.

C_1 = correction factor for the depth of embedment of the footing. For a square, circular or long footing at the ground surface, $C_1 = 1.0$. For a square or circular footing, embedded to depth $D/B = 1$, $C_1 = 0.72$, and for a long footing ($L/B = 10$), embedded to depth $D/B = 1$, $C_1 = 0.88$.

C_2 = correction factor for the thickness of the stratum having E_s properties. In some cases this may be a correction factor for the depth to bedrock. For a depth of soil below the footing to bedrock equal to $2B$, $C_2 = 0.65$, and for a long footing $C_2 = 0.83$.

I_w = stress influence factor that considers foundation shape (circle, square or rectangle) and stiffness (flexible or rigid). Beneath the center of flexible circular, square and long ($L = 10B$) footings, $I_w = 1.00$, 1.12, and 2.50; and for settlements beneath the edge or corner of circular or square footings use 2/3 of these I_w values, and use $^1/_2$ of the I_w value for long footings. For settlements beneath rigid circular, square or long footings use $I_w = 0.85$, 0.85, and 2.10. For a more detailed listing of suggested I_w values please refer to standard geotechnical/foundations texts such as Bowles' books.

q = net bearing pressure applied to the ground at the base of footing.

B = width of the footing.

L = length of the footing.

μ = Poisson's ratio of soil, for sandy soils $\mu = 0.2$ to 0.35.

E_s = soil modulus in units of force per length.^{2}

3.2.6.2 The Schmertmann Method

The Schmertmann method (Schmertmann, 1970; Schmertmann, Hartman and Brown, 1978) is a standard practice technique for calculating settlements in predominately silty, sandy soil with some layers of unsaturated, low-plasticity clayey soil. Several published case history evaluations of the Schmertmann method, including Sargand, Masada, and Abdalla (2003), have determined that is method gives conservative, reasonably reliable estimates of settlement. Bowles, in his third edition text (1982) suggests that the Schmertmann method will give conservative settlement estimates if the soil modulus is constant or increases with depth, but may be not be conservative if the soil modulus is very low toward the bottom of the stress zone of influence.

The Schmertmann method of calculating settlements in sandy soils is frequently recommended for use in calculating settlements of granular soils for commercial projects. I have used this method for years, but prefer to use site-specific CPT data and not correlated q_c values determined from SPT blow counts by use of q_c/N ratios as has become standard practice in many regions of the United States. Foundation settlement is calculated by the Schmertmann method (Schmertmann, 1978) by solving the following equation:

$$\Delta S = C_1 C_2 \left(\Delta q \right) \sum_{1}^{n} \left(\frac{I_z}{E_s} \right) (\Delta z) \tag{3.2.3}$$

$$C_1 = 1.0 - 0.5 \left(\frac{p'_o}{\Delta q} \right) \tag{3.2.4}$$

$$C_2 = 1 + 0.2 \log_{10} \left(\frac{t_{in\ years}}{0.1} \right) \tag{3.2.5}$$

where,

ΔS = foundation settlement, calculated by summing up the settlement of each layer, layer 1, layer 2, layer 3, and so on in the same units as $\Delta z_{1,2,3\ldots}$

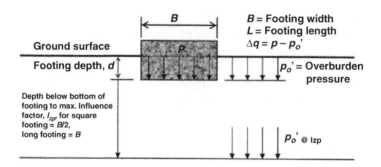

Figure 3.2.6 Definition of terms used in Schmertmann equations (Schmertmann, 1978)

C_1 = correction factor for foundation depth below ground surface

C_2 = correction factor for creep settlement of granular soils. For years this factor for estimating settlement of granular soils versus time was the only estimate of creep settlement of sands that was readily accessible (that I was aware of . . .)

n = number of sublayers of soil, each having a constant cone Penetration resistance, q_c, soil layer data required to depth $2B$ for square footings and depth $4B$ for long footings.

p = pressure exerted on the Footings bearing surface

p'_0 = the overburden vertical effective stress at the footing bearing level prior to placement and loading of the footing, see Figure 3.2.6 below.

$\Delta q = (p - p'_0)$ = foundation pressure increase at bearing surface in the same units of pressure as q_c, where q_c is the cone tip bearing pressure

$p'_{0\ @\ Izp}$ = the overburden vertical effective stress at the depth of the maximum influence factor, I_{zp}, see Figure 3.2.6 below.

I_z = strain influence factor that includes type of footing (square or long) and the depth of influence of foundation stresses generating settlement, see Figure 3.2.7 below.

E_s = soil modulus equal to 2.5 q_c for square footings ($L/B = 1$) and equal to 3.5 q_c for long footings, assume that your footing is a long footing for $L/B \geq 10$, between E_s of 2.5 q_c for L/B of 1 and $E_s = 3.5\ q_c$ for L/B of 10 calculate settlement for both cases and interpolate settlement for your footing length. the settlement for your problem's L/B.

t_{yr} = time in years from the application of foundation loading to the point in time when you want to calculate the footing settlement.

Δz = soils beneath the footing are divided into n representative layers with each layer n having a thickness equal to Δz.

A brief description of the Schmertmann method is given below. For more detailed information, articles by Dr. John Schmertmann (Schmertmann, 1970; Schmertmann, Hartman and Brown, 1978) are recommended.

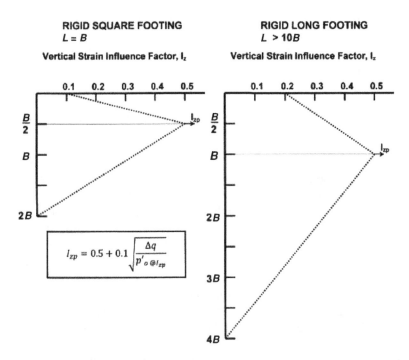

Figure 3.2.7 Definitions of vertical stress distributions for use in Schmertmann settlement analyses (Schmertmann, 1978)

First you have to identify if the footing is a square footing or a long strip footing. If it is a square footing (or a circular footing), you use a triangular stress distribution to a depth equal to two times the footing width, $D = 2B$. If the footing is a long footing with a length to width ratio $L/B \geq 10$, you use the modified triangular stress distribution to a depth equal to four times the footing width. Stress distributions for the square and long strip footings are given below in Figure 3.2.7.

What do you do if your footing's length is between $1B$ and $10B$? You calculate the settlement for both extremes of $1B$ and $10B$, and linearly interpolate the settlement for your actual footing's length between the $1B$ and $10B$ settlement values.

By the way, I have always considered a footing to be long if its L/B ratio is greater than or equal to 6, and if the L/B is greater than or equal to 4, I often go with the 3.5 q_c value. As you will see below, my lack of conservatism in using the 3.5 q_c value for footings having $L/B < 10$ has not been misplaced, because Schmertmann settlement calculations are often larger than measured settlement values.

3.2.6.2.1 Settlement of a Square or Circular Footing

After determining whether your footing is square (or circular), rectangular, or a long strip, you need a cone penetration log that represents the soil stratification from the ground surface to at least $2B$ beneath a square footing and to a depth of at least $4B$ for a long footing.

If the footing is circular, with diameter D, calculate the equivalent width of a square footing by $B = (0.8 * D^2)^{1/2}$, or by other published correlations between square and circular footings.

3.2.6.3 Schmertmann Settlement Calculations – A Few Issues to Consider

Schmertmann did his field testing on Florida sandy soils. I assume that this method is best for normally consolidated, uncemented sands. I also assume that his creep settlement factor C_2 is for normally consolidated sands. In FHWA-TS-78-209 on the bottom of page 50, Dr. Schmertmann said that this method should be used only with "first-loading cases with adequate bearing capacity." By use of the term "first-loading," I assume that Dr. Schmertmann means a normally consolidated sand that has not been preloaded or compacted to artificially densify the material. Dr. Schmertmann also says on page 50 that sands that have be prestrained by loadings that caused prior shear strains will likely have actual settlements less than those calculated by his method. If design engineers expect that sands on their site have been overconsolidated, preloaded, or roller compacted, he says that the increased modulus caused by these factors will not likely increase cone bearing values, q_c, by a similar amount. He suggests that if preloading has occurred that, tentatively, settlement values calculated by his method could be decreased by 50% and still be conservative. I interpret his use of the term "tentatively" to mean that he suggests that you could use the 50% reduction to estimate the reduced amount of settlement caused by preloading or compaction, but that actual field measurements or advanced analysis techniques would be required to confirm the settlement reduction caused by these factors. I have observed similar increased modulus values (i.e., reduced settlements) on sites with highly over-consolidated, aged granular deposits. I recommend that you design and perform large prototype or full-scale load tests to confirm load-settlement behavior of foundations on large or critical projects. Based on these tests you can develop site-specific correlations between q_c and E_s, and if necessary you can develop site specific q_c/N factors for your site.

Given that undisturbed samples of granular soils are difficult to obtain for use in high-quality laboratory testing, I believe that the future of settlement analyses in granular soils lies in the use of field tests such as the cone penetration test and full-scale load tests for determining moderate to large strain values of soil modulus, and in the use of seismic geophysical tests for determination of small strain values of soil modulus.

3.2.7 Janbu Method of Settlement Calculation

Dr. Nilmar Janbu was researching the topic of foundation settlements in Norway long before most of us were out of grade school. Born in 1920 (same year as my mother), he is a geotechnical professor emeritus at the Norwegian Technical University, Trondheim, Norway. I'm not quite sure how to say this, but it is my impression that Dr. Janbu was doing things his own way and developing his own methods of geotechnical analysis at the same time that nearly everyone else in the world was closely following Dr. Karl Terzaghi and his associates in the United States. Dr. Janbu reminds me of Ken Burns, who when asked where did he come up with his historical documentary films like *The Civil* War said something like, "This is what I was doing when you weren't looking." The methods of settlement calculation given below were what Dr. Janbu was doing while the rest of us were not looking. Apparently we in the west weren't too interested in Dr. Janbu's work, because he gave the series of invited lectures that this material comes from at Moscow State University in the Soviet Union from 23–25 May 1967.

Dr. Janbu developed a general soil settlement equation that applies to highly over-consolidated clays, silt-sands, and to soft, normally consolidated clays. His method is appealing because it is based on simple mechanics and it uses the tangent modulus rather than the secant modulus of soil materials. Best of all Dr. Janbu developed one equation that can be used for calculating settlements in all types of geomaterials: sands, silts, clays, and soft rocks. Personally, I've always wondered why we use the secant modulus to calculate the settlement of soils when we all know that the stress–strain curves for most soils are highly non-linear (take another look at Figures 3.2.4 and 3.2.5). If we would divide the soil strata into layers and define the tangent modulus for each layer, we could calculate the strain experienced by each layer, multiply the layer strain by the layer thickness to calculate the layer settlement, and then just add up the settlements for all of the layers to find the total settlement.

The Janbu settlement analysis starts by drawing a profile of the soil below the loaded area and determining the existing vertical effective stress σ_{vo}' at the center of each soil layer. I recommend that your profile extend to at least a depth of $2B$ for square or nearly square footings and to a depth of $4B$ for long footings. You can subdivide the soil into as many sublayers as you desire; presumably the more layers you use the better your answer. I suggest you limit your sublayer thicknesses to 2 feet . . . just a suggestion.

After you have a soil profile with existing vertical effective stresses, I suggest you use the Boussinesq chart for uniform soil profiles or the Westergaard chart for layered soil profiles to determine the change in vertical stress at the center of each soil layer or sublayer. Don't forget to use the appropriate side of Boussinesq/Westergaard charts for either square or long footings, and interpolate results for intermediate rectangular footings. As an alternate, you can use the appropriate Boussinesq/Westergaard equations to determine the change in stress $\Delta\sigma'$ at the center of each layer. Janbu had his own vertical stress distribution method which was developed at the Norwegian Technical University. To avoid undue complication, I prefer using standard Boussinesq/Westergaard equations. If you are interested in Dr. Janbu's complete method, please look up his Russian presentation on the internet.

Next we are going to calculate the foundation load-generated settlement strain ε at the center of each soil layer or sublayer and plot the strains on the depth profile. From the plot of loading-induced strain versus depth, we can calculate the settlement of the loaded footing. To see how to calculate settlement from the strain versus depth plot, let's look at an infinitesimally thin layer dz in the soil profile that experiences a stress-induced strain ε. The vertical compression or incremental settlement ΔS of the thin layer can be expressed as:

$$dS = \varepsilon(dz) \tag{3.2.6}$$

To calculate the total settlement S over the depth of stress increase ($2B$ or $4B$) or the depth of the soil layer to bedrock H, whichever is less, we need to sum up the compressions of each infinitesimal layer making up the entire depth. This sum is expressed as the integral in Equation 3.2.7:

$$S = \int_0^H \varepsilon(dz) \tag{3.2.7}$$

Equation 3.2.7 indicates that the total settlement S of a footing is equal to the area under the plot of the soil strain versus depth diagram (i.e., ε versus z diagram). Dr. Janbu wrote that calculating the area under the ε versus z plot was one of the best tools for determining and seeing the settlement distribution in a soil profile. All you need to do is plot strain values at the center of layers on your soil profile, draw a line through the strain values, and calculate the area under the ε versus z plot to determine the total foundation settlement S.

Now all we need to determine is the stress-induced strain ε at the center of each soil layer or sublayer. Dr. Janbu developed his equations for determining soil strains directly from mechanics principles. He called the soil compression deformation modulus M, and defined this modulus for each incremental stress $d\sigma_v'$ and incremental strain $d\varepsilon$ change as:

$$M = \frac{d\sigma_v'}{d\varepsilon} \tag{3.2.8}$$

From this basic definition of deformation modulus M, he rewrote the equation to express the incremental change in vertical strain $d\varepsilon$ caused by an incremental change in vertical stress $d\sigma_v'$ as:

$$d\varepsilon = \frac{d\sigma_v'}{M} \tag{3.2.9}$$

Then to calculate the total strain ε caused by a change in stress from the initial vertical effective stress σ_{vo}' to the final vertical effective stress $\sigma_{vo}' + \Delta\sigma$ we need to integrate Equation 3.2.9 from the initial effective vertical stress to the final vertical effective stress, as shown in Equation 3.2.10:

$$\varepsilon = \int_{\sigma'_{vo}}^{\sigma'_{vo}+\Delta\sigma} \frac{d\sigma'_v}{M} \tag{3.2.10}$$

Introducing constants m (modulus number), a (stress exponent), and p_a (atmospheric pressure, used to normalize the equation, $p_a \approx 1$ tsf $= 2$ ksf ≈ 1 kg cm$^{-2} = 100$ KPa), Janbu developed an expression to functionally relate vertical effective stress σ_v' to the compression modulus M:

$$M = m\, p_a \left[\frac{\sigma'_v}{p_a}\right]^{1-a} \tag{3.2.11}$$

Equation 3.2.11 is bounded by values of $0 \le a \le 1$, which reasonably covers the range of values encountered for normally consolidated clays ($a = 0$), silts and sands ($a = 0.5$), and over-consolidated clays ($a = 1$).

Substituting M from Equation 3.2.11 into Equation 3.2.10, Janbu obtained his general equation for calculating the strain ε caused by an increase in vertical stress $\Delta\sigma$:

$$\varepsilon = \frac{1}{m\,a} \left[\left(\frac{\sigma'_{vo} + \Delta\sigma}{p_a}\right)^a - \left(\frac{\sigma'_{vo}}{p_a}\right)^a \right] \tag{3.2.12}$$

We are discussing settlement of sandy soils in Section 3.2, so Equation 3.2.12 for sandy soils where $a = 0.5$ becomes:

$$\varepsilon = \frac{2}{m} \left[\left(\frac{\sigma'_{vo} + \Delta\sigma}{p_a}\right)^{0.5} - \left(\frac{\sigma'_{vo}}{p_a}\right)^{0.5} \right] \tag{3.2.13}$$

For a sandy soil deposit with $m = 125$ (a loose sand), $\sigma'_{vo} = 1.20$ tsf, $\Delta\sigma = 1.0$ tsf, calculate the strain ε using Equation 3.2.13: $\varepsilon = 0.0062 = 0.62\%$.

As a guide, Janbu gave m values as follows:

Loose sands, $m = 80 - 150$
Medium sands, $m = 150 - 250$
Dense sands, $m = 250 - 400$

You can calculate your own values of m if you perform a consolidation test on your sand (i.e., a confined compression test in a consolidometer) that covers the range of vertical effective stresses in the zone of influence of your foundations. If you get the compression index C_c from this test and the initial void ratio e_o, the value of m is equal to:

$$m = 2.3 \left(\frac{1 + e_o}{C_c} \right) \tag{3.2.14}$$

For a sandy soil with $C_c = 0.029$ and an initial void ratio $e_o = 0.9$, we calculate an m value from Equation 3.2.14 as: $m = 151$ which is a loose to medium dense sand.

Since it is difficult to obtain high-quality undisturbed samples of silty sands, we would be better off if we could find m values from CPT cone tests. Now here is the rub ... where are the published m values from CPT or SPT tests? They don't exist or possibly they are not well publicized, in either case I'm not aware of them. One answer to the question, "Why don't they exist?" may be included in Janbu's 1967 lecture notes. Toward the end of his second lecture, Janbu said that the M modulus values are apparently primarily caused by shearing stresses. What was his point? I believe that he was referring to the fact that engineers in general believe that consolidation of soils is primarily a compression stress phenomena. In reality the only consolidation settlement loading case that is primarily the result of compression stresses is a large semi-infinite area fill, see Figure 3.2.8. Most column footings can be approximated by point loads and most wall footings can be approximated by line loads, both of which generate shearing stresses in the supporting soils, again please refer to Figure 3.2.8.

This issue about the presence of shearing stresses in the zone of settlement brings up a good point on the nature of stresses generated by foundation loading. If we are going to accurately estimate foundation settlements, we need to use a laboratory test that replicates, as close as possibly, the state of stresses generated by our foundation when we determine the value of M to use in Janbu's equations. If we have a large area loading, the confined compression or consolidation test would be suitable for determining the modulus M, and if we have a square footing a triaxial test would be better suited to determine M.

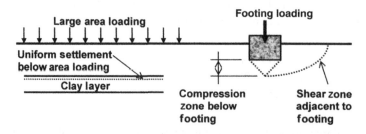

Figure 3.2.8 Comparison of a large area versus a footing settlement

Figure 3.2.9 When settlements are independent, they can be added!

3.2.8 Estimating Settlements – Why Did We Over-Estimate the Settlement?

I mentioned in Section 3.2.1 that in the late 1960s we used to calculate the total settlement, S_{total} using the following Equation 3.2.15:

$$S_{total} = S_{immediate \text{ or elastic}} + S_{consolidation} + S_{secondary \text{ compression or creep}} \qquad (3.2.15)$$

After saying in Section 3.2.1 that we calculated the total settlement using this equation, I said that the calculated settlements using this approach were too large. What I didn't say was why the calculated values were too large. There are several reasons why the settlement calculated by Equation 3.2.15 gives an answer that is too large. I'll go over a few of these issues for you.

3.2.8.1 First Problem – Equation Terms Not Independent

The first problem is the most basic problem in the world of engineering mathematics. The three terms making up this equation are not independent terms, so by definition they cannot be directly added to give a sum. Let me illustrate by a pair of Venn diagrams, Figures 3.2.9 and 3.2.10.

Notice in Figure 3.2.9 that the area of three circles represents the settlement calculated by three separate processes, and as such these three circles do not intersect. The functions that are used to calculate these three areas are independent. Calculating

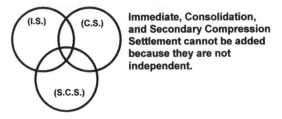

Figure 3.2.10 Adding dependent settlements results in over-estimating total settlement (you're adding same values more than once!)

the complete area by adding the individual areas of each of these three circles, we obtain the total settlement. This is what we were doing by using Equation 3.2.15 above. The problem with this approach is illustrated in Figure 3.2.10. Notice that the circles shown in Figure 3.2.10 overlap or intersect, illustrating that their areas are not independent of each other. To calculate the complete settlement illustrated by the intersecting circles in Figure 3.2.10 we need to count the intersecting areas once not twice. As a result the total settlement represented by the three circular areas shown in Figure 3.2.10 is less than the sum of the areas represented in Figure 3.2.9.

What I am saying above in geotechnical terms is that during incremental loading (i.e., 1/16 tsf, 1/8 tsf, $\frac{1}{4}$ tsf, $\frac{1}{2}$ tsf, 1 tsf etc.) of a consolidation sample, each increment's settlement readings includes an initial elastic settlement when the load is applied, a continuing consolidation settlement through the loading time, and a component of secondary compression or creep during the entire loading time. Although consolidation test results include all three settlement types described in Equation 3.2.15, they are all attributed to consolidation settlement because the consolidation test is assumed to measure only consolidation settlement up to $t_{100\%}$ and only secondary compression or creep settlement after consolidation settlement is complete.

3.2.8.2 Second Problem – Mixing Stress States

Another problem in combining settlements from elastic settlement, consolidation settlement, and secondary compression settlement is mixing stress states that are incompatible and don't represent the state of stresses in the field. The elastic modulus may be determined from triaxial compression testing (don't use the unconfined compression test for determining soil moduli) and the consolidation settlement determined by the consolidation or confined compression test. The triaxial compression test primarily generates shearing stresses in the sample, while a consolidation test is intended to minimize shearing stresses in the sample. The types of stresses generated in these tests are so different that they likely don't model field stresses and as a result can't be added to represent settlement of a field loading case.

3.2.9 Additional Sand Settlement Information – Specific

If you have time and budget for a detailed analysis of granular soil settlement, there are several ways to go. I suggested that you first determine whether your soil is: (1) loose, dry, or collapsible soil with a metastable structure, (2) loose soil with a void ratio above the steady state line (SSL), or (3) medium dense to dense soil with a void ratio below the steady state line (SSL).

If the soil is a metastable structured collapsible soil, you need to determine how much of the soil profile may be wetted, and whether it will be wetted enough to collapse. After you determine how deep significant wetting will penetrate, you then need to determine how much collapse upon wetting will occur (normally given as the percentage of layer height reduction). Without wetting, collapsible soils do not normally experience significant settlement from foundation loading unless the loading is very high and generates partial soil-structure collapse. Given normal foundation loadings (generally less than 8000 pound per square foot), foundation settlements in collapsible soils are determined from collapsible-upon-wetting analyses conducted at expected vertical stresses.

If the soil is a loose dry granular material with a stable structure and a void ratio above the steady state line, it may experience two types of settlement: (1) settlement due to loading stress increases, and (2) settlement due to wetting, which may cause a loss of dry strength (i.e., loss of soil tension stresses/soil suction). Settlement due to foundation loading is caused by increases in soil compression and shearing stresses that cause settlement until the soil reaches its critical void ratio at large shearing strains. If the loose granular soil is at or near saturation and does not have dry strength (apparent cohesion) due to soil tension stresses generated by soil suction, then it will experience settlement due to loading stress increases alone.

If the soil is a dense sand with a stable structure and a void ratio below the steady state line, it will experience generally small elastic settlement due to compression stresses, but it will experience expansion or dilation due to shearing stresses.

Calculating the time rate of creep settlement of a granular soil is a difficult problem to solve. If you need a rough estimate of the time rate of creep settlement, I recommend the Schmertmann Equation 3.2.5 for C_2 discussed above. If you need a highly accurate measurement of the time rate of creep settlement in sandy soil, I recommend a long-term, full-scale load test with controlled soil moisture content (i.e., controlled soil suction), controlled relative humidity, controlled soil temperature, controlled vertical effective stresses, and with accurate continuous measurements of foundation movement. If you pour a full-scale concrete footing, I suggest you monitor foundation concrete movements during concrete curing and soil movements separately.

References

Bowles, J.E. (1968) *Foundation Analysis and Design*, 1st edn, McGraw Hill Inc., 659 pages.

Bowles, J.E. (1982) *Foundation Analysis and Design*, 3rd edn, McGraw Hill Inc., 816 pages.

Bowles, J.E. (1996) *Foundation Analysis and Design*, 5th edn, McGraw Hill Inc., 1024 pages.

Das, B.M. (2006) *Principals of Foundation Engineering*, 6th edn, CL-Engineering, 750 pages.

Samtani, N.C. and Nowatzki, E.A. (2006) *Soils and Foundations Reference Manual, Volumes 1 and 2*, FHWA-NHI–06-089, National Highway Institute, Federal Highway Administration, Washington, D. C., Two volumes 1056 pages.

Sargand, S.M., Masada, T., and Abdalla, B. (2003) Evaluation of cone penetration test-based settlement prediction methods for shallow foundations on cohesionless soils at highway bridge construction sites. *Journal of Geotechnical and Geoenvironmental Engineering*, Geo Institute of the American Society of Civil Engineers, **129**(10), 900–908.

Schmertmann, J.H. (1970) Static cone to compute static settlement over sand. *Journal of the Soil Mechanics and Foundation Division*, American Society of Civil Engineers, **96**(3), 1011–1043.

Schmertmann, J.H., Hartman, J.P., and Brown, P.R. (1978) Improved strain influence factor diagrams. *Journal of the Soil Mechanics and Foundation Division*, American Society of Civil Engineers, **104**(8), 1131–1135.

Schmertmann, J.H. (1978) *FHWA TS-78-209, Guidelines for Cone Penetration Test, Performance and Design*. The United States Department of Transportation, Federal Highway Administration, Offices of Research and Development, Implementation Division (HDV-22), Washington, D.C., 145 pages.

3.3

Self-Weight Settlement of Sandy Soils

3.3.1 Introduction to Collapsible Soils

Various authors use the term "collapsible soil" with differing meanings. In this text, I use the term soil collapse to describe a two-stage settlement upon wetting of a dry granular soil due to its self-weight.

The first stage of collapse is structural collapse of a very loose granular soil with a metastable structure. This collapse is due to loss of particle cementing and macrostructure support upon wetting. After the metastable soil's cemented particle bonds break and its support structure collapses, it is still a very loose soil, and so it may continue to settle by loss of soil suction due to wetting, as described by Fredlund (Tadepalli, Fredlund, and Rahardjo 1992).

Montessa Park located south of Albuquerque, New Mexico is infamous for the presence of highly collapsible soils (Hansen, Booth, and Beckwith, 1989). In Figure 3.3.1, the jail at Montessa Park settled so severely after wetting of collapsible soils that it had to be abandoned and demolished.

In Figure 3.3.2, I am collecting block samples of collapsible soil from an exposure located near the abandoned jail in Montessa Park. I had read the papers written by my colleagues at Sergent, Hauskins and Beckwith, and had discussed theories of collapsible soils extensively with George Beckwith and Bob Booth, but I still needed to study Albuquerque collapsible soils for myself. I had to "see it to solve it."

The second stage of collapse settlement is identical to the first stage of settlement of a loose dry *in situ* sand or poorly compacted, loose, dry granular fill that has a stable loose structure. Again loss or reduction of soil suction in this "second stage" of collapse settlement is from self-weight not from foundation loading, see Figures 3.3.3 and 3.3.4.

Geotechnical Problem Solving, First Edition. John C. Lommler.
© 2012 John Wiley & Sons, Ltd. Published 2012 by John Wiley & Sons, Ltd.

Figure 3.3.1   Collapse settlement damage to building, Montessa Park, New Mexico caused facility to be abandoned six months after construction

Figure 3.3.2   Collecting collapsible soil block samples near Montessa Park

Figure 3.3.3   Trench fill settled after a rain requiring gravel backfill to prevent traffic from dropping into a large hole

Figure 3.3.4   Retaining wall backfill settled upon wetting

3.3.2 Soil with a Metastable Collapsible Structure

Working example

A few years ago, I started a study of highly collapsible soil in Albuquerque, New Mexico, see Figure 3.3.2. The idea for this study came from a 2007 luncheon discussion about collapsible soils that I had with a senior member of our Albuquerque consulting community Benny McMillan. When I came to Albuquerque in 1990, it was my impression that local engineers had studied and figured out the nature of Albuquerque collapsible soils long ago, or at least by the time of the Northwestern University conference in 1989. Benny corrected my understanding; he mentioned that as recently as 1970, local engineers believed that collapsible soils were gypsum-cemented loose sandy soils or very dry, loose silty sands. Upon arrival in Albuquerque, I was told that local collapsible soils were weakly cemented silty sands with small pin holes that had unit weights less than about 95 pounds per cubic foot and moisture contents less than about 5%.

My discussions with Benny got me thinking that local engineers might have missed something while working on collapsible soil consulting projects and if I took a close look at collapsible soils, there might be something new I could learn. First I located a well-documented collapsible soil deposit and obtained some block samples, see Figure 3.3.2.

Upon returning to the laboratory my trusted, best laboratory technician Eric and I obtained unit weights on the undisturbed block samples and obtained values that ranged from 65.7 to 103.6 pounds per cubic foot. Then we destructured (i.e., ground up) the samples, keeping them dry, poured a composite sample into a container, and lightly hand-jiggled the sample container to obtain a loose laboratory sample with a unit weight of 87.9 pounds per cubic foot. With the exception of the one sample at 103.6 pcf, all of the field samples were lighter than the laboratory loose sample. The loosest field sample was 25.2% lighter than the loosely prepared laboratory sample. How can a field sample be so much lighter than a very loosely prepared laboratory sample? The answer is that the field sample has a metastable structure. By metastable, I mean that this soil has some mechanism that holds soil grains in an extra loose configuration, and prevents it from densifying due to natural disturbances. There are published documents that show sketches (not photographs) of grains of sand with cementing material that hold the individual sand particles apart in a loose metastable structure. While we were at it, Eric and I decided to take a close up photograph and a microscopic photograph of our collapsible soil samples to see the cementing described in published materials. Remember my motto is that you have to see it to solve it. Take a look at Figure 3.3.5(a) and (b) to see the results of our collapsible soil photographs.

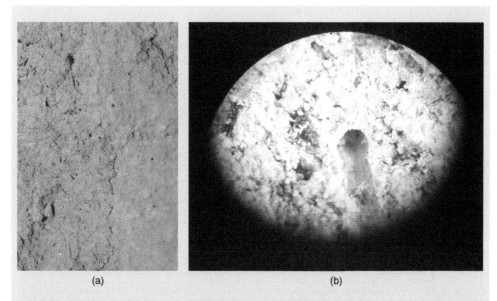

(a)										(b)

Figure 3.3.5 (a) Close-up photograph of collapsible soil with visible pin holes; (b) microphotograph of collapsible soil showing holes are tubes

In the close up photograph in Figure 3.3.5(a) you can see the small pin holes that are known to characterize collapsible soils in much of the Southwestern United States. These pin holes have a diameter of about 0.2 mm. The surprise comes in Figure 3.3.5(b) where you can see in the microphotograph that the pin holes are actually small tubes and not spherical holes in the soil structure. Upon very close inspection you can see the loose structure of the collapsible grains with small amounts of cementing paste at the grain to grain contacts. During my attempts to replicate the soil structure with the fine tubular features in the laboratory, I determined that the cementing paste between grains is high-plasticity smectite clay, with liquid limit values in excess of 100 and PI values ranging from about 60 to 80. This came as a surprise to me because laboratory testing of these collapsible samples for plasticity index came back with PI values from 0 to 5. The high plasticity cementing clay material is apparently present in such small amounts in the collapsible soil sample that it doesn't significantly affect the results of a plasticity index test taken on the entire sample.

We found that wetting of metastable collapsible soils having high-plasticity clay for a particle cementing agent softens the hard dry cementing clay causing it to swell. We found that samples of this collapsible soil placed dry in a consolidometer and placed dry in a Proctor mold with low vertical confining stresses experienced measureable swell as small amounts of water were incrementally added to the samples. It is standard geotechnical practice in New Mexico to place dry *in situ* samples of collapsible soil in a consolidometer and incrementally add loads until the vertical stress approximates the anticipated foundation loading (i.e., 1 or 2 tons per square foot), then the consolidometer is flooded with water and the collapse settlement measured as shown in Figure 3.3.6.

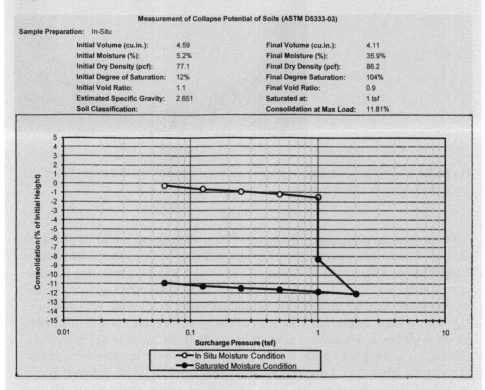

Measurement of Collapse Potential of Soils (ASTM D5333-03)

Sample Preparation: In-Situ

Initial Volume (cu.in.):	4.59	Final Volume (cu.in.):	4.11
Initial Moisture (%):	5.2%	Final Moisture (%):	35.9%
Initial Dry Density (pcf):	77.1	Final Dry Density (pcf):	86.2
Initial Degree of Saturation:	12%	Final Degree Saturation:	104%
Initial Void Ratio:	1.1	Final Void Ratio:	0.9
Estimated Specific Gravity:	2.651	Saturated at:	1 tsf
Soil Classification:		Consolidation at Max Load:	11.81%

Figure 3.3.6 Typical collapse upon wetting test

With this standard procedure, I never knew that collapsible soils could swell when water is slowly added and the vertical stress is 1/16 to 1/8 ton per square foot. These light loadings approximate slab on grade or paving surcharge loadings. Of course if the collapsible soil profile is wetted deeply enough, the collapse settlement will predominate over the initial swelling tendency. The point I'm trying to make here is that assumptions made during standard geotechnical testing procedures may mask significant properties of soils tested.

3.3.3 Collapse Settlement of Dry Loose Sandy Soil with a Stable Structure

As we discussed earlier, loose unsaturated sandy soil has a high dry strength due to soil suction. As loose, dry sandy soils are wetted, they lose dry strength as their matric suction decreases until they reach a minimum strength in the saturated state, see Figure 2.7.5. As these sandy soils lose strength they cannot support their own weight without settling. Wetting of loose dry sandy fill to "settle it down"is an old contractor trick called ponding. Ponding is often used to densify trench backfill in many parts of the United States. The problem with the ponding technique is that it doesn't develop uniform or consistent settlement of trench backfill, and water from trench-fill ponding can soak into adjacent soils causing unintended settlement of nearby utilities or paving.

Placing dry loose fill adjacent to buildings or in sewer trenches that are later paved over is a significant problem for contractors and engineers in many parts of the country. Since these materials are stable while dry, they are often overlooked for many months or even years. Finally when a wet year comes along, or a water line leaks, these soils are wetted and experience significant settlement. It has been my experience that loose, dry uncompacted fills can settle from 2 to 10% of their height upon wetting. The sad part of these wetted fill settlement problems is that the contractors often have compaction test results in their files proving that these fills were compacted to 95+% of specified Proctor maximum density. Further study frequently shows that compaction tests were done intermittently and that fill lift thicknesses of five feet went untested. One case I clearly recall involved 30 to 35 feet of backfill that was not in place on Saturday at 5:00 p.m., but was in place and ready for testing on top by the inspector on Sunday morning. The night shift had been very productive without the help of the compaction technician and project engineers. Testing on top of the 30+ feet of fill produced passing compaction tests. Monday morning they paved over the fill. Two weeks later after a heavy rainfall, the fill settled more than a foot! Subsequent testing of the fill by test borings showed that the fill had SPT blow counts of less than 4 throughout much of its depth. I found out later from a contractor employee (unofficially, since he was a friend of mine) that the fill had been pushed in full depth without compaction effort.

3.3.4 Standard Laboratory Testing of Collapsible Soils

Figure 3.3.6 represents a standard collapse upon wetting test performed in a consolidometer. This test is referred to as D-5333-03, *Standard Test Method for Measurement of Collapse Potential of Soils* by ASTM.

I am often asked the same questions about the interpretation of collapse upon wetting tests. At what level of collapse might I consider this test to be a metastable

Table 3.3.1   Soil properties indicative of collapse potential

Property	"Collapsible"	"Likely Collapsible"	"Possibly Collapsible"	"Not Likely Collapsible"
Dry Density, γ_{dry} pcf	$\gamma_{dry} \leq 80$ pcf	$80 \leq \gamma_{dry} \leq 90$	$90 \leq \gamma_{dry} \leq 103$	$\gamma_{dry} > 103$ pcf
SPT Blow Count, N_{70}	$N \leq 4$	$4 \leq N \leq 10$	$10 \leq N \leq 20$	$N > 20$
Water Content, $w\%$	$w\% \leq 4\%$	$4\% \leq w\% \leq 6\%$	$6\% \leq w\% \leq 8\%$	$w\% > 8\%$
Degree of Saturation, $S_r\%$	10% to 40%	40% to 60%	60% to 80%	$S_r > 80\%$
Soil Fabric	Porous with visible pin holes	—	—	Soil wetted, remolded and re-compacted

soil and when might I consider it to be a loose, dry granular soil with a stable structure? Does sample disturbance cause an increase or a decrease in the measured collapse potential?

Table 3.3.1 below gives some general measures to determine if a soil is a metastable collapsible soil or whether it is a loose, dry granular soil. The values given in Table 3.3.1 are from my experience with collapsible soils in New Mexico, please consider these values as general trends that should be confirmed or modified for geotechnical practice in your local region.

Relative to sample disturbance affects, if the sample was a metastable soil to begin with it will have a decrease in the measured collapse potential if the sample is disturbed during sampling and testing. If the sample is a dense sample with a void ratio below the SSL, disturbance will loosen the soil and likely increase its apparent collapse upon wetting. It is my experience that dry, dense soils that are disturbed during sampling and testing rarely, if ever, have an indicated collapse upon wetting of more than 2%. If laboratory testing of a disturbed dense granular sample indicated that it had some collapse potential, and field sampling indicated that this sample had SPT blow counts over 20, and a dry density over 110 pounds per cubic foot, it is likely that it is not really a collapsible soil.

References

Hansen, L.A., Booth, R.D., and Beckwith, G.H. (1989) Characterization of a Site Underlain by Deep Collapsing Soils. *Foundation Engineering: Current Principles and Practices*, **Volume 1**, Geotechnical Engineering and Construction Divisions of the American Society of Civil Engineers, Evanston, Illinois, Fred H. Kulhawy, Editor, pages 191–208.

Tadepalli, R., Fredlund, D.G., and Rahardjo, H. (1992) Soil collapse and matric suction change. Proceedings of the 7th International Conference on Expansive Soils, Dallas, Texas, **Volume 1**, pp. 286–291.

3.4

Bearing Capacity of Shallow Foundations

3.4.1 Background and History of Bearing Capacity

In the era before Karl Terzaghi's bearing capacity equation was introduced in the United States, engineers went to their local building codes to determine what the allowable bearing pressure was for design of their foundations. If they were working outside of a major metropolitan area, where no local experience or building codes existed, they had to refer to general texts on "foundation soils."

In my library, I have a textbook, *Foundation Soils*, published by the International Textbook Company in 1909 with no author noted, which I find to be weird. This 1909 engineering text suggests that foundation soils should first be examined to determine their character, be it rock, gravel, sand, or clay.

They go on to say, "Where no test of the sustaining power of the soil is made, different soils, excluding mud, at the bottom of the footings shall be deemed to sustain safely the following loads to the superficial foot: soft clay – 1 ton per square foot; ordinary clay and sand together in layers, wet and springy – 2 tons per square foot; loam, clay or fine sand, firm and dry, 3 tons per square foot; very firm, coarse sand, stiff gravel or hard clay – 4 tons per square foot." They strongly suggest performing a full-scale footing load test. They state that a large building or important structure where the designer wants a greater bearing capacity than given in the list above, a load test is to be performed. In the 1909 text, they give a detailed description of a full-scale footing load test to determine the load versus settlement curve of the foundation–soil system.

Before introducing the bearing capacity equation in his 1943 textbook, *Theoretical Soil Mechanics* (Terzaghi, 1943), Terzaghi discusses a footing's load settlement curve and how to determine a footing's failure loading from the load test curve. Then

Geotechnical Problem Solving, First Edition. John C. Lommler.
© 2012 John Wiley & Sons, Ltd. Published 2012 by John Wiley & Sons, Ltd.

Terzaghi goes on to suggest that the bearing capacity equation is a simplified method of estimating the failure loading of a shallow footing. He discusses a general shear failure and a local shear failure of foundation soils, indicating that excessive settlement of loose sands results in a local failure before a general shear failure can be mobilized.

Early geotechnical texts such as *Soil Mechanics in Engineering Practice* by Terzaghi and Peck, first printed in 1948, include charts that give allowable bearing pressures for 1 inch of settlement for footings of varying widths on sands of varying density. The point of this review of early geotechnical publications is that bearing capacity was not limited to a discussion of soil strength, as it often is today. Early pioneers of geotechnical practice knew that performance of foundations was directly related to foundation settlement. A foundation that was load tested and designed for a high allowable bearing pressure may not experience a general shear failure, but if the footing settled excessively, it failed to support the structure properly.

3.4.2 Allowable Bearing Pressure

One engineer told me, "When I was in school, the professor said that you find the soil's cohesion and friction values, plug them into the bearing capacity equation, divide by a factor of safety of three and presto, you have the allowable bearing pressure for foundation design. Simple enough, game-set-match, I'm ready to design my footings."

Another engineer said, "After I got out of school and started working as a staff geotechnical engineer, I never used the bearing capacity equation. We related standard penetration blow counts to approximate allowable bearing pressures, sized footings and calculated footing settlements for various loading cases, what happened to the three N's bearing capacity equation?"

I was in a meeting with an architect and structural engineer of a building that had experienced excessive foundation settlements requiring underpinning. The structural engineer was genuinely confused. He said, "All of my footings were sized for less than the recommended allowable bearing pressure. How could the footings fail if I didn't overload them?"

"It's simple," I told him, "The footings were built on dry, loose fill that was contaminated with clay balls. The fill soil settled when it got wet. When the foundation support soil goes down, the footings go down, no matter what bearing pressures they were designed for. It's settlement that counts, not bearing capacity."

No matter how hard I try to explain the importance of foundation settlement under loading, architects and structural engineers want an allowable bearing capacity value to design their foundations. In all requests for a geotechnical proposal from architects and structural engineers they specifically include in their geotechnical requirements the need for a recommended allowable foundation bearing pressure. It is always an absolute necessity. They even include a list of geotechnical requirements as an appendix A to their contract with the geotechnical engineer, so the geotech has no choice but to provide a value of allowable bearing pressure in the geotechnical report or be in violation of a signed contract.

What to do? First, I strongly recommend that you do not violate your contracts! Secondly, I recommend that the allowable bearing pressure value that you give in your geotechnical report be based on acceptable settlement values. Do not recommend an allowable bearing pressure for footing design based on the ultimate bearing pressure determined by the bearing capacity equation without consideration of settlement. Thirdly, I suggest you talk with your clients about performance of their structures. What settlements are acceptable, and what maximum settlement constitutes their definition of structural failure.

The simplest procedure I know for finding an allowable bearing pressure for design of spread footings on sandy soil (although I also use it for low-plasticity silty clays) is to divide the average blow count N by five (since the footings are not sized at this point, I average N over 5 to 6 feet to start and then after I have a preliminary footing size I repeat the process by averaging over $1B$ depth for square footings and $2B$ for long footings), and interpret the result as kips per square foot for $1/2$ inch settlement.

$$q_{\text{allowable}} = {}^N/_5, \quad \text{kips per square foot} \qquad (3.4.1)$$

Equation 3.4.1 is an experienced-based allowable bearing capacity equation that is similar to early Meyerhof recommendations; see Meyerhof (1956), Equation 7c, page 5. Equation 3.4.1 works reasonably well for footings about 2 to 4 feet wide that bear at depths of about 2 to 3 feet below grade in soils with SPT blow counts of about 5 to 20. This equation comes with an admonition, which is actually an early "rule of thumb," that soils with blow counts below 4 are "un-buildable" (that is, you need to improve the soil or use deep foundations). For the same 2 to 4 foot wide footings bearing on soils with blow counts of 25 to 50, Equation 3.4.1 changes because you divide N by four to get an allowable bearing pressure in kips per square foot for $1/2$ inch settlement, as shown in Equation 3.4.2.

$$q_{\text{allowable}} = {}^N/_4, \quad \text{kips per square foot} \qquad (3.4.2)$$

For footings more than 4 feet wide, there are equations and charts in textbooks that give you allowable bearing pressure values versus SPT blow count N for 1 or $1/2$ inch settlement. For a selection of these charts and graphs, I suggest Meyerhof's 1984 book (Meyerhof, 1984) (although it is quite hard to find), or you could check standard geotechnical texts such as Peck, Hanson, and Thornburn (1974) or Joseph Bowles (2005).

Although dividing the standard penetration blow count N by 4 or 5 is the easiest way I know to find an allowable bearing pressure, the second easiest standard practice procedure is to use a design chart developed by my friend Ralph Peck. Dr. Peck's chart was originally included in the Peck, Hanson, and Thornburn text of 1953 (Peck,

Hanson and Thornburn, 1953). A large, nearly full-page copy of this chart is given on page 159 of T.H. Wu's 1966 textbook (Wu, 1966). Both of these early text books can be purchased used on the Internet. Dr. Peck and his fellow authors revised the bearing capacity chart in their second edition of 1974 on page 309 (Peck, Hanson and Thornburn, 1974) to include footings of varying embedment depths. These bearing pressure charts are well known to be conservative, so I use them with an assumed set-tlement of 0.5 inch rather than their published 1.0 inch settlement value, as suggested by Joseph Bowles in his third edition text (Bowles, 1982).

Joseph Bowles in his fifth edition text (Bowles, 1996) gives similar charts with higher bearing pressures than Peck or Meyerhof. In Bowles' first edition text (Bowles, 1968) for footings 4 feet wide or less, Bowles divides N by 2.5 to get an allowable bearing pressure in kips per square foot for 1.0 inch settlement, which is essentially the same as dividing N by 5 for 0.5 inches of settlement.

All of the SPT-derived bearing pressure relations described above may be used for preliminary design, although modern practice uses advanced analyses to determine final allowable bearing pressure values. Oh, I might remind you that early references to SPT blow count N are likely N_{45} or N_{55} values.

3.4.3 How Structural Engineers Use Allowable Bearing Pressures

When structural engineers design building column footings using the concrete code, ACI-318-05, they use the allowable bearing pressure to size the footing using service loads (unfactored loads that the column is expected to regularly support). For exam-ple, if the service dead load is 150 kips and the service live load is 250 kips, and you recommend a net[1] allowable bearing pressure of 4 kips per square foot, the structural engineer will add 150 plus 250 for a total of 400 kips and divide by 4 kips per square foot to get 100 square feet. Then he will take the square root of 100 to find a footing size of 10 feet by 10 feet square.

After the structural engineer has a footing size, he or she uses load factors of $1.2DL$ and $1.6LL$ to determine the design bearing pressure as follows: 1.2×150 kips $= 180$ kips, 1.6×250 kips $= 400$ kips, and 180 kips plus 400 kips equals 580 kips. The design factored bearing pressure for determining the footing thickness and the reinforcing steel is 580 kips divided by 100 square feet equals 5.8 kips per square foot.

If your ultimate bearing capacity was three times the allowable value (i.e., a factor of safety of 3), then it was equal to 3×4 kips per square foot or 12 kips per square

---

[1] A word about net versus gross bearing pressure is required. Most geotechnical engineers intend that their allowable bearing pressure be used to size the column footing by dividing the service loading by the bearing pressure to determine the required footing area. They assume that soil and concrete weigh "about" the same and that the soil removed to construct the footing weighs essentially the same as the concrete that replaces it in the ground. Structural engineers must be related to accountants because they take every pound into account and assume the recommended allowable bearing pressure is "gross" pressure. They often subtract the weight of the footing and the weight of any surcharge soil loading above the bearing surface from the allowable bearing pressure to determine the net bearing pressure for use in footing design.

foot. If the structural engineer's statistically derived, load-factored design bearing pressure ever actually occurred, you would still have a soil bearing factor of safety of 12 ksf divided by 5.8 ksf equal to a factor of safety of 2.07.

Here is a plan of action for determining the allowable bearing pressure to recommend in your geotechnical report that will provide the structural engineer what he needs for footing design, while still considering the affect of footing settlement on structural performance:

1. After studying the site geology, topography. and history, design a site investigation plan that adequately characterizes the site for its required uses. Remember the graded approach. Such an investigation may include test borings, cone penetration tests, pressure meter tests, model or full-scale load tests, and geophysical tests. Findings of field tests might not match what you expected in your plan, so be flexible and anticipate that plan adjustments will be required. Plan adjustments may require additional field or laboratory tests so you should make your client aware that a second phase of testing may be required if his site holds geologic or manmade surprises, such as fill.

2. Conduct laboratory testing that is designed to provide data required for project specific analyses, again remember the graded approach. Generic testing plans often fall short of providing needed data for your analyses. Use of multiple levels of correlations based on a few SPT, CPT, or PI tests often introduces unacceptable uncertainty into your analyses that are incorrectly interpreted by your client and by other consultants as excessive conservatism.

3. Obtain a list of estimated column and footing loads from the project structural engineer. If he won't give you a table of values, at least get a minimum and a maximum value, and an estimate of how closely spaced these potential differential settlement parts of the structure may be.

4. Prepare several profiles across the site and select a critical profile for your allowable bearing pressure analysis. The critical profile should have the worst boring included, even if you are not sure that it may occur at the location of the largest column or wall loading. When preparing these profiles, make corrections to SPT or CPT data as required by standard practice in your area or refer to federal highway, state highway, US Army Corps of Engineers, or other standard guides, documents, or publications.

5. Determine or estimate the bottom of footing elevation. In most parts of the United States footing bottoms are required to be between two and five feet below the ground surface. Don't forget that your site may be sloped and your footing elevations will vary accordingly. Depths of footings below finished grade are often specified in local building codes. If there is no local building code, footing depths used are often standard practice values set by local engineers and architects.

6. Assuming that the zone of influence of the building footings will be twice the footings' width, approximately five to 15 feet below the bottom of the footing

bearing level, average test boring SPT blow counts in the zone of influence for the worst and best borings on the critical profile.

7. Using the average blow counts corrected for depth and for hammer energy, calculate the foundation preliminary allowable bearing pressure in kips per square foot by using $N_{70}/4$ or $N_{70}/5$ as discussed above.

8. Calculate the approximate sizes of the largest and smallest adjacent footings by dividing their column service loading in kips by the allowable bearing pressure in kips per square foot.

9. Using the footing size and service load-bearing pressure, calculate the settlement of the adjacent large and small footings. See Sections 3.1, 3.2, or 3.3 for methods of calculation of footing settlements. If your project is a standard commercial project supported on granular or unsaturated soils, I would use the Schmertmann method or the Janbu equation, $S = \mu_o \, \mu_1 \, q \, B \, (1 - \mu^2) \, / \, E_s$ (see Section 3.2 or check out Bowles' first edition textbook, pages 91 and 92).

10. Unless you have specific direction from the project structural engineer, I would make sure that the maximum calculated differential settlement is $^1/_2$ inch or less. I know that both geotechnical and structural engineers frequently bandy about a total or differential settlement of 1 inch in discussions, but it is my experience that these 1 inch numbers are not supported by measurements of actual structures.

11. If your calculated differential settlement is not less than $^1/_2$ inch, go back to step 7 and adjust the allowable bearing pressure. Redo the calculation until you have an acceptable differential settlement, or if the differential settlement is excessive for all reasonable footing sizes then develop alternate foundation recommendations, such as soil improvement, a mat, combined footings, micro-piles, augered piles, driven piles, drilled shafts, and so on.

If you insist on using one of the standard Terzaghi, Meyerhof, or Hansen bearing capacity equations, which in some circles is referred to as the three N's equation, you can substitute the allowable bearing pressure in Step 7 by one-third of the ultimate bearing capacity $q_{ultimate}$ calculated by the bearing capacity equation. Bearing capacity equations are covered very thoroughly in almost any geotechnical text book that you may pick up, so I will not repeat that material here. I would add two notes of caution when using bearing capacity equations published in older books. Back in Terzaghi's day, let's say the 1940s and 1950s, it was standard practice to define the footing width as $2B$, that is B was equal to one-half of the footing width. In later textbooks this early convention was changed to make the entire footing width equal to B. Whatever book you use for the bearing capacity equation, please make sure to check the definition of B. My second caution to you is to check values calculated for the third bearing capacity factor N_γ. In earlier bearing capacity references, the values calculated for N_γ were not properly bounded. In other words, calculated values of N_γ were much too large. I suggest you refer to a modern geotechnical textbook or current FHWA publications covering shallow footing analysis and design to make sure your N_γ values are properly limited.

Now that I think about it, there is one additional caution that I would like to mention relative to use of bearing capacity equations. The size of test footings used in development of bearing capacity equations were generally much less than 10 feet (3 meters) wide. In Bowles' 5th edition textbook (Bowles, 1996), he gives corrections for the bearing capacity equation for large footings. If your footing is much larger than 10 feet wide, Bowles recommends that you multiply the third term of the bearing capacity equation by a term r_γ as follows: $0.5\gamma\, B\, N_\gamma s_\gamma d_\gamma r_\gamma$, where γ is the unit weight of bearing soil located beneath the footing, B is the width of the footing, N_γ is the third bearing capacity factor, s_γ is the shape correction factor, d_γ is the depth correction factor, and r_γ is the size correction factor as follows:

$B = 2$ m	2.5 m	3 m	3.5 m	4 m	5 m	10 m	20 m	100 m
$r_\gamma = 1.0$	0.97	0.95	0.93	0.92	0.90	0.82	0.75	0.57

3.4.4 Advanced Bearing Capacity Material

Current geotechnical practice is moving away from allowable stress design (ASD) to load and resistance factor design (LRFD). Most of the credit for this change in the United States can be given to the United States Federal Highway Administration and their insistence that all structures designed for projects using their money after Fall 2007 employ AASHTO LRFD design procedures. I'm not sure, but I've heard that they decided that the change had to be made, even if we didn't have all of the necessary data to mathematically define all of the geotechnical resistance factors required. I agree that you can't wait for everything to be perfect to start work. Consider an artist; does he need a perfect studio and the best art supplies to start his first painting? If he waited to have everything just right, he would never get any artwork done!

3.4.4.1 Bearing Capacity by Geotechnical LRFD Methods

From earlier parts of this chapter, you know that I am not a fan of using bearing capacity alone to design footings. If a footing moves excessively relative to its structural purpose, it has by definition failed. The excessive movement could be vertically, laterally or it could be excessive rotation about any of the three axes: x, y or z. The AASHTO design manual requires calculation of the factored failure or ultimate bearing pressure for a footing and comparison of this limit strength loading value to a factored loading resistance (see Section 5.3). Then AASHTO requires that the designer calculate footing settlements using service loadings and evaluate these movements to the structural requirements of the bridge, wall, and so on. I like this geotechnical LRFD approach because it includes consideration of both factored bearing capacity and settlement. Currently the 2010 AASHTO LRFD Design Guide in Chapter 10 has geotechnical guidance for design of foundations. We will discuss geotechnical LRFD for footings in Chapter 5 Section 5.3.

3.4.4.2 What is Wrong with the Bearing Capacity Equation?

We need to examine the stress state in soil beneath a shallow spread footing to evaluate the bearing capacity equation. First of all, the stresses generated beneath a shallow spread footing are a combination of compressive stresses and shearing stresses. Larger footings generate a greater portion of compressive stresses acting directly beneath the footing than do smaller footings. As a limit state you might consider a pointed pile as an extremely small footing that has little compressive stress capacity at its tip and a majority of its load capacity is carried by shearing stresses along the side of the pile. Larger footings also generate greater amounts of shearing stresses at greater depths, as is illustrated by study of the Boussinesq chart included in most elementary texts. You might be tempted to say, "So what?," but let's take an even closer look.

We know that loose sands act like dense sands when they are subjected to very small confining stresses because they dilate during shearing. The same loose sands act like loose sands when they are subjected to large confining stresses when they are sheared because they compress during shearing. Since the confining stresses acting on the shallow stress bulb of a small footing are small, the footing loading could result in dilation, while the greater confining stresses acting on the deeper stress bulb of a large footing could result in soil compression in the same loose sand deposit. Soils beneath the small footing would dilate at relatively low loadings and then finally compress at larger loadings, making us think the footing had reached an ultimate or limit load capacity, while the larger footing's bearing soils would compress under all loadings suggesting that there were no peak bearing pressure.

The proportion of shearing stresses versus compressive stresses generated beneath our theoretical small and large footings also causes a problem in interpretation. Figure 3.4.1 below illustrates a steady state line with distance from the initial void ratio to the steady state line given as upsilon, $Y = e_i - e_{SSL}$ (see Fellenius and Altaee, 1994). The upsilon parameter is also called the "state parameter" in the literature (Been and Jefferies, 1985). You can see that the distance from the initial void ratio to the SSL is a function of the mean confining stress. Loading induced soil shearing stresses generate a change in void ratio that can cause the material void ratio to reach the critical void ratio and develop steady state shearing. At the steady state line, no additional volume changes due to shearing strains will occur. Compression strains due to vertical compressive stresses do not have a similar limit value, although very high compression stresses do result in grain crushing and eventual molding of the soil into a solid mass. The amount of compression caused by an increase in vertical stresses is a function of Young's Modulus, E, which is dependent on confining stresses. Again the deeper soils have higher confining stresses which result in increased E values.

Given the above discussion, notice that soil stress–strain and volume change properties depend on confining stress. The problem with the bearing capacity equation is that it suggests that soil shear strength parameters are independent of soil confining stresses, and it suggests that soil shear strength is independent of soil strains.

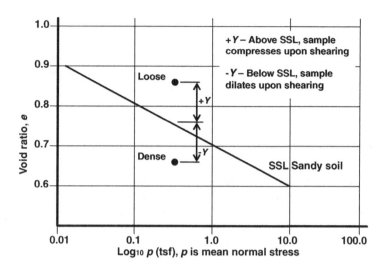

Figure 3.4.1 Definition of the upsilon parameter, $Y = e_{initial} - e_{SSL}$

3.4.4.3 Footing Design Using the Load-Settlement Curve

To design a footing using the bearing capacity equation, values of soil shear strength parameters c and ϕ are required. Standard practice over the years has been to select peak values of vertical compression stress from stress–strain curves, and to plot Mohr's circle from maximum vertical and horizontal stresses on a Mohr–Coulomb plot to determine c and ϕ values. This procedure has some inherent problems. For instance, what if one of the three tests used to draw Mohr's circle has a well-defined peak stress and the other two tests do not have peak stresses. How do you plot stresses from three compression tests when they represent different stress–strain soil performance. I've been aware for some time now, but others apparently consider it a new phenomenon that soils do not actually exhibit peak shear strength (Chu and Wanatowski, 2009). Some say that the observed peak stress in triaxial shear tests is an artifact of the testing method caused by shearing stresses induced between the ends of samples and the top and bottom platens in the triaxial test. Others, such as Dr. Bengt Fellenius (Fellenius, 2008), suggest that small footings only affect soil to shallow depth where the soil (even loose soil) behaves like an over-consolidated soil. When loaded this shallow zone soil dilates and after some strain accumulation it finally contracts, resulting in a stress-deformation curve that indicates a peak or ultimate value. I suggest that you review Holtz and Kovacs' first edition text, *An Introduction to Geotechnical Engineering*, Prentice-Hall 1981, pages 496 to 503 (Holtz and Kovacs, 1981) or their second edition text 2011, pages 545 to 553 (Holtz, Kovacs and Sheahan, 2011).

   Rather than deal with the problems of selecting a peak strength, reduced strength, residual strength, or a strength at a given soil strain to use in their bearing capacity equation analyses, engineers are starting to turn back to the load-settlement curve method (Fellenius, 2008; Briaud, 2007) which is similar to the full scale footing load test methods used prior to 1925. If you think about it, why do we perform load tests on pile foundations and analyze pile load-settlement curves, and then turn around and use the bearing capacity equation for a deep footing? Why not develop load-settlement curves for both types of foundations?

References

Been, K. and Jefferies, M.G. (1985) A state parameter for sands. *Geotechnique*, **35**, 99–112.

Bowles, J.E. (1968) *Foundation Analysis and Design*, 1st edn, McGraw Hill Inc., 659 pages.

Bowles, J.E. (1982) *Foundation Analysis and Design*, 3rd edn, McGraw Hill Inc., 816 pages.

Bowles, J.E. (1996) *Foundation Analysis and Design*, 5th edn, McGraw Hill Inc., 1024 pages.

Briaud, J.L. (2007) Spread footings in sand: load settlement curve approach. *Journal of Geotechnical and Geoenvironmental Engineering*, **133**(8), 905–920.

Chu, J. and Wanatowski, D. (2009) Effect of loading mode on strain softening and instability behavior of sand in plane-strain tests. American Society of Civil Engineers, GeoInstitute, *Journal of Geotechnical and Geoenvironmental Engineering*, 108–120.

Fellenius, B.H. and Altaee, A. (1994) Stress and settlement of footings in sand. Proceedings of the American Society of Civil Engineers, ASCE, Conference on Vertical and Horizontal Deformations for Foundations and Embankments, Geotechnical Special Publication, GSP, No. 40, College Station, TX, June 16 - 18, 1994, Vol. 2 pp. 1760–1773.

Fellenius, B.H. (2008) Foundation Design Approach of Past, Present and Future. Geo-Strata Magazine, November/December 2008, pp. 14–17.

Holtz, R.D. and Kovacs, W.D. (1981) *An Introduction to Geotechnical Engineering*, 1st edn, Prentice-Hall, Inc., Englewood Cliffs, New Jersey, 733 pages.

Holtz, R.D., Kovacs, W.D. and Sheahan, T.C. (2011) *An Introduction to Geotechnical Engineering*, 2nd edn, Pearson, Prentice-Hall, 853 pages.

Meyerhof, G.G. (1956) Penetration tests and bearing capacity of cohesionless soil. *Journal of Soil Mechanics and Foundation Division of the American Society of Civil Engineers*, **82**(SM 1), 1–19.

Peck, R.B., Hanson, W.E., and Thornburn, T.H. (1953) *Foundation Engineering*, 1st edn, John Wiley & Sons, 410 pages.

Peck, R.B., Hanson, W.E., and Thornburn, T.H. (1974) *Foundation Engineering*, 2nd edn, John Wiley & Sons, 514 pages.

Terzaghi, K. (1943) *Theoretical Soil Mechanics*, John Wiley & Sons, Inc., New York, 528 pages.

Terzaghi, K. and Peck, R.B. (1948) *Soil Mechanics in Engineering Practice*, John Wiley & Sons, Inc., New York and Chapman & Hall, Ltd., London, 566 pages.

Wu, T.H. (1966) *Soil Mechanics*, Allyn and Bacon, Inc., Boston, 431 pages.

3.5

Load Capacity of Deep Foundations

3.5.1 Deep Foundations – What Are They?

This is a more difficult question than you might imagine. Back in the nineteenth and early twentieth century, most footings were shallow spread footings that supported building columns and walls. Often when soft soils were encountered in the upper 10-feet of the site, engineers and constructors just dug through the soft material to obtain footing support on harder soil. These footings were called deepened spread footings, and I suppose you could call them "deep foundations." When side walls of the excavations in soft soils became unstable and started collapsing, they used timbers and wood sheeting to support the excavation walls. When the wood sheeting used to support the excavation sides started to push into the excavation, they added wood timbers across the excavations to brace the wood sheeting. If you have ever seen a wood-sheeted, timber-braced excavation it is like chaos in motion with cross members going everywhere. There are photographs of these sheeted/braced excavations in old engineering books, and they are still used in Third World nations to this day. Topics of retaining structures and laterally braced retaining structures are included in Sections 4.2 and 4.3.

Early bridge engineers trying to cross rivers couldn't normally use deep spread footings to support their bridges because they couldn't keep river water out of the required excavations. Some bridge engineers used timber piles which were trees with their limbs cut off. Many early timber piles were driven with bark still on the tree trunk. When iron pipe and I-beams became available, they were driven into the river bed to provide bridge support. Timber piles, pipe piles, and I-beam piles used by early engineers could all be considered to be deep foundations.

Geotechnical Problem Solving, First Edition. John C. Lommler.
© 2012 John Wiley & Sons, Ltd. Published 2012 by John Wiley & Sons, Ltd.

Some early bridge engineers insisted on having massive masonry or concrete footings bearing on bedrock beneath rivers for support of their monumental structures. A good example is the Brooklyn Bridge. They invented a pressurized chamber that looked like a huge diving bell that was used to protect workers that hand-excavated the soil down to bedrock and constructed the massive bridge foundations. Foundations constructed by this pressured chamber method were commonly called caissons. These caissons were deep foundations. By the way, workers who spent their days digging soil in these early pressurized caisson chambers developed the bends and died by the hundreds. Older workers who were overweight and had high blood pressure were resistant to development of nitrogen bubbles in their blood stream during decompression after work in the pub. As a result the image of the big, old tough laborer was born.

Development of mechanical drills in the early 1900s allowed deep drilled holes to be excavated and filled with concrete and they became known as drilled caissons. The term drilled caisson was still prevalent in the 1960s and early 1970s when I started working as a "soils engineer." In the late 1970s in an effort to distinguish between pneumatic chamber caissons and drilled caissons, the name drilled caisson was changed to drilled pier. In many parts of the United States, the term drilled pier is still used. In the late 1990s, bridge engineers complained about confusion when using the term pier to describe drilled piers. The ends of a bridge are called abutments, if the bridge is a one-span bridge it is supported by two abutments. If the bridge is a multiple span bridge, its support requires center structures, which are called piers. So there is the rub, bridge engineers call their center supports "piers" and if we call drilled foundations "drilled piers" and use these drilled piers to support bridge abutments we are faced with a linguistic problem, that is piers supporting abutments. Some bridge engineers gave in and changed the name of their center supports to "bents." Most bridge engineers working in the United States, and especially those at the federal government level, requested (or demanded) a change in the name of our drilled foundations from drilled piers to drilled shafts.

So if you are an engineer who is confused by the name of those deep drilled foundations, let's review:

1. The term drilled caisson is now obsolete, and only appears in older text books. If you have a deep foundation excavation using a pneumatic chamber it may still be called a caisson.
2. The term drilled pier is still used in building construction where they don't have any bridge piers.
3. The term drilled shaft is required in highway work where bridge and roadway engineers have bridge support piers that are not to be confused with foundations. By common usage, the term drilled shaft seems to be becoming the preferred description of a deep drilled foundation. I expect that soon the term drilled shaft will replace the term drilled pier in all types of construction.

So what are deep foundations? Deep foundations are driven piles, drilled piles, drilled shafts, and dozens of variations of piles and shafts developed by deep foundation contractors around the world.

When talking about deep foundations be they piles or drilled shafts, it has been common practice to refer to them as friction piles (or shafts) and end-bearing piles (or shafts). An end-bearing pile is a pile that has been driven or drilled into very stiff or hard clay, a dense sand and gravel, or to bedrock. A friction pile or drilled shaft is a pile that has been drilled to driven into soil materials to develop a specified load capacity without tipping or bearing in hard soils or on bedrock. When analyzing the load capacity of a friction pile, the tip resistance is often neglected.

When analyzing the load capacity of an end-bearing pile or long drilled shaft, the side and tip resistances are normally added. When analyzing the load capacity of a short (length less than or equal to three diameters) drilled shaft or belled drilled shaft bearing on bedrock, the side resistance is often neglected.

If it is standard practice to follow similar guidelines for including or excluding pile side or tip resistance in your practice area, that is fine with me, if you are not sure whether you should consider side resistance, tip resistance or some combination of side and tip resistance in your capacity calculations, I recommend that you perform some pile or drilled-shaft load tests.

A word about drilled-pile and drilled-shaft contractors may help clarify confusion about varying sizes of piles and shafts. When I started work as a geotechnical engineer in the early 1970s, drilled piles used to be limited to an upper diameter of about 18-inches. Today, drilled, augercast piles are up to 60-inches in diameter, and 72-inch diameter augercast piles are not far off. Drilled shafts used to range from 18 inches in diameter up to about 14 feet in diameter, where smaller diameter shafts were drilled by truck rigs and very large diameter shafts were drilled by crane mounted rigs. Today, drill-shaft contractors are drilling foundations down to 12 inches in diameter and some even smaller, while mobile rigs are drilling very large shafts and crane-mounted rigs are nearly extinct. Do you see the trend? Drilled-pile contractors are developing larger and larger pile sizes with the intention of "eating the drilled-shaft contractor's lunch." While on the other hand, the drilled-shaft contractors are drilling smaller and smaller shafts to fight back. Crane-rig manufactures are out of business (my guess). As a result of competition between contractors and manufacturers, there is no clear defining line between diameters of piles and drilled shafts in the twenty-first century. In the end, the deep foundation type to be selected is the one that best fits the site soil conditions, that best fits the needs of the project, and best fits the budget and schedule of the project.

3.5.2 Allowable Load Capacity of Deep Foundations

"When I was in school, the professor said that you add the pile shaft side friction capacity to the end bearing capacity, and you have the total pile or shaft load capacity.

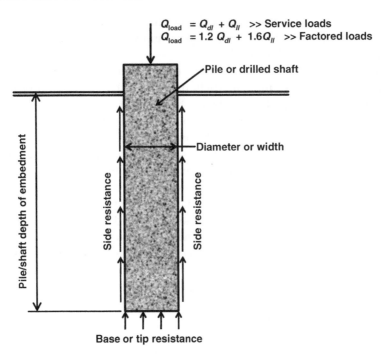

Figure 3.5.1 Pile and drilled shaft load capacity

I think we used to call it the ultimate pile capacity. Then we just divided the total pile capacity by three and we had the allowable pile load capacity. Isn't that all there is too it?"

I hate to equivocate, but my answer is yes and no. Let's go through the pile and drilled-shaft load-capacity problem step by step.

First let's consider the pile/drilled-shaft capacity equations 3.5.1 and 3.5.2, and the pile/drilled shaft illustrated in Figure 3.5.1:

$$Q_{total} = Q_{side\ resistance} + Q_{tip\ resistance} \tag{3.5.1}$$

where

$$Q_{side\ resistance} = \sum_{i=1}^{n} f_s\, P_{ep}\, H_i \tag{3.5.1a}$$

and

$$Q_{\text{tip resistance}} = A_p q_p \qquad (3.5.1b)$$

f_s is pile or shaft side resistance per unit area, that is, psf.
P_{ep} is the effective perimeter of the pile, ft.
H_i is the soil layer thickness of the ith layer, ft.
A_p is the bottom or tip cross sectional area, square ft.
q_p is the bearing capacity of soil below the pile tip, that is, psf.

$$Q_{\text{allowable}} = \frac{Q_{\text{total}}}{3} \qquad (3.5.2)$$

Looking at Equation 3.5.1, it appears to be a simple matter that we are adding the pile resistance developed along the side of the pile by the pile–soil friction to the pile resistance developed below the tip of the pile (Note: To avoid being too wordy, I'm going to use the word "pile" in most instances when I mean "pile and drilled shaft"). Remember your mathematics; can you really add pile side friction to pile end bearing? In math class you learned that you can only add two things if they are from the same set. Two red balls plus two red balls equals four red balls. Two red balls plus two blue balls does not equal four red balls. Back to the pile load, if you load a pile to failure and generate a maximum loading of 2000 kips at 3.5 inches deflection and beyond 3.5 inches deflection the pile loses resistance capacity, you may have a loading case where both the skin friction and the end bearing are in a state of failure, see Figure 3.5.2.

If both pile resistances are in a state of failure, you can add them. But now let's look at Equation 3.5.2, where you divide Q_{total} by three to calculate $Q_{\text{allowable}}$, let's break that calculation in to its parts, see Equation 3.5.3:

$$Q_{\text{allowable}} = \frac{Q_{\text{total}}}{3} = \frac{Q_{\text{side resistance}}}{3} + \frac{Q_{\text{tip resistance}}}{3} \qquad (3.5.3)$$

Equation 3.5.3 says that you can add one-third of the maximum pile side resistance to one-third of the maximum pile tip resistance. Look at Figure 3.5.2, notice that although the maximum (or near maximum) values of side and tip resistance occur at the pile deflection of 3.5 inches, values of one-third of the side resistance and one-third tip resistance *do not* occur at the same pile deflections. If one-third of the pile side resistance and one-third of the pile tip resistance do not occur at the same time during your pile load test, how can you add them? The simple answer is you can't add them? The pile head can only be at one location at a time, so you have to add the values of side residence and tip resistance at a given pile head deflection to obtain a load–deflection curve for the pile.

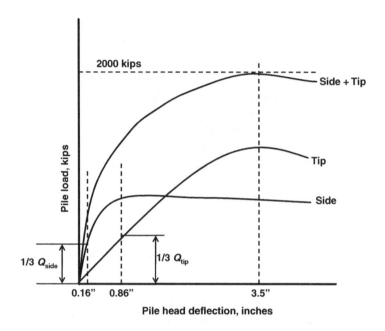

Figure 3.5.2 Pile and drilled shaft load versus head deflection

Now comes the hard part, why did engineers in the past add one-third of the pile side resistance to one third of the pile tip resistance and declare the pile to be OK when this sum exceeded their expected (i.e., service) pile loading. Well first let me say that all engineers did not do this. There were and are a significant numbers of engineers both then and now that ignore pile tip resistance (assuming that they don't have a short end-bearing pile) and use only pile side resistance when considering support of service loadings. Their rationale is that significant pile tip resistance occurs at large deflections (see Figure 3.5.2) and they want service loadings to be supported by piles that deflect about 0.2 to 0.5 inches, which is primarily developed by pile side resistance. These same engineers include the maximum tip resistance for analysis of extreme loading events such as large earthquakes or hurricanes where deflections of their structure foundations may acceptably exceed 2 inches.

Back to the engineers who add one-third of the pile side resistance to one third of the pile tip resistance and declare the pile to be OK when this sum exceeds their service loading. Why do they do this if the method is not deflection compatible? I guess the easiest way to explain this practice is to say that use of a factor of safety of 3 is conservative. Even if the tip resistance were actually zero at service load deflections for piles designed using a factor of safety (FOS) of 3, the pile FOS considering only side resistance would have an actual FOS of about 2 which is still safe. So even if they include the tip resistance in their calculations the answer would still be safe. I'm just the messenger here; I'm not saying that I agree with this approach. This matter of using an FOS of 3.0 and neglecting pile load deflection curves when adding tip and shaft resistances also suggests that reducing pile factors of safety from 3.0 to 2.5 to 2.0, because we have "better SPT or lab data these days" could result in foundation problems in the field.

I recall that several years ago, Professor Bill Lambe (Dr. William Lambe, co-author of the classic 1969 Lambe and Whitman text) asked a group of practicing geotechnical engineers to calculate the uplift resistance of a pile. As I recall none of them came up with the same answer, and there were almost as many analysis methods as there were participating engineers. Maybe I was young and naive at the time, but I was shocked at the differing pile uplift capacity results included in Dr. Lambe's report. So what can I say, different engineers often do things differently.

Now that we have some idea about the issues related to adding pile side resistance to tip resistance, let's consider how pile side and tip resistances are calculated.

3.5.2.1 Pile and Drilled Shaft Side Resistance

When piles are driven into the ground they displace the soil. Some piles such as closed-end pipe piles or timber piles displace large amounts of soil during driving, and other piles, such as H-piles displace much smaller amounts of soil during driving.

Drilled shafts and augered piles remove soil during drilling which can allow excavation side walls to move inward or even collapse into the open hole. How the sidewalls of drilled shafts and augered piles are maintained open has a considerable impact on the side resistance developed. From the top after they are installed, augercast piles and drilled shafts look alike, see Figure 3.5.3. Augercast piles are installed

Figure 3.5.3 30-Inch diameter drilled shaft or augercast pile?

Figure 3.5.4 Augercast pile installation

with long augers that maintain an open hole during drilling, see Figure 3.5.4. Unless casing is used, drilled shafts are installed in unsupported holes with short augers on a Kelly bar extension that has to be withdrawn after each drilling pass, as shown in Figure 3.5.5.

As you might expect, clays and sands generate pile side resistance differently, and clays and sands generate pile side resistance differently for driven piles than for drilled piles. When a driven pile is advanced through a clay layer into a sand layer, the pile drags clay into the sand layer, and when a driven pile comes out of a sand layer into a clay layer the sand comes along with the pile into the clay layer. With all of these different ways that piles can generate side resistance to loading, it is natural to assume that many different methods of calculating pile side resistance have been developed. We need to break these pile resistance cases up for reasonable consideration. I propose to first show you how to calculate pile side resistance and pile tip resistance, and then we will classify piles as driven or drilled, and consider side friction and end-bearing load resistance calculations for each pile type in clayey and in sandy soils.

Figure 3.5.5 Drilled shaft installation – auger on Kelly bar

3.5.2.2 Calculating Side Resistance of Piles and Shafts

Classically in soil mechanics the Mohr–Coulomb equation, $s = c + \sigma\ (\tan\ \phi)$, has been used to estimate the shearing strength of soils. When determining the shearing resistance between the side of a vertical pile and the surrounding soil a similar equation was adapted:

$$f_s = c_{\text{adhesion}} + \sigma_{\text{horizontal}} \tan\delta = c_a + K\sigma_{\text{vertical}}\tan\delta \qquad (3.5.4)$$

where,

  $c_a =$ adhesion between pile and soil in force per unit area, that is, psf
  $K =$ lateral earth pressure coefficient $= \sigma_h\ /\ \sigma_v$ used to relate the vertical soil stress
      to the lateral soil stress acting at the pile surface, Note: This lateral pressure
      value is modified from in situ values of K by the installation of the pile.
  $\sigma_v =$ vertical soil stress adjacent to the pile where f_s is being calculated
  $\delta =$ friction angle between the pile material and the adjacent soil.

  To solve Equation 3.5.4, we need to determine a value of the soil adhesion on the pile sides. A common method of determining pile–soil adhesion is to assume that the soil adhesion is a portion of the soil cohesion: $c_{\text{adhesion}} = \alpha\ c$, where α is the factor used to determine the appropriate portion of cohesion that is acting on the pile sides. This method of calculating pile side resistance to loading is called the Alpha method.
  If we assume that the soil friction angle $\phi = 0$ and that all of the soil shear strength comes from soil cohesion, then the equation for pile side resistance reduces to

Equation 3.5.5.

$$Q_{\text{side resistance}} = \sum_{i=1}^{n} c_{\text{adhesion}} P_{ep} H_i = \sum_{i=1}^{n} \alpha c P_{ep} H_i \qquad (3.5.5)$$

where,

$c_{\text{adhesion}} = \alpha c$ pile–soil adhesion equals soil cohesion times α
$\quad P_{ep} =$ the effective pile perimeter.

The Alpha method as represented by Equation 3.5.5 is typically used to determine pile side resistance in clayey soils.

To calculate the pile side friction resistance f_s in sandy soil, the adhesion term is assumed to be zero, and we need to determine the value of the second term in Equation 3.5.4. There are several methods commonly used to calculate f_s for sandy soils including the Nordlund method (Nordlund, 1963) and the Beta (β) method. The Beta method is also called the effective stress method in some references (Hannigan *et al.*, 2006).

The Nordlund method (Nordlund, 1963) is a semi-empirical method based of field observations of piles. Nordlund developed his method in 1963 and updated it in 1979 (Nordlund, 1979). This method includes consideration of pile taper and differences in pile materials in the generation of pile side resistance to loading. Nordlund's method includes consideration of pile side resistance and pile tip resistance. Since Nordlund's calculation of tip resistance is equivalent or similar to the standard method discussed below, I use Nordlund's pile side resistance method, as discussed below and ignore his tip resistance calculation. Nordlund's equation for calculating the pile side resistance to loading is given in Equation 3.5.6.

$$Q_{\text{side resistance}} = \sum_{z=0}^{z=D} K_\delta C_f \sigma_z' \frac{\sin(\delta + \omega)}{\cos \omega} P_{ep} \Delta z \qquad (3.5.6)$$

where,

$\quad z =$ depth to center of stratum layer
$\quad D =$ pile length embedded in soil layers
$\quad K_\delta =$ coefficient of lateral earth pressure at depth z
$\quad C_f =$ correction factor for lateral earth pressure when $\delta \neq 0$
$\quad \sigma_z' =$ effective overburden pressure at depth z
$\quad \delta =$ pile–soil friction angle
$\quad \omega =$ pile taper angle, measured from vertical
$\quad \phi =$ Friction angle of soil adjacent to pile at depth z
$\quad P_{ep} =$ effective pile perimeter at depth z
$\quad \Delta z =$ Length of pile segment at depth z.

Nordlund does not limit the value of pile side friction by specifying an upper limit of side friction. Most guides and building codes do specific upper limits on pile skin friction f_s. For example, the 1996 FHWA specification requires an upper limit on pile or shaft skin friction of 4.0 kips per square foot for sandy soils and 5.5 kips per square foot for clayey soils. If you don't have code-specified upper limits on f_s, I recommend that you study results of full-scale load tests conducted in your region to start a library of f_s values. You should soon be able to set an upper bound based on field test results.

Nordlund uses a very interesting approach to calculating the pile side friction. He uses the volume of soil displaced by the driven pile during driving in cubic feet per foot of pile to determine a ratio of pile–soil side friction to the soil friction angle, δ/ϕ. The greater the volume of soil displaced by the pile per foot, the greater the soil–pile friction. Of course since Nordlund discusses soil volume displaced by the pile, he must be talking about driven piles since drilled piles and shafts don't displace soil when they are installed.

To use the Nordlund method you need to determine values of δ, K_δ, and C_f. Nordlund presents up to 10 pages of tables to find these values. In the 2006 FHWA Pile Manual (Hannigan et al., 2006), they give a step by step procedure for calculating pile side resistance by the Nordlund method, including all of the required graphs. On the internet I found a free publication from the New York State Department of Transportation that gives a complete set of Nordlund charts.

Why didn't I give you a copy of the Nordlund charts? First I didn't want to fill up this book with charts that are presented nicely elsewhere. Second I want to encourage you to explore geotechnical reference material like the FHWA pile manual. Thirdly I freely admit that I am friends with Frank Rausche and Garland Likins. I mention this not because I'm selling material written by Frank and Garland, I recommend the 2006 FHWA manual because it includes their work. Frank Rausche and Garland Likins have spent their entire careers studying driven piles and anything about piles written by them is worthy of study.

3.5.2.3 Calculating Tip Resistance of Piles and Shafts

How do you think that engineers developed an equation for calculating the tip or end-load resistance of piles? Would you use what you are familiar with or would you come up with a whole new equation? It is not surprising that standard practice has adopted the Terzaghi bearing capacity equation for use in calculating the bearing capacity of soil or rock beneath a pile tip, see Equation 3.5.7.

$$q_p = cN_c + qN_q + \frac{1}{2}\gamma BN_\gamma \tag{3.5.7}$$

N_c, N_q and N_γ are bearing capacity factors that depend on the soil friction angle ϕ. These bearing capacity factors are similar to Terzaghi's bearing-capacity factors, although pile factors are modified for use with deep, pile foundations. The terms c, q and γ are cohesion, vertical stress and unit weight of soil or rock beneath the tip of the pile, and B is the diameter or effective width of the pile tip.

Maybe it is just me, but standard practice seems to mix up drained shear strength and undrained shear strength when analyzing piles and drilled shafts. Commonly undrained strength analyses are called total stress analyses. For the total stress analysis, ϕ is assumed equal to zero and c is assumed equal to the undrained shear strength $c = s_u = \frac{1}{2} q_u$, where q_u is determined by the unconfined compression test. This total stress method for calculating pile tip resistance is commonly used for clayey soils or rock beneath the pile tip. It doesn't seem to matter whether the clayey soil is saturated or unsaturated, when the unconfined compression test is used for determining s_u by this total stress method.

Using the total stress method, resulting values of $N_q = 1$ and $N_\gamma = 0$, and the value of $N_c = 9$ is commonly used for a deep foundation, although I recall seeing reported values of N_c ranging from 6 to 16. Considering that Terzaghi's shallow foundation equation uses $N_c = 5.7$ for c only ($\phi = 0$) soils, I believe that using N_c values less than 9 is too conservative unless you anticipate soil softening to occur below the pile tip.

Using the total stress analysis method described above, the equation for calculating the pile tip load resistance is condensed to Equation 3.5.8.

$$Q_{\text{tip resistance}} = A_p S_u N_c \qquad (3.5.8)$$

When soils below the pile tip are drained during loading the analysis used for determining the pile tip resistance is referred to as the effective stress analysis. As you might imagine, this method is commonly used for determining pile tip resistance for granular soils. When using the effective stress analysis method, the first and third terms (i.e., $\frac{1}{2}\gamma B N_\gamma$ and $c N_c$) are assumed to be small second-order terms and the equation reduces to Equation 3.5.9.

$$Q_{\text{tip resistance}} = A_p \eta q' N_q' \qquad (3.5.9)$$

Use of the prime on q' and N_q' refer to effective stresses, and of course the effective friction angle ϕ' is used to determine N_q'. For all methods except the Vesić and Thurman methods discussed below, assume the correction factor $\eta = 1.0$.

Some engineers (Coyle and Castello, 1981) have compared the results of pile load tests in sand with predicted pile tip capacity from Equation 3.5.9 and have concluded that standard shallow foundation values of N_q determined by Terzaghi's bearing capacity equation are suitable for calculating pile tip resistance. Although Coyle and Castello said that Terzaghi's method was "favored," they did conclude that none of the tested N_q equations correctly predicted the actual N_q values measured.

There are several methods of adjusting tip resistance of piles bearing on granular soils, some methods adjust N_q by limiting ϕ to not less than 30 degrees and not more than 36 or 40 or 43.75 degrees. Please check the method you are using carefully to confirm what limits on ϕ are used with specific equations. Other methods such as those proposed by Thurman (Thurman, 1964) and by Vesić (Vesić, 1975) use correction

Table 3.5.1   Values of deep bearing-capacity factor N_q and correction factors η

Method and factor, N_q or η	$\phi = 30°$	$\phi = 35°$	$\phi = 40°$
Terzaghi, N_q	20	42	100
Meyerhof, N_q for $d/B = 0$	26	32	105
Meyerhof, N_q for $d/B = 16$	60	150	330
Mathias and Cribbs, *Driven Manual*, Table 7-1b, N_q	30	65	170
Thurman, $\eta_{d/B}$, $d/B = 20$	0.58	0.68	0.73
Thurman, $\eta_{d/B}$, $d/B = 30$	0.52	0.66	0.73
Thurman, $\eta_{d/B}$, $d/B = 45$	0.49	0.65	0.73
Vesić, $\eta_{Ko} = \frac{1+2K_o}{3}$, $K_o = 1 - \sin\phi$	0.67	0.62	0.57

factors (I'll call pile correction factors η) to modify the tip resistance computed by Equation 3.5.9. Thurman corrects Equation 3.5.9 for pile tip depth by using d/B ratios, where d is the depth of the pile tip and B is the width or diameter of the pile tip. Table 3.5.1 lists several proposed N_q and η factors.

Vesić corrects Equation 3.5.9 for average confining stress at the pile tip by using the lateral earth pressure coefficient K_o to calculate a dimensionless correction factor, η, see Equation 3.5.10. The at-rest earth pressure coefficient changes as a soil becomes more highly over-consolidated, although not quite as directly as the OCR (over-consolidation ratio). In fact K_o and OCR are inverse relations, as OCR and soil shear strength go up, values of K_o go down.

$$\eta = \frac{1 + 2K_o}{3} \tag{3.5.10}$$

There are many references that suggest values for N_q and for the depth correction factor η. Bowles' 3rd edition text (Bowles, 1982), Meyerhof (Meyerhof, 1976), FHWA program Driven (Mathias and Cribbs, 1998) includes a User's Manual FHWA-SA-98-074 that includes recommended values of N_q and Thurman's term α for use with Equation 3.5.7. There are several new pile and drilled shaft design manuals for LRFD analysis and designs that have been recently issued by the FHWA. We'll discuss LRFD design of piles and drilled shafts in Chapter 5.

3.5.2.4 Driven Piles in Clayey Soils

To calculate the side resistance of a pile in clay, we use Equation 3.5.5, called the Alpha method. To calculate the tip resistance in clay, we use Equation 3.5.8. If you are driving piles in soft clay, the pile penetration often disturbs and remolds the clay during driving and generates excess pore water pressures, which reduce the clay's shearing resistance and adhesion. With time the excess pore water pressures dissipate and the remolded clay gains strength. In clayey soils dissipation of excess pore pressures and gain in strength can take two or more weeks to occur. Some clays called thixotropic clays regain their structure and strength with time. When soils adjacent to driven piles gain strength with time, the gain in pile capacity is called setup.

Piles do not always experience setup. Some hard clays, shales, and dense silts dilate and generate negative pore water pressures or suction when penetrated by a driven pile. With time, water soaks into the disturbed dilated clay around the perimeter of the driven pile and the clay softens. In dense silts, the material doesn't soften, but loss of soil suction with time reduces effective confining stresses resulting in loss of strength.

Piles driven into soils that soften or lose suction with time also lose some of their load capacity, which is called pile relaxation. It is common for engineers to look for pile setup and forget about pile relaxation.

OK, let's take a look at the Alpha method for calculating side resistance developed by driven piles in clayey soils.

Since driven piles push through the soil strata as they are driven, they can drag along some soil from one soil stratum down into a lower stratum, causing a blurring or mixing of soil adhesion conditions for the pile as it moves from stratum to stratum. Tomlinson developed a widely used and repeated method (Tomlinson, 1980) for dealing with pile-soil adhesion in layered soils. Tomlinson's method uses several cases to develop pile–soil alpha adhesion factors: Case 1 – all clay soil from top to tip of pile, Case 2 – soft clay soil over stiffer clay soil where the pile penetrates into the lower stiff clay, and Case 3 – sand or sandy gravel over clay soil where the pile penetrates into the lower stiff clay. Tomlinson defines the depth of the pile below the top of the clay layer as D and the diameter or width of the pile as B. Case 1, where we have all clay soil from top to tip of pile, the depth D is measured from the ground surface. If the clay-bearing layer is under soft clay (Case 2) or under a sand layer (Case 3), the depth D starts not at the ground surface, but at the bottom of the soft clay or sand layer (i.e., top of the bearing layer). In Table 3.5.2, listed values of alpha are given for the three cases; notice that alpha also varies with the ratio of the pile length L to the pile diameter B. Tomlinson gave values for $L/B = 10$ and $L/B > 40$. For copies of Tomlinson's plots of alpha values, you can refer to the Driven user's manual (Mathias and Cribbs, 1998) or the 2006 FHWA Pile manual (Hannigan, et al., 2006). The Driven User's manual also gives digitized values of Tomlinson's alpha values that are useful when generating your own spreadsheets.

Driven pile tip resistance in cohesive soils is calculated by Equation 3.5.8. As I discussed above, pile tip resistance if often neglected unless the pile is considered to be an end-bearing pile. In cases where pile tip resistance is used in the calculation of total pile load resistance in cohesive soils, it is often limited to about 60 to 80 kips per square foot. FHWA and others allow you to use greater values of tip-bearing resistance if you have full-scale pile load test results to confirm higher values of tip resistance. The maximum tip recommendation of 80 kips per square foot unless you have a load test seems reasonable to me.

Table 3.5.2 Pile–soil adhesion values α for clayey soil profiles (Tomlinson, 1980)

Tomlinson pile–clay-side resistance factor, α (Tomlinson, 1980)						
	Case 1 – all clay from ground surface to pile tip		Case 2 – soft clay layer over clay bearing layer		Case 3 – sand layer over clay bearing layer, $D < 10B$, $\alpha = 1.0$	
Undrained shear strength clay (ksf)	$L = 10B$	$L > 40B$	$L = 10B$	$L > 20B$	$L = 20B$	$L > 40B$
0.5	0.98	1.00	0.50	0.83	1.00	1.00
1.0	0.92	1.00	0.40	0.75	1.00	0.93
1.5	0.82	1.00	0.33	0.70	1.00	0.82
2.0	0.69	0.90	0.29	0.67	0.95	0.67
2.5	0.44	0.70	0.27	0.65	0.83	0.53
3.0	0.30	0.50	0.24	0.63	0.75	0.41

3.5.2.5 Driven Piles in Sandy Soils

Driving piles into loose sandy soils can densify the sand causing pile capacity to be greater than indicated by correlations to SPT or CPT data. This is especially true when piles are driven in closely spaced groups. I have seen piles in a group driven in a sequence from the outside of the group to the inside of the group to help densify loose sandy soils. By the time the pile driver got to piles in the center of the group, the sand in the center had densified so much that the piles in the center of the group could not be driven, in other words they encountered refusal driving blow counts and started to experience damage, while tens of feet from predicted depth.

Although driven piles in sandy soils can experience setup and relaxation, these effects are not normally as significant in sandy soils as they are in clayey soils.

To estimate side friction and end bearing of driven piles in sandy soils, I recommend using Equations 3.5.6 and 3.5.9.

Do I recommend designing final pile depths by using the equations given above? No I don't. What I do is estimate pile lengths given above, then specify that load test piles be driven, while monitored by a pile driving analyzer (PDA). After static load testing, I correlate static load test results with each test pile's PDA driving results and refine the project pile driving criteria. During driving of production piles I recommend testing as many piles as the project, client and budget allows by use of the PDA.

3.5.2.6 Drilled Piles and Drilled Shafts – General

Driving piles into clayey soil pushes the soil aside, which can remold it as described above. Drilling piles or drilled shafts in clayey soils removes soil which causes some stress relief, which reduces side wall to soil friction, see Figures 3.5.4 and Figures 3.5.5.

Concreting of drilled piles and shafts causes varying degrees of penetration of concrete or grout into the sidewall soils of the pile excavation. Generally it is believed that pressure injected, augercast piles penetrate grout further into the sidewalls of the excavation than drilled shafts, which are filled by free fall or Tremie concreting methods.

When drilled shafts are installed below the water table or under artesian water conditions, they are frequently filled with bentonite or polymer slurry to prevent cave-in or blow-in of the excavated hole. The effect of remaining bentonite slurry coating on drilled shaft side walls seems to be a reduction of soil-wall friction, while the effect of polymer slurry remaining on side walls seems to result in no detectable change in soil-wall friction.

Even though the auger used to drill a shaft may be 36 inches in diameter, the resulting hole may not be 36 inches in diameter. In soft clays the hole may squeeze and result in a shaft diameter smaller than 36 inches. Shaft hole squeezing can be dealt with by using steel casing or slurry during drilling and concreting of the drilled shaft. Augercast piles use the auger flight to support the hole and if the pile is concreted by use of an adequate grout pressure and controlled rate of pulling the auger, squeezing can be handled.

On the other hand, after drilling a shaft with a 36-inch diameter auger in granular soils or layered sands and clays, the resulting shaft diameter can be much larger than 36-inches. This is called "over-drilling." Over-drilling can be caused by minor cave-ins of clean sand, gravel or soft wet zones that occur along the sides of an uncased hole. If minor cave-ins were the only cause of over-drilling, two drillers working side by side (at a safe spacing distance) on a site installing drilled shafts might be expected to achieve the same volume of concrete in adjacent drilled shafts, in my experience, this is not the case. Driller technique has a significant impact on the amount of over-drilling achieved on completed drilled shafts.

I have observed two drillers on a site in New Mexico drilling adjacent rows of shafts achieve 25% over-drill on one row while the adjacent row had 60% over-drill. The difference was basically caused by driller technique. To drill a drilled shaft, the driller has to repeatedly enter the hole to auger soil and then remove the soil laden auger from the hole to dump the excavated soil. One driller carefully inserted and removed his auger with little contact with the hole side walls, while the other driller banged into the sidewalls consistently both during entering and leaving of the hole.

Let's say that these over-drilled shafts were designed to be 48 inches in diameter and 50 feet deep. The intended volume of these shafts was 23.3 cubic yards per shaft, and the intended perimeter of these shafts was 12.6 feet. Drilled shafts on the row with 25% over-drill had volumes of 29.1 cubic yards per hole with an effective shaft perimeter of 15.7 feet. Drilled shafts on the row with 60% over-drill had volumes of 37.3 cubic yards with an effective shaft perimeter of 20.1 feet. These shaft differences caused by over-drilling can cause problems. As you might expect, the cost of shaft concrete used was more than expected for the 60% over-drilled holes and this led to contention between the owner's representative and the drilled-shaft contractor. What you might not have considered is the difference between the installed drilled shafts and the shafts assumed during design. Assume for a moment that we neglect the ends of these shafts, and assume that the shaft load capacity and settlement potential is directly related to the perimeter area of the shafts. If we accept these assumptions for the sake of discussion, and assume that the calculated shaft service load total settlements were 0.75 inches and maximum differential settlements between shafts were 0.25 inches, what is the effect of over-drilling shafts? The nominal 48-inch diameter shafts with 25% over-drilling had shaft load capacities that were 25% greater than anticipated, and they had service load settlements of 0.60 inches. The shafts with 60% over-drilling had shaft load capacities that were 60% greater than anticipated, and they had service load settlements of 0.47 inches. The maximum differential settlement between a lightly loaded column on a shaft with 60% over-drilling and a heavily loaded column on a shaft with 25% over-drilling is 0.29 inches. Even though the installed shafts were larger than anticipated, the differential settlement was not less, it increased. This is because the original differential settlement calculation assumed that the lightly loaded column had a loading that was 67% of the heavily loaded column and that the adjacent drilled shafts supporting these columns were identical. Now that the heavy column can be supported by a smaller shaft and the light column supported by a larger shaft, the difference in settlements is increased.

By the way, when you do full-scale drilled shaft load tests, it is a good idea to find out the amount of concrete actually used to construct the test shaft, calculate the amount of over-drilling, and adjust your estimated shaft capacity from the load based on the nominal shaft diameter to the load capacity generated by the actual shaft diameter.

Geotechnical engineers are often accused of being "over-conservative," especially in drilled-shaft load capacity determinations. Little do the accusers know of the differences between the designs' nominal shaft dimensions versus the actual installed-shaft dimensions.

3.5.2.7 Drilled Piles and Shafts in Clayey Soils

I will primarily focus on drilled shafts, because I treat drilled large diameter piles in clayey soils essentially the same as drilled shafts. Clayey soils are primarily classified as GC, SC, CL, CH, and MH soils. We used to consider clayey soils as soft, medium stiff, stiff, and hard clays for the purposes of drilled shaft design. Today for the purposes of drilled-shaft design, we subdivide clayey soils into two groups: (1) cohesive soils, and (2) cohesive intermediate geomaterials (IGMs). Intermediate geomaterials are those materials that classify between soils and rocks. We struggled for decades with classification and analysis of properties of very hard clayey soils and soft rocks, because these materials were not clays and they were not truly rocks. I'm pleased that we finally have a classification of geomaterials that have properties between those of soils and rocks. The classification of clayey (i.e., cohesive) soil, cohesive IGM and rock is based solely on compressive strength measured by the unconfined compression test.

Cohesive soils are those materials having unconfined compression test strengths of less than 10 000 pounds per square foot (5 tons per square foot). We used to classify clays with compressive strengths greater than 4 tons per square foot as hard. I guess that means today the classification of hard clay only extends from compressive strengths of 4 to 5 tons per square foot.

Cohesive IGM are those materials that have unconfined compression strengths between 10 000 and 100 000 pounds per square foot, and rocks have unconfined compression strengths greater than 100 000 pounds per square foot (694.4 pounds per square inch). We used to classify soft shale as having an unconfined compressive strength of 1000 psi, and shales having compressive strengths less than 1000 psi as clay shales. The new dividing line of about 700 psi seems in line with past practice.

For cohesive, clayey soils, we have to consider short-term load resistance by using the undrained shear strength, and we have to consider longer-term load resistance by using effective stress, drained soil shear strength parameters.

Let's start with the cases of undrained loading (which is normally during construction) to determine the drilled shaft load capacity in cohesive soils.

The method used to calculate the vertical load capacity developed by side friction of shafts in cohesive soils is called the Alpha method as we mentioned earlier. The equation for calculating shaft side resistance is as follows:

$$Q_{\text{side resistance}} = \sum_{i=1}^{n} c_{\text{adhesion}} \pi DH_i = \sum_{i=1}^{n} \pi DH_i (\alpha S_u)_i \qquad (3.5.11)$$

This is the same equation we use for side resistance in driven piles, except we only have circular cross-sections for drilled shafts, so we know the perimeter (P_{ep}) is equal to πD, and it is common geotechnical practice to use the term S_u for undrained shear strength rather than c or c_u (don't be confused, the terms S_u and c_u are both referring to undrained shear strength). But what about the alpha term (α); is it the same for driven piles and drilled shafts? Alpha in both cases refers to a ratio between the pile–soil side adhesion and the clay's undrained shear strength. But driven piles push the clayey soil aside, remolding soft clays and dilating stiff clays, while drilled shafts unload the clayey soil as the shaft hole is drilled. In theory, the alpha term for driven piles should be different than the alpha term for drilled shafts. So as you might expect the guidance for selecting an alpha value for drilled shafts is different than the Tomlinson method that we used for piles.

I use guide documents for analysis of driven piles and drilled shafts that are developed and published by the US Federal Highway Administration (FHWA), because most of the recent research done on piles and drilled shafts has been done by the

FHWA through their National Highway Institute (NHI). I also use Electric Power Research Institute (EPRI) guidance on drilled shafts, especially for rock sockets. I have already mentioned the 2006 FHWA document on piles, so now I'd like to introduce you to the FHWA's latest manual on drilled shafts, *Drilled Shafts: Construction Procedures and LRFD Design Methods*, publication number FHWA-NHI-10-016, May 2010 (Brown, Turner, and Castelli, 2010). This manual gives the latest guidance on selecting alpha values for drilled shafts as follows:

$\alpha = 0$ between the ground surface and a depth of 5 feet, or to the depth of seasonal moisture fluctuation whichever is greater. I use 5 feet or two times the diameter of the drilled shaft, whichever is greater. The depth of seasonal moisture fluctuation is a somewhat controversial topic that I prefer to avoid when analyzing drilled shafts, unless the site has highly expansive soils.

$\alpha = 0.55$ along portions of the drilled shaft below five feet where $\frac{S_u}{p_a} \leq 1.5$

$\alpha = 0.55 - 0.1 \left(\frac{S_u}{p_a} - 1.5 \right)$ on the remaining portions of the shaft where $1.5 \leq \frac{S_u}{p_a} \leq$ 2.5, the term p_a is atmospheric pressure used to normalize these equations, atmospheric pressure equals 2116 psf or 14.7 psi and should be expressed in the same units as S_u.

Some publications have recommended that the side resistance along the bottom five feet of drilled shafts be neglected. One well-respected reference recommending that the resistance on the bottom 5 feet of shaft be neglected is the AASHTO 1999 drilled shaft manual (Reese and O'Neill, 1999). I never agreed with this recommendation. I know that Reese and O'Neill had computer models that predicted tension stresses adjacent to the bottom one diameter (often taken as 5 feet) of shaft, and that these tension stresses would reduce cohesive resistance. My experience with full-scale, instrumented drilled-shaft load tests has been that the predicted loss of cohesive side friction resistance does not occur in the field, so I do not neglect the load capacity generated along the bottom 5 feet of shaft. Well what do you know, the latest 2010 version of the FHWA drilled-shaft manual agrees with me! On page 13–17 (i.e., Chapter 13, page 17), the new 2010 FHWA manual says that, "... this recommendation (i.e., Reese and O'Neill's) is not supported by field load test data and the authors of this version (i.e., of the FHWA Drilled Shaft Manual) recommend that side resistance should not be neglected over the bottom one diameter." Am I right? Are the authors of the new 2010 manual right, or were Reese and O'Neill right? My suggestion to you is to do full-scale drilled-shaft load testing on your project when possible, add some instrumentation to the shaft above its tip, and check the developed side resistance for yourself.

The method used to calculate the vertical load capacity developed by tip or base resistance of shafts in cohesive soils doesn't have a name. Who knows, the side resistance methods for clays and sands are the Alpha and Beta methods, while the tip resistance is calculated by the no-name method. Using total stress analysis (i.e., undrained loading), the method presented by Reese and O'Neill in their 1999 Drilled Shaft manual (Reese and O'Neill, 1999) is still used for clayey soils. This equation is:

$$q_{\text{shaft tip}} = N_c^* s_u \qquad (3.5.12)$$

Where, N_c^* is the bearing capacity factor for deep foundations and s_u is the average undrained shear strength of the clayey soil in a zone within two shaft diameters below the bottom of the shaft tip. Equation 3.5.12 should only be used for drilled shafts that have a length below grade greater than three times the shaft diameter, but then if your shaft was much less than three diameters deep it would be a shallow rather than a deep foundation. N_c^* used in Equation 3.5.12 is equal to 9.0 when the shaft tip is bearing on clays with undrained shear strengths of 2000 pounds per square foot or more. But then I can't imagine why you would want to bear the base of a drilled shaft on clayey soil with undrained shear strength of less than 2000 psf. If your clay is that soft, why not drill deeper?

My friend and next door neighbor Paul died yesterday morning. I went to his wake last night, and it got me thinking. Paul was 63 years old and had been retired from work at the airlines for three years. I'm 65 years old and still working as a geotechnical consulting engineer every day; no part-time or slow downs. You can imagine why Paul's death got me thinking. My long-time bank teller and coin-collecting friend told me day before yesterday that she is retiring from the bank next week. Several of my friends retired at 62 years old in the past five years. What am I doing working so hard? Maybe I should stay home and watch Oprah every afternoon. Oh, I forgot, Oprah is retiring from her TV show after 25 years next week! When thinking about retirement, I always come back to the advice that Ralph Peck gave me. Ralph said, "John, you have got plenty of time to retire after you are 85." Ralph was not kidding. He worked on consulting projects and spoke at engineering conferences up until he was past 90 years old, until failing eye sight forced him to stop working. Ralph and I discussed geotechnical engineering every time we met. He told me that engineers from my generation must keep on working until we mentor and help as many young engineers as we can. Engineering is a necessary and honorable profession that supports human kind.

Now that I think of it, my bank-teller friend said, "John, I'm retiring from a job. You have a profession, a career, not a job. Why would you want to retire from the life you love?" I wouldn't.

3.5.2.8 Drilled Piles and Shafts in Sandy Soils

The method used to calculate the vertical load capacity developed by side friction on shafts in cohesionless, sandy soil is called the Beta (β) method.

The Beta method calculates the side resistance in cohesionless soils as the frictional resistance generated along the outside surface of the drilled shaft. The frictional resistance is calculated as shown below in Equation 3.5.13:

$$f_s = \sigma'_v K \tan \delta \tag{3.5.13}$$

where,

σ'_v = vertical effective stress at the center of stratum thickness Δz

K = coefficient of lateral earth pressure, $\frac{\sigma'_{horizontal}}{\sigma'_{vertical}}$

δ = pile-soil interface friction angle.

For convenience, the product of K times tangent δ is called the shaft side resistance coefficient β, and the equation for shaft side resistance reduces to:

$$f_s = \sigma'_v \beta \tag{3.5.14}$$

Most references refer to the Beta method as if it were one method that was understood by all engineers. This is not the case. There are several methods of determining values of beta for use in design. Some methods use curve fitting to match results of load tests to back calculated values of beta (Reese and O'Neill, 1999). Other methods use OCR, K and δ values to determine appropriate values of Beta. I prefer methods that use site specific soil parameters to those that use curve fitting to match load tests performed elsewhere. I have used methods outlined in Bowles' 5th Edition Chapter 19 to calculate beta values ($K \tan\delta$ in Bowles' text), but currently I prefer the Chen and Kulhawy method (Chen and Kulhawy, 2002) outlined in the FHWA 2010 Drilled Shafts Manual.

To determine a value of β, we need to determine an appropriate value of the lateral earth pressure coefficient K. If we assume that there is no stress relaxation in the soil around the shaft due to drilling then it would be reasonable to assume that K equals the "at-rest" earth pressure coefficient K_o, or that $K/K_o = 1$. For normal sandy soils that are not cemented or aged, the value of K_o increases with the soil's overconsolidation ratio, OCR. The higher the value of the OCR the higher the value of K_o, although we know that the upper bound value of K_o would be equal to or less than the passive earth pressure coefficient, K_p. Why? Well the theory is that a soil that has reached a state of passive earth pressure has achieved an upper bounding state of soil failure in compression. So if the soil is in a state of failure when it has an earth pressure coefficient of K_p, it can't exceed failure stresses, because it has failed. The value of $K = K_o$ is approximated by the Mayne and Kulhawy equation (Mayne and Kulhawy, 1982) which is:

$$K_o = \left(1 - \sin \phi'\right) OCR^{\sin \phi'} \leq K_p \tag{3.5.15}$$

where

$$OCR = \frac{\sigma'_p}{\sigma'_v} \tag{3.5.16}$$

where,

σ'_p = plainvertical effective preconsolidation pressure at the center of the stratum
σ'_v = vertical effective stress at the center of the stratum
K_p = passive earth pressure coefficient = $\tan^2\left(45^o + \frac{\phi'}{2}\right)$.

So determining a value of K_o using the Mayne and Kulhawy equation comes down to determining the vertical effective preconsolidation pressure of a sandy soil. We use the consolidation test to find the preconsolidation pressure of a clayey soil, but that method doesn't work too well in granular soils where it is difficult enough to get an undisturbed sample not less trim that sample for a consolidation test. So for practical purposes a value of σ'_p for use in finding a value of OCR is obtained from correlation to the standard penetration test blow count, N_{60}. The 2010 FHWA manual suggests using another of Mayne's equations (Mayne, 2007):

$$\frac{\sigma'_p}{p_a} \approx 0.47\,(N_{60})^m \tag{3.5.17}$$

where p_a is atmospheric pressure used to normalize the equation, $p_a = 2116$ pounds per square foot, and $m = 0.6$ for clean quart sands and 0.8 for silty sands and sandy silts. For gravelly soils, Equation 3.5.18 by Kulhawy and Chen (Kulhawy and Chen, 2007) is recommended by the 2010 FHWA manual:

$$\frac{\sigma'_p}{p_a} = 0.15 N_{60} \tag{3.5.18}$$

Since $\beta = K \tan \delta$ and we assumed that $K = K_o$, we can substitute values from equations given above to calculate an approximate value of β that is based on K_o, OCR and the vertical effective stress as follows:

$$\beta \approx \left(1 - \sin\phi'\right)\left(\frac{\sigma'_p}{\sigma'_v}\right)^{\sin\phi'}\tan\phi' \leq K_p \tan\phi' \tag{3.5.19}$$

The FHWA 2010 Drilled Shaft Manual suggests that Equation 3.5.19 should not be used for vertical effective stresses less than about 900 pounds per square foot, or a depth of about 7.5 feet when soil moist unit weight is about 120 pounds per cubic foot. This limitation doesn't mean much to me, because I ignore side resistance developed by the upper 5 feet or two diameters of shaft anyway.

Now all we have left is drilled-shaft tip resistance in sandy/cohesionless soils. The tip resistance of a drilled shaft is calculated in standard practice by using the equation:

$$q_{tip}\,(ksf) = 1.2N_{60} \tag{3.5.20}$$

The 2010 FHWA Drilled Shaft manual limits tip resistance calculated by Equation 3.5.20 to a blow count of 50, and a maximum tip resistance of 60 kips per square foot. Earlier FHWA guidance (Reese and O'Neill, 1988) allowed blow counts up to 75 with a maximum tip resistance of 90 kips per square foot. All of the standard guides that I'm aware of allow you to exceed these maximum limit values of tip resistance in design if you have results of full-scale load tests that confirm higher tip resistances.

Working example

About 10 years ago, I worked on two highly instrumented drilled shaft tests in Albuquerque, New Mexico. One test failed in tip loading at just over 120 kips per square foot and the second test failed at over 220 kips per square foot. Both test shafts were tipped in the same soil formation, located about 2 miles apart. Test boring blow count data indicated that a maximum value of tip resistance of 90 kips per square foot would be suitable for design. Use of load test data allowed project designers to achieve savings of approximately $1 000 000 by reducing diameters of drilled shafts. Did you ever notice that savings quoted on large projects are always at least a million dollars!

3.5.3 Case Histories and Full-Scale Load Tests

The use of equations to calculate load resistance of driven and drilled piles and drilled shafts is standard geotechnical practice in designing deep foundations for buildings and bridges. Given the considerable uncertainty involved in subsurface conditions and in foundation installation techniques, I always recommend performing full-scale load tests to confirm pile/shaft load capacity, reduce uncertainty in design and allow a lower factor of safety (in ASD design) or a larger resistance factor (in LRFD design) that is based on analysis of load test data and not on gut feelings.

What we need in geotechnical engineering practice and in geotechnical research is an extensive library of well-documented case histories that measure side and tip resistance of piles and drilled shafts. We also need good documentation of pile head and tip deflections associated with development of side and tip resistance. It is no coincidence that the Ralph B. Peck Lecture and Award is for documentation and presentation of case histories. Ralph believed that case histories are very important to the advancement of geotechnical engineering practice and knowledge. I couldn't agree more.

But why do we have to go to the trouble and considerable expense to perform full-scale load tests? Why can't we take some half-inch diameter pipe that is 40 inches long and simulate a half-foot diameter pipe pile that is 40 feet long? Over the years I have seen dozens of geotechnical papers that report model pile tests using small-scale piles installed in laboratory test chambers with scaled-down pile load tests. What's wrong with these mini-load tests? Small-scale pile load tests may sound like a good idea but they don't work. "Why not?" you're probably asking.

Well the simple answer is that the scaling of pile performance from a small half-inch diameter pipe to a full-scale 6-inch diameter pile is not correctly modeled by the ratio of pile sizes. To better explain this answer, pile performance is related to the vertical soil effective stress at each element of pile length. If you recall from Section 2.5, we discussed the effect the mean confining stress has on the shear strength and volume change properties of a soil. Dense sands dilate and loose sands compress during shearing tests, only if they are tested at moderate confining stresses. If dense sands are tested at high confining stresses they can act like loose sands and compress when sheared, and loose sands tested at very low confining stresses act like dense sands because they dilate when sheared. If you are modeling a 40-foot-long pile or drilled shaft, the actual pile has a vertical stress (assuming a moist unit weight of 120 pounds per cubic foot) of 1200 psf at 10 feet, 2400 psf at 20 feet, 3600 psf at 30 feet, and 4800 psf at 40 feet. If we try to model this 40-foot-long pile by a 40-inch long pile, the vertical stress is 100 psf at 10 inches, 200 psf at 20 inches, 300 psf at 30 inches, and 400 psf at 40 inches. Soil performance during shearing depends on the actual value of confining stresses acting, so assuming the sandy soil in both the test chamber and in the field have exactly the same moist unit weight (or relative density) and the exact same soil fabric, the model pile soils sheared at vertical confining stresses of 100 to 400 psf will likely all dilate, while soil at the 10-foot depth of the real pile may slightly dilate, but soils at 20, 30, and 40 feet will likely compress when sheared, see Figure 3.5.6. The result of this model versus actual pile mismatch is that the model pile shearing stresses, apparent friction angle, and measured pile performance will not be representative of the real-world pile.

If you would like to see what modeling parameters to use for your laboratory scale pile load tests, I suggest that you read Bengt Fellenius' paper on physical modeling in sandy soils (Altaee and Felenius, 1994). I'll give you a hint. The size of your model pile (in a 1g gravity environment) has to be very close to full-scale for your laboratory test results to be useful. I'm asked sometimes by students and practicing engineers, why Ph.D.s at major universities perform centrifuge testing. Well now you know the answer, the only way to increase the gradient (from 1 g to higher multiples of gravity) of vertical stresses acting adjacent to a model pile, wall or footing is to increase the gravity forces acting on the soil used to build the model. The Altaee–Fellenius paper also describes how to plan centrifuge testing for equal changes in state parameter Y upsilon, so that tested samples will perform the same in the field as in the laboratory,

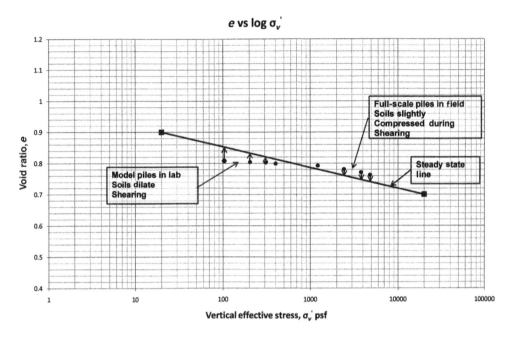

Figure 3.5.6 Small laboratory piles versus full-scale field piles

that is by adjusting the test so that the void ratio changes during shearing and the resulting dilation and compression when sheared is similar in both the test model and the full-scale pile. The vertical distance between the initial void ratio and the SSL in Figure 3.5.6 is equal to the upsilon parameter.

References

Altaee, A. and Felenius, B.H. (1994) Physical modeling in sand. *Canadian Geotechnical Journal*, **31**, 420–431.

Bowles, J.E. (1982 and 1996) *Foundation Analysis and Design*, 5th edn, McGraw Hill Inc., 816 and 1024 pages.

Brown, D.A., Turner, J.P., and Castelli, R.J. (2010) *Drilled Shafts: Construction Procedures and LRFD Design Methods.* NHI Course No. 132014, FHWA NHI-10-016, Geotechnical Engineering Circular No. 10, 970 pages.

Chen, Y.J. and Kulhawy, F.H. (2002) *Evaluation of Drained Axial Capacity for Drilled Shafts*, (eds M.W. O'Neill and F.C. Townsend), American Society of Civil Engineers, Reston, VA, Geo-Institute, Geotechnical Special Publication No. 116, Deep Foundations 2002, pp. 1200–1214.

Coyle, H.M. and Castello, R.R. (1981) New design correlations for piles in sand. *Journal of the Geotechnical Engineering Division*, The American Society of Civil Engineers, **107**(GT 7), 965–986.

Hannigan, P.J., Goble, G.G., Likins, G.E., and Rausche, F. (2006) *Design and Construction of Driven*

Pile Foundations – Volumes I and II, Report Number FHWA-HI-05-042, U.S. Department of Transportation, Federal Highway Administration, 968 pages.

Kulhawy, F.H. and Chen, J.-R. (2007) Discussion of 'Drilled Shaft Side Resistance in Gravelly Soils' by Kyle M. Rollins, Robert J. Clayton, Rodney C. Mikesell, and Bradford C. Blaise, *Journal of Geotechnical and Geoenvironmental Engineering, American Society of Civil Engineers*, **133**(10), 1325–1328.

Mathias, D. and Cribbs, M. (1998) *Driven 1.0: A Microsoft Windows^{TM} Based Program for Determining Ultimate Vertical Static Pile Capacity*. Users Manual, Report Number FHWA-SA-98-074, U.S. Department of Transportation, Federal Highway Administration, 112 pages.

Mayne, P.W. and Kulhawy, F.H. (1982) Ko-OCR relationships in soil. *Journal of the Geotechnical Engineering Division*, American Society of Civil Engineers, **108**(GT6), 851–872.

Mayne, P.W. (2007) In-situ test calibrations for evaluating soil parameters, in *Characterisation and Engineering Properties of Natural Soils II*, Proceedings of Singapore Workshop (eds Tan, T.S., Phoon, K.K., Hight, D.W., and Leroueil, S.), Taylor & Francis Group, London, pp. 1601–1652.

Meyerhof, G.G. (1976) Bearing capacity and settlement of pile foundations. The Eleventh Terzaghi Lecture. *Journal of Geotechnical Engineering Division, American Society of Civil Engineers*, **102**(GT3), 195–228.

Nordlund, R.L. (1963) Bearing capacity of piles in cohesionless soils. The American Society of Civil Engineers. *Journal of the Soil Mechanics and Foundations Division*, **SM3**, 1–35.

Nordlund, R.L. (1979) *Point Bearing and Shaft Friction of Piles in Sand*. University of Missouri-Rolla, 5th Annual Short Course on the Fundamentals of Deep Foundation Design.

Reese, L.C. and O'Neill, M.W. (1999) *Drilled Shafts: Construction Procedures and Design Methods*, Publication No. FHWA-IF-99-025, Federal Highway Administration, Washington, D.C., 758 pages.

Thurman, A.G. (1964) Computed *Load Capacity and Movement of Friction and End-Bearing Piles Embedded in Uniform and Stratified Soil*. Ph.D. Thesis, Carnegie Institute of Technology.

Tomlinson, M.J. (1980) *Foundation Design and Construction*, 4th edn, Pitman Advanced Publishing Program.

Tomlinson, M.J. (1994) *Pile Design and Construction Practice*, 4th edn, E & FN Spon, London, 411 pages.

Vesić, A.S. (1975) *Principles of Pile Foundation Design*. Soil Mechanics Series No. 38, School of Engineering, Duke University, Durham, N.C., 48 pages + figures.

3.6

Laterally Loaded Piles and Shafts

3.6.1 Introduction of Laterally Loaded Piles and Shafts

In the 1950s and 1960s laterally loaded piles were analyzed primarily by three methods. The first method specified the allowable lateral loading on vertical piles or poles. These allowable lateral loadings were based on local load test experience. For example, a 12-inch diameter timber pile driven into sand or medium clay had an allowable lateral loading of 1500 pounds per pile; if soft clay was present the allowable lateral force was often limited to 1000 pounds per pile or less. To resist a 15 000 pound lateral loading by vertical piles driven into medium stiff clay, you needed 10 piles.

The second method was based on the assumption that lateral loading on piles was similar to vertical loading on spread footings. The upper 5 to 10 feet of the pile was assumed to resist lateral forces (some codes allowed up to 15 feet of pile), and so the area resisting lateral forces was assumed to be equal to the width of the pile times 5 or 10 feet, see Figure 3.6.1. Building codes gave values of allowable lateral bearing pressure or allowable lateral loading per foot of pile for soils described as loose, dense, soft, or hard. The 2000 International Building Code gave allowable lateral bearing pressures for isolated poles supporting buildings as 400 pounds per square foot for sandy gravel, 300 pounds per square foot for silty sand, and 200 pounds per square foot for silty clay. All you had to do to check the design of a pile or pole foundation for lateral loading was to divide the lateral pile force by the resisting pile area and compare the resulting stress to the allowable lateral pile pressure based on your assumed soil type. For years I thought that these building-code allowable lateral bearing pressure values were intended for spread footings only. It was after a structural engineer pointed out the paragraph on isolated flagpoles, signs, and

Geotechnical Problem Solving, First Edition. John C. Lommler.
© 2012 John Wiley & Sons, Ltd. Published 2012 by John Wiley & Sons, Ltd.

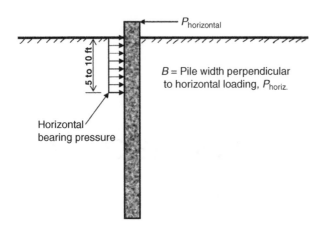

Figure 3.6.1 Laterally loaded pile with horizontal bearing pressure

pole-supported buildings that I learned that these values were actually being used for design of laterally loaded piles.

Notice that I used the words "assume or assumption" several times in the above description of the second lateral pile analysis method. Bridge engineers who design piles to resist lateral loadings on a regular basis didn't particularly like either of the building-code-specified allowable lateral force or the allowable lateral bearing pressure concepts. Unlike building pile foundations that are in or adjacent to heated spaces, bridge piles are exposed to freeze–thaw conditions so bridge designers concluded that the upper 3 to 4 feet of their piles would not develop reliable lateral resistance. Bridge piles are often subjected to erosion from streams and rivers which remove near-surface soils, exposing the upper portion of these piles and reducing lateral pile resistance. Since available 10- to 12-inch diameter bridge piles did not have large diameters, bridge engineers in the 1950s and 1960s assumed that their piles did not have any bending resistance. To resist lateral loadings bridge engineers used battered piles, that is they used piles installed on an incline to vertical as shown in Figure 3.6.2.

Notice that the battered pile's lateral resistance is calculated by using vector analysis from statics. The ratio of pile forces is taken as equal to the ratio of pile dimensions, as shown in Figure 3.6.2 for a 5 on 12 battered pile.

Geotechnical engineers and structural engineers in the 1960s were not satisfied with the use of allowable lateral pile force, allowable lateral pile bearing pressure and battered pile methods to resist lateral forces. Using earth pressure theories developed by Terzaghi, Wayne C. Teng of Teng and Associates in Chicago wrote a geotechnical text in the early 1960s (Teng, 1962) that used soil mechanics to analyze laterally loaded piles. Teng's work was with sheet piles walls that were used for lake-front construction and retaining of deep building foundation excavations. The simplified lateral earth pressure approach used by Teng to design laterally loaded sheet piles is

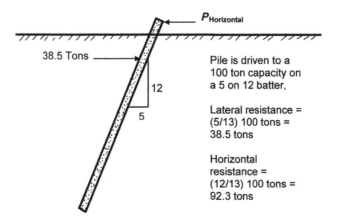

Figure 3.6.2 Laterally loaded battered pile

shown in Figure 3.6.3. This method assumed that the pile was a rigid structure and that active and passive pressures are present on the loaded and unloaded sides of the pile.

Active earth pressure develops at relatively small lateral pile deflections and passive earth pressure develops at relatively large lateral pile deflections, so the lateral pressures shown on both sides of the pile in Figure 3.6.3 couldn't act at the same time at the same point on the laterally loaded pile. Besides, all piles are not rigid, so lateral bending of the laterally loaded pile would also modify the earth pressures generated

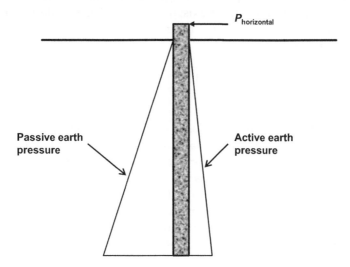

Figure 3.6.3 Lateral loaded pile with active and passive soil pressures

on opposite sides of the laterally loaded pile. Teng's methods were widely distributed by reproduction in the United States Steel Sheet Pile Design Manuals (USS, 1969 and 1975) and in the US Navy's NAVFAC manuals. Teng's methods with some modifications are still used by present-day designers and are included in several lateral pile design computer codes.

The main problem with all of the methods discussed above is the lack of strain compatibility between the laterally loaded pile and the soil. This is where structural engineers come into the laterally loaded pile analysis history. A famous structural mechanics engineer named Stephen P. Timoshenko (1878–1972) (I recommend you look up Timoshenko's book on advanced strength of materials, it is the second volume of a two volume set (Timoshenko, 1956)) presented the equations for deflection, shear and bending moment of a beam on elastic (i.e., soil) foundation. Up until 1930, you would have had to read German or be a German engineer to study the works of Winkler (1867) or Zimmermann (1888). Timoshenko used a term he called *the modulus of the foundation* which denotes the foundation reaction that we now call the modulus of subgrade reaction. Timoshenko's modulus of subgrade reaction is the foundation reaction per unit length of the beam when the beam deflection is equal to one unit. Timoshenko's beam was a line, and its subgrade reactions were one-dimensional springs. Reviewing Timoshenko's book, he emphasized that the springs beneath his beam on an elastic foundation were capable of exerting both upward and downward forces. In other words, Timoshenko's subgrade could exert tension stresses on the beam, which is not likely with real-world soils. The beam's subgrade reactions are sometimes referred to as Winkler springs (Winkler, 1867). Since Timoshenko's beam was one-dimensional and his subgrade material was linearly elastic, the modulus of subgrade reaction used in his equations was a series of linear springs and his subgrade modulus k has units of pounds per inch length per inch deflection. Given the length of Timoshenko's beam is along the x-axis, and vertical beam deflection is along the y-axis, and the axis of bending is about the z-axis, Timoshenko's equation of a beam on elastic foundation is:

$$EI_z \frac{\mathrm{d}^4 y}{\mathrm{d}x^4} = -ky \tag{3.6.1}$$

Turning Timoshenko's beam on an elastic foundation up into a vertical orientation several engineers, notably Reese and Matlock (Matlock and Reese, 1960), in the middle to late 1950s solved the equations for laterally loaded pile foundations. The main advantages of these numerical solutions were beam–soil deflection compatibility and the development of non-linear soil springs by use of p–y curves. We know that soil materials are not elastic, so the development of non-elastic stress–deflection relationships was a big advancement in modeling lateral deflection of piles in real soils. The main problem with these laterally loaded pile numerical solutions was timing. They were developed before the personal computer was invented. To use the Reese and Matlock method, you had to have access to a mainframe computer or use cumbersome graphs. Thanks to Federal Highway Administration support, the program COM624 was developed for use on mainframe and early personal computers, and not soon after a commercial version called L-Pile was developed as a PC application and put on the market in 1986. From the time I got my first IBM XT personal computer to the present day, I have used COM624 and then the successor versions of the L-Pile program to analyze laterally loaded piles and drilled shafts.

But what if you don't want to perform a computer analysis or if you want to check the L-Pile program by a hand calculation method, are you out of luck? No you are not out of luck; you have an alternate method available for hand calculation of laterally loaded piles. The hand calculation method that I use is Broms' method (Broms, 1965). Broms' method is discussed in the FHWA 2006 pile manual (Hannigan *et al.*, 2006). The New York State Department of Transportation has developed a step by step procedure for the Broms method that is available on their website. The FHWA also has a free download of their manual FHWA-HI-97-013 (Hannigan *et al.*, 1998) that includes the Broms method with all charts, tables and figures required to use this method. A few things to consider when using the Broms method:

1. Broms' method is a hand calculation method for analyzing single vertical laterally loaded piles using charts and graphs to determine applicable parameters.
2. You can calculate the pile head deflection and loading, but you cannot calculate pile deflections at any other point along the pile. If pile deflection criteria are project critical, another method should be used.
3. Broms' method ignores axial pile loading.
4. Broms' method calculates the maximum soil resistance to lateral loading.
5. Broms' method calculates the pile's maximum bending moment induced by lateral loading.
6. Broms' method considers both free- and fixed-head piles. Free head means that the top of the pile is completely free to rotate during lateral loading. Fixed head means that the pile's top remains vertical, and that a bending moment is generated by the pile cap that restrains the pile head from rotating during lateral loading.
7. Broms' assumes only one soil type at a time; either the soil is all cohesive or all cohesionless. Layered soils are *not* considered.
8. FHWA-HI-97-013 indicates that Broms' method over-estimates the lateral load resistance for long, fixed-head piles in sands.

3.6.2 L-Pile Program Use – A Few Pointers

When asked by my young geotechnical engineers if there is any bit of general advice I can give them about the L-Pile program, I always say, "Remember that L-Pile was developed by structural engineers, so it is a structural engineering program with all of the conventions that go along with beam analyses." Nearly every time I give this advice, I get the same response … a blank stare. What I consistently forget is that geotechnical engineers these days don't start out as structural engineers. Ralph Peck started as a structural engineer; I started as a structural engineer, and my friend Bob Meyers started as a structural engineer. Ralph has passed away, I am 65 years old, and my friend Bob is 57 or 58 years old (sorry Bob, I forgot). The point I'm making is that nearly every geotechnical engineer under 55 years of age has specialized from the beginning of their career in geotechnical or geological engineering. If I'm about to tell you some things that you already know well, consider this a refresher and please humor me.

The model of a vertical beam-pile in structural engineering is a one-dimensional line. It has no width or diameter like a real pile. All loading on the beam is a point load, a concentrated moment, a line load, or a variation on a triangular loading. All loadings are one-dimensional. The line loading has only length; it has no width, so it has units of pounds per lineal foot along the length of the beam, not pounds per square foot. If the model beam-pile has a free head, it can have a shear, but not a moment, at the head because it is free to rotate. If the beam-pile has a fixed head, it can have both a shear and a moment at the head because it is restrained against rotation. At every point of zero shear along the beam-pile, a maximum or minimum value of bending moment occurs. The equivalent point of fixity of a beam-column is not the point where the pile lateral deflection becomes zero, nor is it the point where the pile rotation is zero. The equivalent point of fixity is the point along the length of the pile where a fixed cantilevered pile would have the same deflection at the pile top when analyzed by the structural (above grade analysis) analysis program as the deflection computed by the (below grade) L-Pile program. This means that both computer programs (the structural engineer's bridge program and the geotechnical engineer's L-Pile program) have considered compatible stiffness when calculating lateral deflection of the top of pile. OK, I've adequately vented on structural versus geotechnical issues.

The L-Pile program is intended for analysis of a single pile or drilled shaft that is far removed from other piles, that is, no lateral pile interaction is assumed in the analysis. Lateral pile interaction factors are given in FHWA guidance documents, but they are considered by many engineers to be overly conservative. To include the pile interaction affect in an L-Pile analysis, the simplest technique is to use the "p–y modifiers." For example, if the resisting soil in front of a laterally loaded pile is stressed by other piles in the group, the FHWA might suggest using a group reduction factor of 0.8 for this pile, I reduce the lateral soil resistance by using a p-modifier of 0.8. By use of a p-modifier in the L-Pile program, we reduce the reaction generated by pile deflection y as shown in Figure 3.6.4.

If you have a group of piles to analyze, you can roughly estimate group effects using the L-Pile program by analyzing individual piles using group reduction factors, but I don't recommend it. If you can afford the investment, I recommend you use the GROUP program, which was developed by the same engineers who wrote L-Pile. A word of caution, please read the GROUP manual carefully. There are several pile

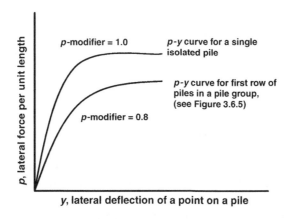

Figure 3.6.4   Pile-soil-pile group interaction, *p*-modifier

group interactions to consider. Geometry and stiffness of piles (lateral and vertical stiffness) making up a pile group affect the distribution of lateral and vertical loading in individual piles, the GROUP program considers these stiffness effects. You might be fooled to believe that the result of the GROUP program considers all of the factors required to provide specific "real" group reduction factors. Oh, I wish that were the case.

The GROUP program does not consider pile-to-soil-to-pile interaction affects. That is, the program still doesn't help you with the soils that are stressed by more than one pile. The pile group designer has to evaluate what *p*-modifier terms to use to include these pile–soil–pile interaction effects, as shown in Figure 3.6.5. I know this is difficult, but pile group interactions are one of the most difficult topics in geotechnical engineering.

At this point you have some choices, you can use guidance on group shadowing affects in FHWA manuals, you can read the group analysis material in Poulos and Davis book (Poulos and Davis, 1980), or you can perform a huge full-scale pile group load test. If you decide to do the load test, please let me know, because I am interested in the results of such tests.

What references do I recommend? I know that some geotechnical engineers don't like Bowles' books (Bowles, five editions, 1968–1996), but then he was a structural engineer and his material on beams on elastic foundations and laterally loaded piles are good background information. In addition to carefully reading of the L-Pile and GROUP manuals, I recommend that you purchase and study a copy of Poulos and Davis' book (Poulos and Davis, 1980). If you start to look over Poulos and Davis' pile analysis book you will notice that their pile group analyses are based on elasticity principles, and I expect you might comment that soils are not elastic. True enough! With all of the complexity and multiple levels of variables involved in pile group analyses, if you use group reduction factors developed from elasticity principles you will be better off than guessing or better off than using factors that are not specific to your pile group. Then again, if you are up for a full-scale pile group load test, I'm game!

3.6.3 L-Pile Soil Input Parameters

When deciding what soil parameters to use for input into L-Pile, remember that clayey soils are c only, sandy soils are ϕ only, and soils with both c-ϕ parameters used to characterize the soil shear strength are called silts. I know, your c-ϕ soil is a clayey silt or a silty clay or a sandy silty clay, no matter, if you want L-Pile to consider both the c and the ϕ properties of your soil, you will have to call it a silt. It's not geotechnical reality, it's just a computer convention used in this program.

I work mostly with silty sandy soil in New Mexico, so I use the API sand option in the L-Pile program, and I use the L-Pile Technical Manual figure that shows modulus of subgrade reaction k_s (pounds per cubic inch) versus relative density versus angle of internal friction ϕ' to select a suitable value of k_s.

Now is as good a time as any to discuss the modulus of subgrade reaction, k_s. First of all the units of k_s are *not* pounds per cubic inch. Pounds per cubic inch are units for density or unit weight, not modulus! The units of k_s are pounds per square inch per inch deflection. When the pile applies a lateral pressure to the soil it causes a lateral deflection, the pressure applied to the soil divided by the resulting deflection of the soil is modulus of subgrade reaction for use in L-Pile. OK, I'm a purist, but I think that use of the units pounds per cubic inch for k_s is just short-hand laziness!

I have given several lectures over the years on what modulus of subgrade reaction is and is not. I normally start out by saying that $k_s = p/\delta$ where p is the applied stress and δ is the displacement caused by application of p. I also say that k_s is not a constant, because it varies with stress level and it also varies with the width b of the applied stress (where b can be the footing width, the pile diameter, etc.). Since k_s varies with stress level, it varies with depth z for deep foundations. The variation of subgrade modulus with foundation width and depth has led to two alternate definitions of soil subgrade modulus. To account for footing width, another subgrade modulus K_s in some references and k'_s in other references is defined as $K_s = k'_s = k_s\, b$, where K_s or k'_s has units of pounds per square inch and b has units of inches. The third alternate modulus of subgrade reaction that takes both foundation width and depth into account is most commonly called n_h because it is applied most frequently to horizontal loadings where the horizontal modulus k_h is required. The equation for $k_{s\,horizontal} = k_h = n_h\,(z/b)$, where n_h has units of pounds per square inch per inch deflection or (it hurts) pounds per cubic inch.

When it comes to describing soil moduli in general, and modulus of subgrade reaction in specific, I am a fan of a short paper written by Jean-Louis Briaud. Dr. Briaud's paper is called *Introduction to Soil Moduli* and I highly recommend it (Briaud, 2001). I got a copy of this paper on Dr. Briaud's web page at the Texas A&M University website. Dr. Briaud summarizes the modulus (i.e., coefficient) of subgrade reaction as follows:

> . . . the coefficient of subgrade reaction is not a soil property and depends on the size of the loaded area. Therefore, if a coefficient of subgrade reaction k is derived from load tests on a footing or a pile of a certain dimension, the value of k cannot be used directly for other footing or pile sizes.

3.6.4 Lateral Pile/Shaft Group Reduction Factors

It seems as if every time I turn around, there is a new published set of recommended lateral pile/shaft group reduction factors expressed as *p*-modifiers (i.e., for use in the L-PILE and GROUP programs). Remember my discussion of complexity in Section 1.2, and the saying hanging on my office wall, "You have to see it to solve it."

I expect that the reason we see so many changes in recommended lateral pile/shaft group reduction factors is that there are too many unidentified variables impacting the results of lateral pile/shaft group reduction factor determinations. Let's take a look at some of the lateral pile/shaft group reduction factor recommendations from AASHTO/FHWA publications and see if we can identify where the hidden variables might be lurking. As you will see, the guidance and testing of laterally loaded piles and shafts seems to blend together, with many references to the words "piles" and "shafts" being used interchangeably in the literature. This is not too much of a problem if all of the piles and shafts are installed by drilling. A point to keep in mind is that driven displacement piles densify granular soils between piles and drilled piles/shafts loosen granular soils between piles. Driven piles should have higher values of *p*-multipliers than drilled piles for the same pile group installed at the same site. Many reported pile group reduction recommendations seem to ignore this fact. I suggest reading NCHRP Report 461 (Brown *et al.*, 2001) for an interesting report on a full-scale lateral group load test on driven and drilled piles at Taiwan High Speed Rail Authority near the city of Chaiyi, Taiwan.

Tables 3.6.1 and 3.6.2 illustrate lateral pile group reduction factors given in the 1996 AASHTO ASD Design Manual. Table 3.6.3 gives similar recommendations for laterally loaded pile groups group reduction factors (GRF) from the AASHTO LRFD manuals published in 2007 and 2010. Best I can tell the data presented in Tables 3.6.1, 3.6.2, and 3.6.3 was intended for use with drilled piles and shafts, and would be conservative if used for driven pile groups.

Table 3.6.1 Load perpendicular to line of piles, group reduction factors from 1996 AASHTO ASD design specification

Soil Type	Pile Spacing in "Pile" Diameters	Lateral Load Direction to Line of Piles	Lateral Group Reduction Factor
Clay	> 2.5B	Perpendicular	1.00
Clay	< 2.5B	Perpendicular	Analysis Req'd
Sand	> 2.5B	Perpendicular	1.00
Sand	< 2.5B	Perpendicular	Analysis Req'd

Table 3.6.2 Load parallel to line of piles, GRF 1996 ASD design specification

Soil Type, Clay or Sand	Pile Spacing in "Pile" Diameters	Lateral Load Direction to Line of Piles	Group Reduction Factor
Both	8B	Parallel	1.00
Both	6B	Parallel	0.70
Both	4B	Parallel	0.40
Both	3B	Parallel	0.25

One of the first things I notice when comparing lateral pile group reduction factors from earlier AASHTO Bridge Design Specification manuals that use allowable stress design (ASD) with the 2007 and 2010 load and resistance factor design (LRFD) group reduction factors is the use of soil descriptions in earlier guides and the lack of soil descriptions in the newer design guides. I have always been a bit skeptical of geotechnical guidance that divided all soil types into clay versus sand, and I expected that the new LRFD guidance would have provided more specific soil descriptions

Table 3.6.3 FHWA 2007/2010 group reduction factors for lateral loading

Load Direction	Pile spacing	p-Multiplier Row 1	p-Multiplier Row 2	p-Multiplier Row 3+
Parallel	3B	0.7/0.8	0.5/0.4	0.35/0.3
Parallel	5B	1.0/1.0	0.85/0.85	0.7/0.7
Perpendicular	3B	0.7/0.8	—	—
Perpendicular	5B	1.0/1.0	—	—

when giving group reduction factors. I was surprised, they not only didn't give more soil information, they dropped reference to soil types altogether! Why do you think they eliminated soil descriptions when specifying lateral group reduction factors? My guess is that they were trying to reduce complexity. Like I said in Section 1.2 engineering practice is becoming more and more complex with passing time, and the latest research into pile/shaft group reduction factors is no exception. The entire AASHTO 2007 Design Guide is 3.5 inches thick. The 2010 Design Guide is two volumes and the second volume by itself is 3.5 inches thick. I believe the authors of AASHTO's 2010 Design Guide were thinking of their primary audience, which is bridge design engineers, not geotechnical or foundation engineers, when deciding what material to include in the new manual. If every design topic from aluminum to timber structural design were covered in increasing detail in each issue of AASHTO's manual, it would become a 20 volume encyclopedia of bridge design.

The next difference that I noticed when comparing older ASD guidance for lateral loading perpendicular to a line of piles/shafts is the minimum spacing to achieve a group reduction factor (GRF) of 1.0. As shown in Table 3.6.1, older ASD design guidance used a group reduction factor of 1.0 (that is no reduction) when piles were spaced more than $2.5B$ apart (center to center spacing equal to 2.5 diameters). Both the 2007 and the 2010 LRFD design guidance requires shaft spacing of $5.0B$ or more to get a group reduction factor of 1.0. The 2007 LRFD guidance reduced the GRF to 0.7 for shaft spacing of $3B$ (which I believe is excessive reduction) and the 2010 LRFD guidance adjusted the GRF to 0.8 for shaft spacing of $3B$. It has been my understanding that the guidance of $2.5B$ spacing given in the 1996 AASHTO ASD pile design guidance for perpendicular loading to a line of shafts resulting in a GRF of 1.0 went back to the early 1950s and was based on bench scale testing of anchor blocks at Princeton University (Merkin, 1951).

I have always assumed that computer modeling was used to double the spacing requirement from $2.5B$ to $5.0B$ to achieve a GRF of 1.0, and I have been on the lookout for a publication or reference that would define the assumptions used in the model runs. In the 2010 AASHTO manual a reference is given for this GRF change from earlier GRF values and the reference is the 2006 FHWA Pile Manual (Hannigan et al., 2006). That got me thinking, so I looked up the GRF guidance in both the FHWA 2006 pile manual and the 2010 Drilled Shaft Manual. As I expected, they don't spare us the pain of complexity in these geotechnical-based references.

In the 2006 AASHTO Pile Manual (Hannigan et al., 2006) on page 9–153 in Table 9-19, the authors report group reduction factors for 10 cases, including laterally loaded pile groups in stiff clays, medium clays, clayey silt, loose sands, medium dense sand, and very dense sand. None of the 10 cases indicate whether the piles were driven displacement piles or drilled piles. Eight of the reported 10 cases are at pile spacings of $3B$, while the remaining two cases are at $5B$ spacings. Five of the tests are full-scale field studies, four tests are in the centrifuge, and one of the tests is a scale model with cyclic loading. Knowing the problems with scale models at $1g$ testing, I would be somewhat suspicious of the scale-model test results. Group reduction factors

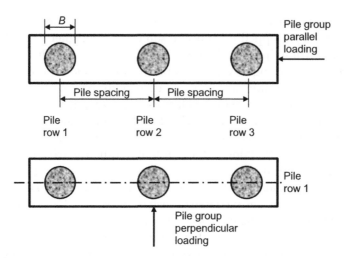

Figure 3.6.5   FHWA 2007/2010 guidelines for group reduction factors

reported in the 2006 AASHTO Pile Manual for the first row, second row, and third or higher rows (see Figure 3.6.5) do vary with soil type, as summarized in Table 3.6.4 below:

Table 3.6.4   Pile group reduction factors from the 2006 AASHTO Pile Manual

Soil Type	Pile Center to Center Spacing	Pile GRF Row 1, 2, and 3+	Test Reference
Stiff clay	3B	0.70, 0.50, 0.40	Brown, Reese, and O'Neill (1987)
Clayey silt	3B	0.60, 0.40, 0.40	Rollins, Peterson, and Weaver (1998)
Loose fine sand	3B	0.80, 0.70, 0.30	Ruesta and Townsend (1997)
Loose medium sand	3B	0.65, 0.45, 0.35	McVay et al. (1995)
Medium dense sand	3B	0.80, 0.40, 0.30	McVay et al. (1995)
Very dense sand	3B	0.80, 0.40, 0.30	Brown et al. (1988)
Loose medium sand	5B	1.0, 0.85, 0.70	McVay et al. (1995)
Medium dense sand	5B	1.0, 0.85, 0.70	McVay et al. (1995)

It looks like the GRF recommendations for 5B pile spacing included in the 2010 AASHTO Design Manual is based on loose to medium dense sand tested by McVay et al., in 1995. It looks like the GRF recommendations for 3B spacing included in the AASHTO 2010 Design Manual are based on medium dense to very dense sand tested by McVay et al., in 1995 and by Brown et al., in 1988. I see enough variations in GRFs for 3B spacing in Table 3.6.4, that is, varying from 0.65 to 0.80 for the first row of piles

Table 3.6.5 Shaft group reduction factors, *p*-multipliers for each row

Shaft Spacing, center to center	3 Diameters	4 Diameters	5 Diameters	Greater than 6 diameters
First row	0.7	0.85	1.0	1.0
Second row	0.5	0.65	0.85	1.0
Third and higher rows	0.35	0.50	0.70	1.0

that I would like to see many more tests to determine soil-specific group reduction factors.

What group reduction factors are recommended in the 2010 AASHTO Drilled Shaft Manual (Brown, Turner, and Castelli, 2010)? They say that it is important to consider group reduction factors when drilled shafts are spaced less than about six diameters apart in any direction. This seems like reasonable advice since older design guides recommended a GRF of 1.0 for shafts spaced five diameters apart in clays and eight diameters apart in sands.

The AASHTO 2010 Drilled Shaft Manual covers similar group reduction *p*-multipliers as discussed in the 2006 Driven Pile Manual, in fact they reference the same full-scale and centrifuge studies described in the pile manual. This might make the reader believe that there is no difference between driven piles and drilled shafts. But later in the text of the drilled shaft material, they explain that *there is a difference* in the impact on group reduction factors between soil types and pile/shaft installation techniques. Reduction factors recommended in the 2010 Drilled Shaft Manual are given above in Table 3.6.5.

Although it doesn't say specifically in the 2010 Drilled Shaft Manual, I believe the factors listed in Table 3.6.5 are for loadings parallel to the piles as shown in Figure 3.6.5. Notice that for spacings of three and five diameters, the reduction factors given in Table 3.6.5 are identical to those given in the AASHTO 2007 Design Manual for Piles, see Table 3.6.3.

I have always thought that the lateral group reduction factors included in Table 3.6.2 were conservative. I first recall seeing these values in the 1996 AASHTO Design Guide, and was confused why these factors were so much smaller (that is a much greater reduction of lateral resistance) than those included in Reese and O'Neill's Drilled Shaft Manual of 1988 (Reese and O'Neill, 1988). Upon review of my 1988 manual, the problem became clearer, or at least I began to understand the difference. In the 1988 FHWA Drilled Shaft Manual Reese and O'Neill recommended a minimum shaft spacing of 3*B* and a group reduction factor of 0.67 for 3*B* shaft center-to-center spacing for sands and clays. They recommended 6*B* spacing for both sands and clays to achieve a GRF of 1.0 and they allowed you to linearly interpolate between shaft spacings of 3*B* and 6*B* to find intermediate GRFs. *But then I realized the big difference* *between the 1996 and the 1988 recommendations, in 1988 the given group reduction factors were for vertical loadings only.* Group reduction factor research for drilled shafts was just beginning in the 1980s. Up until that time engineers generally assumed drilled shafts had large load capacities and that they would replace pile groups. Since single drilled shafts were replacing pile groups they were assumed to be widely spaced. Group reduction factor research was initially conducted for vertically loaded shaft groups, and not for laterally loaded shaft groups. The 1988 Drilled Shaft Manual does not include recommendations of group reduction factors for laterally loaded drilled shaft groups, which explains why the 1996 lateral GRF values seemed so restrictive to me, I'd never seen them before!

As shown in Table 3.6.3, a spacing of 5*B* with no regard to soil type gives a reduction factor of 1.0 for the first row of piles, 0.85 for the second row of piles, and 0.70 for the third row and all higher numbered rows, as illustrated in Figure 3.6.5. This change back to something closer to earlier recommendations makes me wonder if they have more data or a change of mind or both.

Thinking about the difference in installation techniques, it doesn't seem right that the 2010 AASHTO Design Guide gives the same GRF for drilled shafts as augercast piles, and it surely doesn't seem right that the GRF for driven displacement piles is the same that given for augered piles. If, as I mentioned earlier, you check the 2001 NCHRP Report 461, *Static and Dynamic Lateral Loading of Pile Groups*, you will see that they report higher group reduction factors (i.e., less reduction of lateral resistance) for driven displacement piles versus smaller GRFs for augered piles (i.e., more reduction of lateral resistance). This makes sense to me.

It doesn't seem right to neglect soil types when considering group reduction factors for lateral loading, and it doesn't seem right that densification of loose soils by installation of driven displacement piles is not considered in determining group reduction factors.

My suggestion to you the reader is to focus on specific driven pile and drilled shaft manuals and reports for guidance in selecting lateral group reduction factors, and use the AASHTO LRFD Design Manual as general guidance when applying the geotechnical LRFD design methods.

References

Bowles, J.E. (1996) *Foundation Analysis and Design*, 5th edn, McGraw Hill Inc., 1024 pages.

Briaud, J.-L. (2001) *Introduction to Soil Moduli*, Geotechnical News, vol. 19(2), June 2001, BiTech Publishers Ltd, Richmond, B.C., Canada, pp. 54–58.

Broms, B.B. (1965) Design of laterally loaded piles. *Journal of the Soil Mechanics and Foundation Division*, The American Society of Civil Engineers, **91**(SM 3), pp. 79–99.

Brown, D.A., Reese, L.C., and O'Neill, M.W. (1987) Cyclic lateral loading of a large-scale pile group in sand. American Society of Civil Engineers. *Journal of Geotechnical Engineering*, **113**(11), 1326–1343.

Brown, D.A., Morrison, C., and Reese, L.C. (1988) Lateral load behavior of pile group in sand. American Society of Civil Engineers. *Journal of Geotechnical Engineering*, **114**(11), 1261–1276.

Brown, D.A., O'Neill, M.W., Hoit, M., McVay, M., El Naggar, M.H., Chakraborty, S. (2001) *Static and Dynamic Lateral Loading of Pile Groups*, NCHRP Report 461, National Academy Press, Washington, D.C., 59 pages and 6 Appendices.

Brown, D.A., Turner, J.P., and Castelli, R.J. (2010) *Drilled Shafts: Construction Procedures and LRFD Design Methods*, NHI Course No. 132014, FHWA NHI-10-016, Geotechnical Engineering Circular No. 10, 970 pages.

Hannigan, P.J., Goble, G.G., Likins, G.E., and Rausche, F. (2006) *Design and Construction of Driven Pile Foundations* – Volumes I and II, Report Number FHWA-NHI-05-042, U.S. Department of Transportation, Federal Highway Administration, 968 pages.

Hannigan, P.J, Goble, G.G., Thendean, G., Likins, G.E. and Rausche, F. (1998) *Design and Construction of Driven Pile Foundations* — Volumes I and II, Report Number FHWA-HI-97-013, U.S. Department of Transportation, Federal Highway Administration, 828 pages.

Matlock, H. and Reese, L.C. (1960) Generalized solutions for laterally loaded piles. *Journal of the Soil Mechanics and Foundation Division*, The American Society of Civil Engineers, **86**(SM 5), 63–91.

Merkin, T.A. (1951) Small-scale Model Tests of Dead-man Anchorages. Master's Thesis, Princeton University, May 18, 1951, number of pages unknown. Photographs and data from Merkin's work are included in Tschebotarioff, G. P., 1973, *Foundations, Retaining and Earth Structures*, 2nd edn, McGraw Hill Book Company, New York, pp. 538–539.

McVay, M., Casper, R., and Shang, T.-I. (1995) Lateral response of three-row groups in loose to dense sands at 3D and 5D pile spacing. American Society of Civil Engineers, *Journal of Geotechnical Engineering*, **121**(5), 436–441.

Poulos, H.G. and Davis, E.H. (1980) *Pile Foundation Analysis and Design*, John Wiley and Sons, pages 397.

Reese, L.C. and O'Neill, M.W. (1988) *Drilled Shafts: Construction Procedures and Design Methods"*, Federal Highway Administration Publication No. FHWA-HI-88-042 and ADSC: The International Association of Foundation Drilling Publication No. ADSC-TL-4, 564 pages.

Rollins, K.M., Peterson, K.T., and Weaver, T.J. (1998) Lateral load behavior of full-scale pile group in clay. American Society of Civil Engineers, *Journal of Geotechnical and Geoenvironmental Engineering*, **124**(6), 468–478.

Ruesta, P.F. and Townsend, F.C. (1997) Evaluation of laterally loaded pile group at Roosevelt Bridge. American Society of Civil Engineers. *Journal of Geotechnical and Geoenvironmental Engineering*, **123**(12), 1153–1161.

Teng, W.C. (1962) *Foundation Design*, Civil Engineering and Engineering Mechanics Series (eds Newmark, N.M. and Hall, W.J.), Prentice-Hall International, Inc., 466 pages.

Timoshenko, S.P. (1956) *Strength of Materials, Part II, Advanced Theory and Problems*, 3rd edn, D. Van Nostrand Company, p. 572.

United States Steel Corporation (no author given) (1969 and 1975), *Steel Sheet Piling Design Manual*, printed by USS, 132 pages (1969) and 132 pages (1975).

Winkler, E. (1867) *Die Lehre v. d. Elastizität u. Festigkeit*, Prag, 182 pages.

Zimmermann, A. (1888) *Die Berechnung des Eisenbahn-Oberbaues*, Berlin, pages unknown.

4

Retaining Structures – Lateral Loads

4.1

Lateral Earth Pressure

4.1.1 Lateral Earth Pressure Introduction

I heard a structural engineer bragging, "Active earth pressure, passive earth pressure, 'at-rest' earth pressure, who needs them? I use the Rankine active earth pressure and it works fine for me, why do I need anything else?"

Are you confused about which earth pressure to use, or which earth pressure theory to use, Rankine or Coulomb? What about intermediate earth pressures between active and passive? Have you ever heard of prestressing the soil to reduce lateral strains developed during lateral loading? How do you include seismic lateral stresses into determination of the design loading for a retaining wall? Do you use an equivalent fluid pressure of 30 pounds per cubic foot for all of your retaining wall designs?

Determination of the lateral earth pressure is a problem that must consider the lateral soil strains that develop in the soil mass. Calculation of lateral earth pressures by exclusive use of active earth pressure coefficients or by use of an equivalent fluid pressure of 30 pounds per cubic foot can lead to problems.

4.1.2 Lateral Earth Pressure – The Problem

Selecting the correct lateral earth pressure for use in retaining structure design is another perennial issue of confusion with engineers. I receive on average about one call per month about problems involved with determining the appropriate lateral earth pressure.

As we discussed in Section 2.6, lateral earth pressure equations were developed by considering limit or bounding states of lateral tension and lateral compression failure in soils. To calculate lateral earth pressures we have to consider strains involved in developing limit states of stress in the retained soil mass. Development of active earth pressure requires movements of the retaining structure away from the soil that

Geotechnical Problem Solving, First Edition. John C. Lommler.
© 2012 John Wiley & Sons, Ltd. Published 2012 by John Wiley & Sons, Ltd.

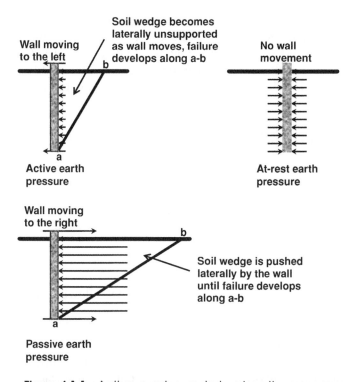

Figure 4.1.1 Active, passive and at-rest earth pressures

are large enough to cause the retained soil mass to experience a shearing failure due to lateral extension (please refer to Figure 4.1.1). Passive earth pressure requires movements of the retaining structure toward the soil that are large enough to cause the retained soil mass to experience a shearing failure due to lateral compression (again, please refer to Figure 4.1.1). "At-rest" earth pressure implies no movements of the retaining structure, that is zero lateral soil strain. Without soil movements occurring, the lateral earth pressure cannot reduce from the "at-rest" state to the active state of stress, nor can it increase from the "at-rest" state to the passive state of stress. Since earth pressure equations determine only three states of soil lateral stress, an upper bound (passive earth pressure), a lower bound (active earth pressure), and an intermediate state (at-rest earth pressure), we cannot determine the lateral earth pressure for intermediate strain states without resorting to rather complex computer analyses.

4.1.3 Coulomb Earth Pressure Equations

The Rankine and Coulomb earth pressure equations are commonly used to calculate lateral earth pressures. The Coulomb equation (Coulomb, 1776) is the more general equation in that it considers all of the applicable factors and as a result it is more complex than the Rankine equation (Rankine, 1857). The Coulomb equation for active lateral earth pressure is given below as Equation 4.1.1. The Coulomb active lateral earth pressure coefficient is K_a. This equation was later modified by A. L. Bell (Bell, 1915) to include cohesion so that both soil shear strength terms c and ϕ are considered.

$$\text{Coulomb active earth pressure, } P_a = \frac{1}{2}\gamma H^2 K_a \qquad (4.1.1)$$

where

$$K_a = \cfrac{\sin^2(\alpha + \phi)}{\sin^2\alpha \sin(\alpha - \delta)\left[1 + \sqrt{\cfrac{\sin(\phi + \delta)\sin(\phi - \beta)}{\sin(\alpha - \delta)\sin(\alpha + \beta)}}\right]^2}$$

The Coulomb earth pressure equation includes the following assumptions and factors:

- Coulomb assumed that the soil is isotropic and homogeneous.
- Coulomb assumed that the soil's failure surface was a flat inclined plane, and that the surface of the backfill behind the wall was a flat planar surface.
- Shearing resistance along the soil failure plane was assumed to be uniformly distributed (even though friction forces at each point along the failure plane are a function of the weight of the soil directly above each point on the failure surface, $\tau_{\text{shearing stress}} = \sigma'_{\text{vertical}} \tan\phi$
- The friction angle of backfill soil is ϕ. The original Coulomb equation is for sandy soils with ϕ only. Soil cohesion was not considered by Coulomb.
- The total unit weight of backfill soil is γ_{total}. The original Coulomb equation assumes only one type of soil. Layered backfill soil and/or soil with properties that vary with depth were not considered.
- The slope angle of the backfill soil is β. The Coulomb equation allows a rising slope $(+\beta)$, a horizontal slope $(\beta = 0)$, and a falling slope $(-\beta)$ behind the retaining wall, see Figure 4.1.2.
- The angle of the back of the retaining wall is α. The angle α is an acute angle that is less than or equal to 90 degrees.
- The friction angle between the backfill soil and the back of the retaining wall is δ. The soil–wall friction angle δ may be positive or negative with respect to a perpendicular line to the back of the retaining wall, see Figure 4.1.2.

The Coulomb equation for passive earth pressures is given below as Equation 4.1.2. The Coulomb active and passive earth pressure equation terms are the same, the only

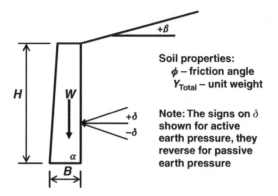

Figure 4.1.2 Coulomb equation terms defined

difference is that some of the signs ($+$ or $-$) change.

$$\text{Coulomb passive earth pressure, } P_p = \frac{1}{2}\gamma H^2 K_p \qquad (4.1.2)$$

where

$$K_p = \frac{\sin^2(\alpha - \phi)}{\sin^2 \alpha \sin(\alpha + \delta)\left[1 - \sqrt{\dfrac{\sin(\phi + \delta)\sin(\phi + \beta)}{\sin(\alpha + \delta)\sin(\alpha + \beta)}}\right]^2}$$

Besides the fact that soil is not homogeneous and isotropic and that soil failure planes are curved and not flat, as assumed by Coulomb, the biggest drawback encountered when solving the Coulomb equations for the active or passive earth pressure coefficient is the complexity of the equations themselves. Over the years I have encountered students that needed to solve the Coulomb active or passive earth pressure equations several times to get two answers that match. As a matter of course, to avoid calculation errors, I solve the Coulomb equation twice to make sure that I haven't made a calculator mistake. If tables of Coulomb earth pressure coefficients weren't so convenient, I'd probably program them on a spreadsheet.

Most engineers I know use the tables of Coulomb earth pressure coefficients published in text books rather than spend time calculating the Coulomb active or passive values. This especially hold true for young engineers taking the professional engineers (PE) examination.

The two biggest advantages of using the Coulomb equation are: (1) the ability of the Coulomb equation to solve for retaining wall earth pressures with rising and falling back slopes, and (2) the ability of the Coulomb equation to include the friction angle between the soil backfill and the back side of the retaining wall, δ.

4.1.4 Rankine Earth Pressure Equations

The Rankine earth pressure equation was developed in a method similar to the Coulomb equation, with the exception that Rankine assumed that the soil–wall interface friction angle is zero. This simplifies the Rankine equation, but it is still rather complex when the slope angle β is included. Relative to using the Rankine equation with slope angle β, you can't use a negative value of β in the Rankine equations. Using a negative slope angle appears to work when you substitute $-\beta$ in the Rankine equations, but don't let it fool you, *it doesn't work*. Most engineers use the Coulomb equation if they need to include $-\beta$ (i.e., falling slope behind the wall) or if there is a significant interface wall–soil friction. If we assume a horizontal backfill behind the retaining wall and assume that the interface soil-wall friction angle is zero, the Rankine equations for active and passive earth pressure are simplified even further to equations that practicing engineers seem to prefer, see Equations 4.1.3 and 4.1.4 below:

$$\text{Rankine active earth pressure, } P_a = \frac{1}{2}\gamma H^2 K_a \qquad (4.1.3)$$

where $K_a = \tan^2\left(45 - \phi/2\right)$

$$\text{Rankine passive earth pressure, } P_p = \frac{1}{2}\gamma H^2 K_p \qquad (4.1.4)$$

where, $K_p = \tan^2\left(45 + \phi/2\right)$.

Now you might be wondering how practicing engineers include earth pressure from a sloped fill. One way is to assume the slope behind the wall is a surcharge loading on the wall, the other way is to take table values of K_a and K_p from text books that include sloped backfill. I guess my question is, if you are going to use table values of earth pressure coefficients rather than calculate them, why not just use Coulomb's equation?

4.1.5 Including Cohesion into Active and Passive Earth Pressures

Both Coulomb and Rankine neglected to include soil cohesion into their earth pressure equations. As I mentioned above, Bell modified the earth pressure equations to include the soil's cohesive shear strength component, C. You can use either Coulomb or Rankine lateral earth pressure coefficients in the following equations which include cohesion: $P_a = \frac{1}{2}\gamma H^2 K_a - 2CH\sqrt{K_a}$ and $P_p = \frac{1}{2}\gamma H^2 K_p + 2CH\sqrt{K_p}$, where C is cohesion in the same force and length units as the in unit weight γ term, and H is the wall height in the same length units as in the unit weight γ and cohesion C terms.

4.1.6 Equivalent Fluid Pressure

Before I started my geotechnical career, I was a structural engineer, and it hasn't been so long ago that I can't remember that as a structural engineer I did not want to hear about active and passive earth pressure coefficients. What I did want for use in my retaining wall design calculations was the equivalent fluid pressure. The equivalent fluid pressure is the unit weight that a true fluid would have to generate to develop the same lateral earth pressure as the soil backfill. The equivalent fluid pressure (EFP) of a soil in an active state of stress is equal to the moist unit weight of the soil, γ_{moist}, times the active earth pressure coefficient K_a. For example, if a soil has a moist unit weight of 120 pounds per cubic foot and an angle of internal friction of 30 degrees, using the Rankine equation, it has an active earth pressure coefficient equal to 0.333. The EFP of this example soil would be 120 pounds per cubic foot times 0.333 or 40 pounds per cubic foot. The structural engineer can calculate lateral forces on his retaining wall assuming that a fluid having a unit weight of 40 pounds per cubic foot is the backfill material.

We can use the same equivalent fluid pressure concept for passive and "at-rest" earth pressures. Using the same friction angle of 30 degrees, the passive earth pressure coefficient, K_p is equal to 3.0 and the "at-rest" earth pressure coefficient is equal to $1 - \sin 30° = 0.50$. Using these coefficients, the passive state EFP would be 360 pounds per cubic foot, and the "at-rest" state EFP would be 60 pounds per cubic foot.

4.1.7 Lateral Earth Pressures for Wet Soil versus Submerged Soil

What if your soil is subjected to a water source and becomes wetted to nearly 100% saturation, but it is not submerged in water, how do you calculate its lateral earth pressure? The difference between a submerged soil and one that is nearly saturated is that the wet, nearly saturated soil supports its own weight, including the pore water weight on its soil structure. Any soil that supports its total weight on its soil structure develops effective stresses $[\sigma'_{vertical\ effective\ stress} = (\gamma_{total\ wet\ unit\ weight})(h)]$ to support itself. To calculate the active earth pressure in this case, we multiply the vertical effective stress at depth h by the active earth pressure coefficient. Remember, in this case I am assuming that the soil is not submerged and does not experience buoyant uplift forces.

Next, how do you calculate the lateral earth pressure for a submerged soil? A submerged soil does not experience full effective stresses due to the buoyant effect of water. The soil is buoyed up by the water it displaces resulting in a reduction of effective stresses generated in the soil structure (or soil skeleton as some call it). Let's take the same soil with the wet total unit weight of 120 pounds per cubic foot and an active earth pressure coefficient of 0.333. If we are looking at the soil as a continuum material and not on the particle level, the soil and the water of the sample make up the soil material, and the soil's buoyant weight is equal to $(120–62.4) = 57.6$ pounds per cubic foot. The equivalent fluid density of this submerged soil would be equal to 0.333 times 57.6 pcf or 19.2 pounds per cubic foot. Since water has an active earth pressure coefficient equal to 1.0 (fluid materials with zero shear strength have equal fluid pressures in all directions), the total lateral pressure would be calculated by using an EFP equal to $19.2 + 62.4 = 81.6$ pounds per cubic foot.

This calculated submerged EFP of 81.6 pounds per cubic foot may be suitable for clayey soils where much of the water is molecularly attached to clay particles and the soil mass acts as a continuum. What about the submerged lateral earth pressure generated by a sandy soil that acts more like a particulate material than a continuum? If we consider that the water in a saturated, submerged sand sample is in hydraulic communication with all of the water, causing buoyancy, it would be reasonable to consider each sand grain as a solid particle that experiences buoyancy. In this case, it would be more reasonable to compute the buoyancy of the sand grains alone without the soil's pore fluid. The vertical effective stress would be calculated by using the dry unit weight of the sand γ_{dry} minus the unit weight of water that buoys the soil particles. The dry unit weight γ_{dry} of the sand is calculated by $(\gamma_{wet})/(1 + wc)$, where the water content is expressed as a decimal. Assume the example sand's water content is 10%, so its dry unit weight would be equal to 120 pcf / 1.1 = 109.1 pounds per cubic foot. The EFP of the particulate sandy material would be $(109.1 - 62.4)(0.333) + (62.4)(1.0) = 15.6$ pcf + 62.4 pcf = 78.0 pounds per cubic foot. You can see that in this case the continuum model of the soil results in an equivalent fluid pressure of 81.6 pounds per cubic foot, which is a bit higher (4.6% higher) than the particulate model of the soil's EFP of 78.0 pounds per cubic foot. For conservatism, many practicing engineers assume that the buoyant unit weight of the soil is equal to its moist or wet unit weight minus the unit weight of water.

Don't forget that you use the lateral earth pressure coefficient K times the soil's vertical effective stress, and use water's lateral earth pressure coefficient of 1.0 times the unit weight of water, then add the soil and water lateral pressures for submerged, buoyant soil's total lateral pressure.

4.1.8 Friction between Retained Fill and Wall – Curved Failure Surfaces

From the very beginning of earth pressure calculations, Coulomb recognized that magnitude and direction of the earth pressure resultant force on the back of a retaining wall was affected by the wall friction. Friction on the back of a wall resisting an active earth pressure wedge acts upward $(+\delta)$ because it is resisting the earth wedge's movement downward, see Figure 4.1.2. When a wall is moving toward the soil mass, generating passive earth pressures the resultant wall force (wall force on the soil) is downward $(-\delta)$ because it is resisting the earth wedge's movement upward, see Figure 4.1.2. The effect of wall friction on the resulting wall force is to rotate it downward for active stresses and rotate it upwards for passive stresses causing the soil failure surface to be curved rather than straight, as Coulomb had assumed. The difference between the resulting curved failure surface and the assumed planar surface for active stress is considered insignificant. The difference between the resulting curved failure surface and the assumed planar surface for passive stress is assumed to be reasonably close (within 10%) for $\delta \leq \Phi/3$ but has been reported to be significantly different for wall friction angles, $\delta > \Phi/3$ (Caquot and Kerisel, 1948). Tables for correction factors for passive earth pressures are included in the NAVFAC manuals and in the National Highway Institute Reference Manual, *Earth Retaining Structures*, from Course No. 13 236 Module 6. Both NAVFAC and NHI manuals are available on the internet for downloading. Now for a word from your mentor, I have done all of the calculations required by FHWA to incorporate the log spiral method into calculation of passive earth pressures on many projects. If you take a flat backfill, and a vertical wall with wall–soil friction angle of zero, you get the same or nearly the same value for the passive earth pressure coefficient as the much simpler Rankine method. Unless I am forced into the Coulomb log spiral calculations, I prefer to use the simplified Rankine method to calculate the passive earth pressure. If I don't have enough passive resistance to satisfy sliding stability of my wall, then I go to the Coulomb method with a friction angle less than or equal to one-third of the backfill soil friction angle, phi.

A word of caution: The affect of wall friction on the active earth pressure is small (as I mentioned above) *except* when the wall settles significantly relative to the backfill soil. If the wall settles relative to the backfill, the direction of the wall-to-soil resultant force changes from upward (i.e., $+\delta$) to downward (i.e., $-\delta$) which can greatly increase the overturning moment of the earth pressure on the wall by increasing the lever arm distance from the toe of wall to the earth pressure force vector. (In the previous sentence, I am talking about the force that the wall exerts on the soil backfill. The force shown in Figure 4.1.2 is the force that the backfill soil exerts on the wall. These forces are equal and opposite.) I discuss the issue of wall settlement effects on lateral earth pressure in some detail in Section 4.2; see Figure 4.2.5(d).

4.1.9 Seismic Earth Pressure

During an earthquake the ground shakes back and forth, and sometimes up and down, and causes increased lateral forces on retaining walls. Just as we found with static lateral earth pressures, the magnitude of lateral earth pressures on a retaining wall in an earthquake are a function of the wall movement. When designing normal retaining walls (i.e., walls that are not critical to the life function of facilities, or walls that hold up nuclear power plants!), engineers do not analyze the walls for all of the complex motions and loadings exerted on the wall during an earthquake. Rather they use simplified methods to calculate estimated forces generated by the earthquake and design their retaining wall for those forces.

4.1.9.1 Seismically Loaded Flexible Retaining Walls

When retaining walls are flexible and designed for active and passive earth pressures, it is common practice to use the Mononobe–Okabe method (Mononobe and Matsuo,

1929; Okabe, 1926) to calculate seismic pressures that are added to static earth pressures. To use the Mononobe–Okabe method you have to use Coulomb earth pressure equations because M–O was developed from the Coulomb equation.

Considering earthquake motion, I imagine the soil moving first to the right and then to the left in a repeating motion, first one way then the other. Since a retaining wall has active earth pressures from the retained soil pushing one way and passive earth pressures on the retaining wall foundation pushing back in the opposite direction, earthquake motions add seismic forces to the active earth pressure force and subtract seismic forces from the passive earth pressure force, making the wall less stable. When the earthquake motion is moving in the opposite direction, seismic forces make the wall more stable momentarily until the earth motion reverses. Designing for the more critical case, we design our retaining walls for the direction of earth motion where the seismic force adds to the active earth pressure force and subtracts from the passive earth pressure force.

The Mononobe–Okabe equations for lateral active and passive earth pressure, including the seismic forces, are given below as Equations 4.1.5 and 4.1.6. In these equations P_{AE} (the active earth pressure resultant force) and P_{PE} (the passive earth pressure resultant force) include both the static and seismic earth pressure affects. To separate the seismic and the static component forces from these equations, you have to calculate the Coulomb static values of active (P_A) and passive (P_P) earth pressure resultant forces and subtract them from the P_{AE} and P_{PE} values. The seismic components of lateral earth pressure in the Mononobe–Okabe equations are ΔP_{AE} and ΔP_{PE}. Horizontal and vertical earthquake accelerations are given as pseudostatic values that are decimal values of gravitational acceleration g (32.2 ft s^{-2} or 9.806 65 m s^{-2}), such that the horizontal acceleration, $a_h = k_h\,g$ and the vertical acceleration $a_v = k_v\,g$.

$$P_{AE} = \frac{1}{2}K_{AE}\gamma H^2 (1 - k_v) = P_A + \Delta P_{AE} \tag{4.1.5}$$

$$K_{AE} = \frac{\cos^2(\phi - \theta - \psi)}{\cos\psi \cos^2\theta \cos(\delta + \theta + \psi)\left[1 + \sqrt{\dfrac{\sin(\delta + \phi)\sin(\phi - \beta - \psi)}{\cos(\delta + \theta + \psi)\cos(\beta - \theta)}}\right]^2} \tag{4.1.5a}$$

$$\psi = \tan^{-1}\left[k_h/(1 - k_v)\right] \tag{4.1.5b}$$

$$P_{PE} = \frac{1}{2}K_{PE}\gamma H^2 (1 - k_v) = P_P - \Delta P_{PE} \tag{4.1.6}$$

$$K_{PE} = \frac{\cos^2(\phi + \theta - \psi)}{\cos\psi \cos^2\theta \cos(\delta - \theta + \psi)\left[1 + \sqrt{\dfrac{\sin(\delta + \phi)\sin(\phi + \beta - \psi)}{\cos(\delta - \theta + \psi)\cos(\beta - \theta)}}\right]^2} \tag{4.1.6a}$$

I noted from review of ASCE 4-86 that they have printed an error in the Mononobe–Okabe equation for the active lateral seismic case in the denominator of the terms under the square root. The second cosine should be cosine $(i - \beta)$ using their terminology *not* cosine $(i + \beta)$. Using my terminology in Equation 4.1.5a above this term is correctly given as cosine $(\beta - \theta)$. I mention this printing error to emphasize my earlier point that you should always check two or more references to confirm complicated equations. Printing errors are not uncommon, and they provide no excuse for you making a mistake in your engineering calculations. If there is a printing error in this book (and I hope there are none), it is your obligation to find them, and if you feel obliged to share your discovery, send me an email with the location of the error and your suggested correction.

In Steve Kramer's book (Kramer, 1996) he includes an additional complication to the Mononobe–Okabe equation that involves correcting the orientation of the active and passive failure planes to seismic values from the static values based on the Coulomb equation. Before I go further, I highly recommend Dr. Kramer's book. I have used mine so frequently that I have split the binding. Given all of the simplifications of seismic conditions by modeling them as pseudostatic forces, and the uncertainty included in terms used in Mononobe–Okabe's method, I can't justify directing my staff engineers to perform the additional computations involved in modifying the size of the soil failure wedge located behind a seismically loaded retaining wall by adjusting α_{AE} (see page 479 in Kramer's text).

Originally Mononobe–Okabe's equations were presented with a single force, either the active P_{AE} or the passive P_{PE} acting at one third of the wall height above the base, just as the Coulomb equation gives for static loadings. In their paper given to the 1970 ASCE Conference on *Lateral Stress and Retaining Structures at Cornell University*, H. Bolton Seed and Robert Whitman presented a paper which shows that although the static component of earth pressure does act at one-third the height of the wall, the seismic component acts higher, between $0.5\,H$ and $0.67\,H$. Current practice suggests that the resultant of seismic lateral earth pressure be taken at $0.6\,H$ above the base of wall.

Seed and Whitman give several graphical solutions to the Mononobe–Okabe equations in their 1970 ASCE Specialty Conference paper (Seed and Whitman, 1970). They also give an approximate solution for the seismic component of lateral earth pressure for flexible walls with the following assumptions:

1. The vertical seismic acceleration coefficient k_v is equal to zero.
2. The horizontal seismic acceleration coefficient k_h is less than or equal to 0.35.
3. Back of the retaining wall is vertical and the soil backfill surface is horizontal.
4. The backfill soil's internal friction angle $\phi \approx 35$ degrees.

The Seed and Whitman's approximate solution is reported to be in close agreement with more exact Wood's solution (Wood, 1973), which is given below as Equation 4.1.7.

$$\Delta P_{AE} = \frac{3}{8}\gamma H^2 k_h \tag{4.1.7}$$

4.1.9.2 Seismically Loaded Rigid Retaining Walls

When retaining walls do not move laterally a sufficient amount to generate active (or passive) earth pressures, we use the "at-rest" earth pressure coefficient, K_0 for static loading. If a seismically loaded wall is rigid and doesn't move adequately to generate Coulomb's active earth pressure, then we can't use the Mononobe–Okabe method, which is based on the Coulomb earth pressure equation. John H. Wood's Ph.D. thesis at California Institute of Technology has been used in several publications to calculate seismic lateral earth pressures on rigid structures. Wood's thesis has been published as EERL Report 73-05 (Wood, 1973) and is available on the internet for study. Wood clearly states in his work that the seismic forces on relatively rigid buried structures are significantly higher than those calculated by the Mononobe–Okabe equations. A good summary of Wood's work is included in Kramer's book, Chapter 11.

4.1.10 Suggested Further Reading

For further study of the Mononobe–Okabe and the Wood methods of calculating seismic lateral earth pressures on retaining walls, I recommend Steven Kramer's book, Chapter 11 (Kramer, 1996), and the ASCE Standard 4-86 Section 3.5.3 (ASCE, 1986). If you can find a copy in the library or on the internet, I recommend reading Seed and Whitman's paper (Seed and Whitman, 1970) from the ASCE 1970 Specialty Conference on Lateral Stress and Earth Retaining Structures. By searching the internet for EERL Report 73-05, you can find a copy of John Wood's 1973 work for further study.

References

ASCE (1986) ASCE 4-86. *Seismic Analysis of Safety-Related Nuclear Structures, Section 3.5.3 Earth-Retaining Walls in the Standard and in the Commentary*, American Society of Civil Engineers.

Bell, A.L. (1915) *The Lateral Pressure and Resistance of Clay, and the Supporting Power of Clay Foundations, in the volume "A Century of Soil Mechanics,"* Institution of Civil Engineers, London, pp. 93–134.

Caquot, A. and Kerisel, F. (1948) *Tables for the Calculation of Passive Pressures, Active Pressure and Bearing Capacity of Foundations*, Gautheir-Villars, Paris, France.

Coulomb, C.A. (1776) Essai sur une application des règles de Maximus et Minimis à Quelques Problèmes de Statique, Relatifs à l' Architecture, Mémoires de Mathématique et de Physique. Présentés a l' Académie Royale des Sciences, par Divers Savans, et lûs dans ses Assemblées, Paris, Vol. 7, pp. 143–167.

Kramer, S.L. (1996) *Geotechnical Earthquake Engineering, Chapter 11 — Seismic Design of Retaining Walls*, Prentice Hall, Upper Saddle River, New Jersey, pp. 406–505.

Mononobe, N. and Matsuo, H. (1929) On the Determination of Earth Pressures During Earthquakes. Proceedings of the Second World Conference on Earthquake Engineering, Tokyo, Japan.

Okabe, S. (1924) General theory of earth pressure and seismic stability of retaining wall and dam. *Journal of the Japanese Society of Civil Engineers*, **10**(6),1277–1323 [in Japanese].

Okabe, S. (1926) General theory of earth pressure and laboratory testings on seismic stability of retaining walls. *Journal of the Japanese Society of Civil Engineers*, **12**(1), 123–134 [in Japanese].

Rankine, W.J.M. (1857) On the stability of loose earth. *Proceedings of the Royal Society, London*, **VIII**,185–187.

Seed, H.B. and Whitman, R.V. (1970) Design of earth retaining structures for dynamic loads. American Society of Civil Engineers Specialty Conference on Lateral Stresses in the Ground and Design of Earth-Retaining Structures, Held at Cornell University, Ithaca, New York on June 22-24, 1970, pp. 103–148.

Wood, J.H. (1973) Earthquake-induced Soil Pressures on Structures. EERL Report 73-05, California Institute of Technology, Pasadena, California, 327 pages with appendices.

4.2

Retaining Walls – Gravity, Cantilevered, MSE, Sheet Piles, and Soldier Piles

4.2.1 Introduction to Retaining Walls

How often do you go out into your yard and try to push a boulder across the lawn up to the side of your driveway? Not often I imagine. In fact I doubt if any of you have tried to move boulders by hand. Who does? Sculptors and landscapers, that's who move boulders for a living. While touring a sculptural installation in a park with the artist, my friend Carl Floyd, he told me how he had to teach students to move and position boulders safely on this project. When they started, the students had no idea of how to move a boulder without seriously injuring themselves. By the end of the project, the students were moving and positioning large boulders with ease. The project is called "All People's Park" and it is located near Lake Erie east of Cleveland in Lake County, Ohio.

How did Carl Floyd teach boulder moving? Carl told me that he had to familiarize students with the center of gravity and the balance point of the boulders, and how to use a lever to rock them over and into position. While Carl was telling me about training his students, I imagined early builders and engineers building rock retaining walls around castles and fortifications during war time. Early engineers and builders had a feel for what it took to move a larger boulder. They developed a feel for gravity much like Carl's students. So it is logical that early retaining walls were gravity retaining walls made of rocks and masonry and unreinforced concrete like that shown in Figure 4.2.1(a). So long as the gravity weight of the wall resisted the lateral earth forces, and the wall didn't slide out or punch into the earth, the wall was stable.

Geotechnical Problem Solving, First Edition. John C. Lommler.
© 2012 John Wiley & Sons, Ltd. Published 2012 by John Wiley & Sons, Ltd.

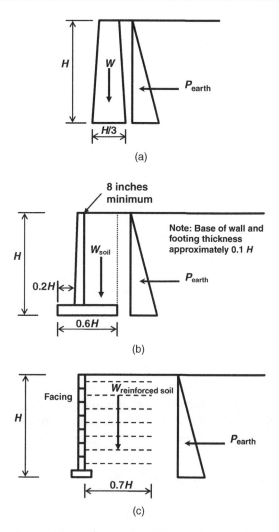

Figure 4.2.1 (a) Gravity walls generate resistance to overturning earth pressure by their weight; (b) Cantilevered retaining wall, weight of earth above wall footing resists overturning earth pressure loading; (c) mechanically stabilized earth (MSE) Retaining wall soil resists earth pressure, facing not required for resistance to earth loading P_{earth}

How stable the wall was relative to some measure of safety was of little or no concern to the wall's builders. Early builders likely had no idea of what the earth pressure force P_{earth} was equal to, nor did they know where the resultant P_{earth} acted on the back of the retaining wall. What early builders did know were ratios of wall height to wall base width for stable walls retaining "normal" soils. They learned that wall ratios of width to height (B/H) between 0.3 and 0.4 were stable by experience. What experience am I talking about? They built walls that failed. After early builders had a few failures, they learned what works and what does not work. Failures of walls were quite common before civil engineering earth pressure principles were invented. If a wall fell down, early builders just rebuilt it with a wider base. Of course this assumes that the early builder had a patron that was patient and wealthy. If you had too many failures, you could lose your head!

In 1776, as we discussed in Section 4.1, Coulomb (Coulomb, 1776) broke through the guess–hope–rebuild era of retaining wall construction by developing his theory of lateral earth pressure. For decades or perhaps centuries, walls were designed as gravity retaining walls, where the weight of the walls' cut stone (or uncut stone) and mortar resisted lateral earth pressure forces. The history of development of retaining wall designs and retaining wall construction is a story of transfer of lateral earth forces from total support by the wall to partial support by the soil material, and finally to total support of lateral forces by the soil itself (with the help of reinforcement). The economic theory behind transferring lateral forces from concrete to soil is that the less mortar, stone, and concrete required in wall design and construction, the cheaper the wall will be to build. Gravity retaining walls use the wall's material strength and weight to resist lateral earth pressures, see Figure 4.2.1 (a).

Cantilevered retaining walls use soil weight over the heel (i.e., back) of the wall footing to help resist soil lateral overturning moment, but the wall's strength and weight are still required to transfer forces into a stable soil/wall structure, see Figure 4.2.1 (b).

Reinforced soil walls, often called MSE (mechanically stabilized earth) walls use soil strength supplemented by tension reinforcement to resist lateral soil forces. MSE wall facings are primarily for appearances, although they do provide resistance to erosion of reinforced backfill soils. MSE wall facings do not resist a significant portion of lateral forces generated by retained soils, see Figure 4.2.1 (c).

4.2.2 Design of Gravity Retaining Walls

Gravity retaining walls were originally built of stones, masonry, unreinforced concrete, masonry units filled with concrete, or cut stone walls either with or without mortared joints. Mortared rock walls used as retaining walls or used as basement walls for support of eighteenth and nineteenth century homes are often referred to as rubble walls.

Gravity retaining walls today are having resurgence in popularity due to the use of bin walls, large precast concrete block walls, and battered, interlocking masonry walls. All you need to do to find suppliers of these patented wall systems is go to a geotechnical conference, read a civil engineering magazine, or search "retaining wall" on the internet.

Gravity walls are analyzed for static stability by using equations of static equilibrium:

$$\sum M_o = 0; \quad \sum F_{horizontal} = 0; \quad \sum F_{vertical} = 0 \qquad (4.2.1)$$

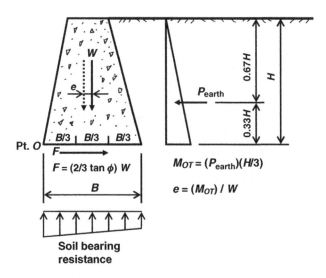

Figure 4.2.2   Analysis of a gravity retaining wall

Gravity retaining walls are designed to have no tensile stresses in the wall or in foundation soils beneath the wall. For initial proportioning, the base of a gravity retaining wall should be taken as approximately 30 to 40% of the height of the wall (0.3 to 0.4 H). After calculating the lateral earth forces from retained earth and surcharge loadings, we need to check that the resultant of wall weight and lateral forces pass through the center third of the retaining wall's foundation, see Figure 4.2.2. If the resulting force on the wall passes through the middle third of the retaining wall foundation, the retaining wall bearing stresses will all be compression. If the resulting force is outside the middle third of the foundation base, the retaining wall will try to develop foundation soil tension stresses (which it essentially cannot do) and as a result the footing will lift off the bearing area. To take care of this problem of foundation tension stresses, the area of the wall base is analyzed by calculating a uniform foundation stress on a reduced bearing area whose width B' is derived from the actual physical foundation width B by use of Equation 4.2.2:

$$B' = B - 2e \tag{4.2.2}$$

This reduction equation to account for uplift moments on the base of a retaining wall footing is often called the Meyerhof equation. By use of the Meyerhof equation to reduce the footing area and increase the resulting footing bearing stress, the need for considering triangular bearing stress distributions on the bottom of a retaining wall footing is eliminated. Structural engineers seem to be unaware of the Meyerhof equation, because I often encounter structural calculations where they calculate foundation bearing stresses by use of the familiar mechanics equation: $\frac{P}{A} \pm \frac{Mc}{I}$.

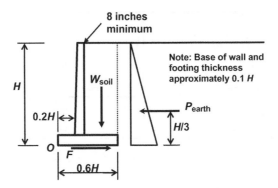

Figure 4.2.3 General stability analysis of cantilever retaining wall

Cantilevered retaining walls are analyzed using equations of statics similarly to gravity retaining walls with the inclusion of the weight of soil above the wall's heel as shown in Figure 4.2.3.

MSE retaining walls are analyzed for external stability by use of statics as shown in Figure 4.2.4. After the wall is proportioned for external stability, it has to be designed for adequate reinforcement of the soil backfill.

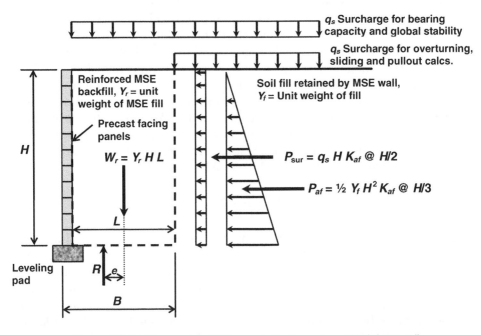

Figure 4.2.4 General stability analysis of an mse retaining wall

4.2.3 Issues with Static Equilibrium Analyses of Walls

What's missing from the static equilibrium analysis of the retaining walls illustrated in Figures 4.2.2, 4.2.3, and 4.2.4? Come on, you know this one. Whenever you apply a stress there is a strain. All of the loadings applied to a retaining wall generate deflections and settlements. Let's start our discussion with deflections generated by the resultant active earth pressure P_{earth}. The retaining wall has to move laterally to allow generation of active earth pressure. Recall from our discussion in Section 2.6 that the active earth pressure is a limit state that results from the tension failure of the soil adjacent to the retaining wall. For discussion purposes, I normally assume that a retaining wall has to move approximately 0.003 times the height of the wall for active earth pressure to be generated. That would be a lateral movement of half an inch for a 15-foot-high wall.

If the wall is infinitely rigid, as assumed by some analysts, or if it is a basement wall supported top and bottom, or if it is a stiff wall socketted into bedrock, or if it is a rigid box structure, and so on, then P_{earth} is commonly assumed to be equal to $0.5\,\gamma$ $(H^2)\,K_0$, using the "at-rest" earth pressure coefficient K_0. This is because we assume that a rigid wall will not move far enough laterally to allow generation of the active earth pressure.

The previous discussion of lateral wall movements is about as far as most discussions of lateral earth pressure go into the topic. Remember I told you there were complicating factors in geotechnical engineering? . . . yes there is more to consider.

Rankine earth pressure theory (Rankine, 1857) assumes that there is no friction between the backfill and the back of the retaining wall, and Coulomb considers friction between the backfill and the wall. The assumption of no friction is conservative so long as the active earth pressure soil wedge is moving downward relative to the wall. If the active soil wedge is moving downward, then the resultant earth pressure generated by the horizontal component and the vertical friction component (ignored by Rankine and included by Coulomb) has a line of action that reduces the overturning moment lever arm, check Figure 4.2.5(b). If we are designing for an overturning moment that is greater than the actual overturning moment, the result is conservatism, as I mention above. If the retaining wall is founded on hard soil or bedrock, there is little or no settlement of the wall and the relative motion between the active wedge and the back of the wall is as assumed (soil wedge down relative to back of the retaining wall). The active wedge moving downward exerts a downward force on the back of the retaining wall.

What if the retaining wall moves downward relative to the active soil wedge? Does settlement of a retaining wall matter? I am asked now and again if the settlement of a retaining wall that is not supporting a building or a bridge needs to be considered. Who cares if it settles, who would know if the wall is not supporting a structure? Settlement of a retaining wall does matter, and the reason that it matters is increased overturning moments due to wall settlement, see Figure 4.2.5(d). When the wall moves down more than the active soil wedge, the result is that the active force is

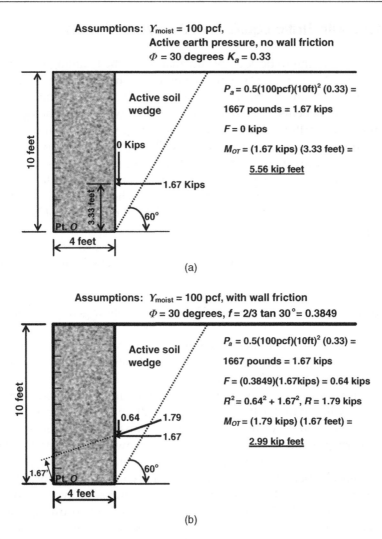

Figure 4.2.5   (a) Retaining wall example, active pressure, no wall friction; (b) Retaining wall example, active pressure with wall friction; (c) Retaining wall example, "at-rest" pressure, no wall friction; (d) Retaining wall example, active pressure, wall friction acting upward due to excessive wall settlement

acting upward rather than downward. When you add the horizontal earth pressure force to the friction force acting upward, the resultant force has a longer lever arm and as a result has a larger overturning moment than assumed. When a retaining wall settles relative to the active soil wedge, the direction of the friction force between the back of wall and the active soil wedge acts upward rather than downward, the resulting increase in overturning moment is not conservative.

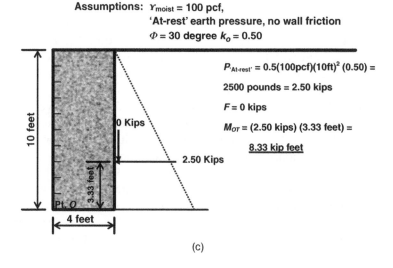

Assumptions: γ_{moist} = 100 pcf,
'At-rest' earth pressure, no wall friction
Φ = 30 degree k_o = 0.50

$P_{At\text{-}rest'}$ = 0.5(100pcf)(10ft)^2 (0.50) =

2500 pounds = 2.50 kips

F = 0 kips

M_{OT} = (2.50 kips) (3.33 feet) =

8.33 kip feet

0 Kips

2.50 Kips

10 feet

3.33 feet

Pt. O

4 feet

(c)

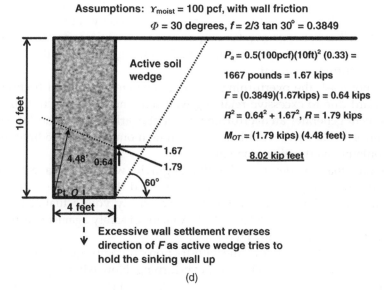

Assumptions: γ_{moist} = 100 pcf, with wall friction
Φ = 30 degrees, f = 2/3 tan 30° = 0.3849

Active soil wedge

P_a = 0.5(100pcf)(10ft)^2 (0.33) =

1667 pounds = 1.67 kips

F = (0.3849)(1.67kips) = 0.64 kips

R^2 = 0.64^2 + 1.67^2, R = 1.79 kips

M_{OT} = (1.79 kips) (4.48 feet) =

8.02 kip feet

1.67

1.79

4.48′ 0.64

60°

10 feet

Pt. O

4 feet

Excessive wall settlement reverses direction of F as active wedge tries to hold the sinking wall up

(d)

Figure 4.2.5 (*Continued*)

For years I've heard stories of dire consequences of walls that settled excessively falling over. Maybe you have heard similar stories, but I wonder have you ever tried to check out the affect of wall settlement on wall overturning moments yourself? It's not that hard to do, so let's give it a try. Assume that a 10-foot tall gravity retaining wall is 4-feet thick with a horizontal backfill that has an internal friction angle of 30 degrees and a moist unit weight of 100 pounds per cubic foot. Consider four cases: (1) active earth pressure P_{active} with no wall friction (Figure 4.2.5a), (2) active earth pressure with wall friction F acting downward equal to $(2/3 \tan \phi) P_{active}$ (Figure 4.2.5b), (3) "at-rest" earth pressure P_o with no wall friction (Figure 4.2.5c) (By the way, how could you have wall friction in the "at-rest" earth pressure condition?), and (4) active earth pressure with wall friction F acting upward equal to $(2/3 \tan \phi) P_{active}$ (Figure 4.2.5d). See Figures 4.2.5(a)–(d) for all four of these cases.

Note from the calculations included with Figures 4.2.5, that the overturning moment for active earth pressure acting without wall friction (as assumed by Rankine) is equal to 5.57 kip ft. The active earth pressure acting with wall friction acting downward (as assumed in standard earth pressure calculations when the wall does not settle excessively) generates an overturning moment of 2.99 kip feet. The lesson I take away from comparing these two values is that the assumption of no wall friction is conservative, resulting in an overturning moment that is nearly twice as much as the moment that occurs when full wall friction is generated (i.e., the no-friction Rankine overturning moment is 1.86 times greater than the with-friction Coulomb overturning moment).

If we assume the retaining wall is rigid and doesn't move and that wall–soil friction doesn't develop, then the overturning moment generated by "at-rest" earth pressure is 8.33 kip feet. The "at-rest" overturning moment is 1.50 times greater than the no-friction Rankine overturning moment, and 2.79 times greater than the with-friction Coulomb overturning moment.

Last but not least, the overturning moment generated with wall friction force F acting upward due to excessive retaining wall settlement is equal to 8.02 kip feet. Wall settlement generates an overturning moment nearly equal to the "at-rest" case, and 1.44 times the no-friction Rankine overturning moment, and 2.68 times the with-friction Coulomb overturning moment.

Let's assume that you designed this 10-foot gravity wall to have a factor of safety against overturning equal to 1.50 and that you used the no-friction Rankine earth pressure in your design. If your wall settled excessively, your actual factor of safety would be equal to $(1.50)(5.57/8.02) = 1.04$. A factor of safety against overturning for a permanent wall of 1.04 is much too small.

Next let's assume that you used the with-friction Coulomb equation to design your wall for a factor of safety of 1.50 against overturning. Now what happens if the wall settles excessively: FOS $= (1.50)(2.99/8.02) = 0.56$. A factor of safety of 0.56 is not good. I suggest that this might explain overturning failures of some retaining walls on soft soil sites. This may also give some insight into why many engineers in the United States prefer the Rankine equation over the Coulomb equation. Personally, I like the Coulomb equation due to its general applicability to various retaining-wall configurations, but I caution you to remember the example given above and always check wall settlements when using the Coulomb equation.

Speaking of settlements, how much retaining-wall settlement is required for the wall-friction force to reverse and cause the earth pressure resultant to act upward as shown in Figure 4.2.5(d)? I do not know the answer to this question, because I have never seen results of a controlled experiment that measures the variables needed to answer it. Until better information is available, I assume that full soil–wall friction is developed when the active soil wedge deflects approximately 0.5 inches relative to the back of the retaining wall. If the wall settles 1 inch *after* full active earth pressure is developed, the wall–soil friction direction should be reversed. Not knowing how long it takes for active earth pressure to develop, I assume that the wall has settled 1 inch during this unknown time period. So my assumption is that a retaining wall has to settle 2 or more inches (total settlement) for the earth pressure resultant to point upward, as shown in Figure 4.2.5(d). The amount of settlement required for this reversal to occur may depend on other variables, such as wall height, wall length, wall type, backfill soil type, foundation soil type, or construction sequence, and if it does you will have to evaluate these factors. If you figure this problem out, please let me know.

4.2.4 Design of Cantilevered Retaining Walls

The design of cantilevered retaining walls has a geotechnical component and a significant structural design component. I often think of cantilevered retaining walls as the structural engineer's answer to retaining soils. The problem with cantilevered retaining walls is an economics problem. Thirty or forty years ago, I recall designing dozens of cantilevered retaining walls for commercial and highway projects. In the past 10 years, I can recall working on three cantilevered retaining walls. In modern times, there are too many alternative wall systems that are more economical than cantilevered walls.

We need to digress from the topic of cantilevered walls for a moment. There are two basic types of retaining walls used on projects these days. They are referred to as top-down and bottom-up walls. Bottom-up walls are walls that you have to either dig a big excavation down to the wall's foundation level before starting construction, or you have to build the wall and then place and compact all of the backfill soil behind the wall. Gravity, cantilevered, and MSE retaining walls are all considered to be bottom up walls.

Top-down walls are walls that you install from the ground surface and then excavate the soil out from in front of the wall. Top down walls include sheet-pile walls, soldier-pile walls, tangent-drilled shaft walls, and slurry-trench walls. Why would you ever want to start building a wall from the top down? Common reasons for using a top-down wall involve nearby obstructions such as highway traffic lanes that cannot be closed, a neighbor's building that must remain (undamaged), a multiple-level structure with a deep basement, or presence of very soft wet soils that will flow into an open excavation if not continuously supported during excavation are a few examples. I might add that top-down walls are nearly always much more expensive to construct than bottom- up walls of similar height. Top-down walls are used because they are required by project constraints, not because they are economical.

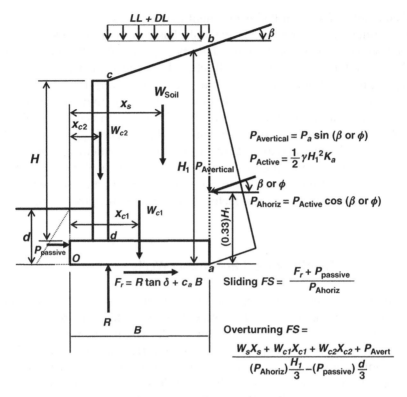

Figure 4.2.6 Detailed analysis of a cantilevered wall

Back to cantilevered walls. I'll discuss the geotechnical considerations of cantilevered wall design here and suggest that you refer to a concrete design textbook or to Bowles' 5th Edition text (Bowles, 1996) for the reinforced concrete design aspects of cantilevered wall design.

Geotechnical analysis of a cantilevered retaining wall is again primarily a static equilibrium analysis, where Equations 4.2.1 are used. Issues arise in cantilevered retaining wall design relative to which forces to include in the analysis and what directions to apply these forces; please refer to Figure 4.2.6.

The soil retained behind the wall in Figure 4.2.6 is sloped upward. If this backfill soil were level, the resultant earth pressure force P_{earth} would be horizontal. Since the soil behind the wall slopes upward, the resultant earth pressure force is taken to be parallel to the soil surface, as shown in Figure 4.2.6. In both the horizontal and sloping upward fill configurations, the Rankine earth pressure method may be used. If the soil backfill slopes downward, you cannot use the Rankine method, and will have to switch to the Coulomb method (which I will discuss shortly).

Notice in Figure 4.2.6 that the lateral earth pressure resultant P_{earth} is taken at the dotted line that projects above the back heel (point a) of the wall. The soil above the wall's foundation slab and between the back of the stem wall and dotted line a–b is considered to be a part of the retaining wall system. This retained soil provides soil–wall resistance to earth pressure overturning generated by P_{earth}. The active earth pressure wedge is considered to develop outside of line a–b with no friction generated along a–b for the Rankine analysis and the resultant inclined at the slope angle β. For the Coulomb analysis friction is generated along line a–b. To use the Coulomb analysis, the wall–soil friction angle δ is taken equal to the soil friction angle ϕ, because the friction generated along line a–b is soil-to-soil friction and not soil-to-concrete friction.

Cantilevered retaining wall stability (assuming the wall itself is structurally stable) analysis includes: (1) sliding along the base of the wall footing; (2) overturning about the toe at point O as shown in Figure 4.2.6; (3) bearing capacity analysis of the foundation support soils, and (4) global stability analysis of the wall system, which is actually a slope-stability analysis to determine if a landslide could move the entire slope and wall system.

Sliding stability analysis of a cantilevered retaining wall is basically a comparison between the sliding resistance forces F_r plus $P_{passive}$ and the driving forces $P_{surcharge}$ and P_{active} and shown in Figure 4.2.6. Seems easy enough doesn't it? But wait a moment; the passive soil resistance at the toe of wall footing, $P_{passive}$ might be reduced by erosion or human excavation of soil for maintenance or repair of utilities. The density and shear strength of the soil at the toe of wall may be loosened and softened by annual frost heaving, thus reducing the magnitude of $P_{passive}$. Given these concerns, many engineers, myself included, ignore the sliding passive soil resistance $P_{passive}$ when calculating the factor of safety against wall sliding. If you were certain that the soil below a fixed depth in front of the wall footing would remain or not be disturbed by frost action, you could include passive resistance of soil below that depth.

What about the sliding friction resistance, $F_r = R\,(\tan\,\delta) + (c_{base\ adhesion})\,B$? Are there any complications in calculating the sliding friction resistance? You bet there are! First, consider the resultant force R which acts perpendicular to the base of the wall foundation. If the surcharge loadings include any live load component, it may not be present during some portion of the lifetime of the wall. The live loading components need to be considered for bearing capacity FOS calculations, when they exist. Live loadings should not be included for sliding and overturning resistance calculations, because they increase the wall's resistance to lateral earth pressure loading, and they (i.e., the live loadings) are not always present to resist these loadings.

The base of wall to soil adhesion is often taken as 0.6 to 0.8 of the foundation soil cohesion, so long as the soil is low plasticity clay. If the wall foundation bears on a cohesive soil of medium to high plasticity, that is a PI greater than 20, the base of wall adhesion is often ignored or severely reduced because the potential for shear strength loss due to wetting and swelling are too great. Speaking of clayey soils, the factor of safety against sliding is taken as 1.5 when foundation soils are granular materials, and it is often increased to 2.0 when foundation soils are cohesive materials.

For bearing capacity analyses, cantilevered retaining walls can be evaluated in one of two ways. As I mentioned above, structural engineers frequently calculate a triangular or trapezoidal bearing pressure distribution beneath the cantilevered retaining wall's footing by using the equation: soil pressure $= P/A \pm Mc/I$. Everything is great using this strength of materials equation until the Mc/I term gets large enough to require soil tension to be developed, then you have to recalculate the size of the resisting area assuming tension is not generated by the foundation to soil interface. I prefer the Meyerhof method which uses a reduced footing width to B', see Equation 4.2.2 above, and the Terzaghi or Meyerhof bearing capacity equation (Bowles, 1996).

4.2.5 Design of MSE Retaining Walls

Design of MSE retaining walls is similar to the design of a gravity retaining wall except MSE walls are designed in two steps. In the first step, we assume that the MSE structure is internally stable, and we analyze it for stability using the same equations of static equilibrium as used for a gravity retaining wall. Then in the second step, we analyze the reinforcements for internal stability to make sure that the assumption of step one is achieved. I think of it as being similar to structural design of a building. First you analyze the overall structure for stability and member loadings, then you analyze the individual members to make sure that they are not over loaded and that their connections have adequate strength. Take a look at Figure 4.2.7 for the forces used to analyze an MSE wall supporting a bridge foundation. For a detailed description of stability analysis of MSE walls, I recommend you obtain a copy of the US Department of Transportation Publication No. FHWA-NHI-00-043, *Mechanically Stabilized Earth Walls and Reinforced Soil Slopes, Design & Construction* (Elias *et al.*, 2001).

I know the MSE wall analysis indicated in Figure 4.2.7 looks complicated, this drawing and an accompanying spreadsheet took an E.I.T (Jon Schermerhorn) and myself more than two days to develop. Analysis of this MSE wall is nearly identical to the gravity and cantilevered wall analyses discussed above in that we use equations of static equilibrium to check the wall for sliding, overturning and bearing pressure. When solving the problem illustrated in Figure 4.2.7, you have to be particularly careful to make sure that you use the appropriate vertical live loadings for cases where it acts as a driving force and don't use it when it acts as a resisting force.

Personally when it comes to analyzing MSE walls, I wonder if the flexibility of the MSE wall versus the rigidity of the gravity retaining wall makes a difference in the distribution of lateral and vertical stresses. In the cases of the gravity and cantilevered retaining walls, we assume that the walls are a rigid body in our application of equations of static equilibrium. I believe that in most cases an MSE wall is at best a semi-rigid structure, and it is more likely a flexible structure. Does this flexibility inconsistency in the MSE wall analysis make a difference? Maybe it does, but like most engineers working on highway projects, Jon and I did not consider wall flexibility

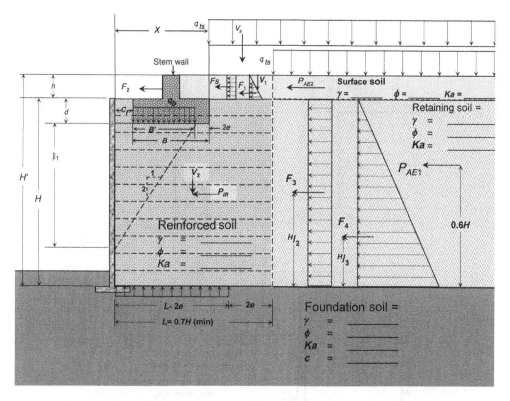

Figure 4.2.7   Detailed analysis of MSE wall supporting a bridge footing

because we were following the MSE wall guidance provided by FHWA-NHI-00-043 (Elias *et al.*, 2001). Sometimes you consider issues, other times you do what you are told. Many engineers are concerned about prescriptive codes in general. I'm not too concerned with prescriptive codes if their design analysis methods make sense. I'm only concerned when a code requires me to do something that I'm fairly sure is incorrect for my site-specific case.

4.2.6 Design of Sheet-Pile Walls

The design of sheet-pile walls requires an analysis that is somewhat different than gravity or cantilevered walls because you don't know how long to embed the sheet piles to generate a stable structure. After you determine how much sheet pile embedment is required for stability, then you can check the sheet piles for bending strength. Figure 4.2.8(a) shows the forces acting on a sheet pile wall. This figure is commonly given in references for analyzing lateral earth pressures acting on a sheet-pile wall.

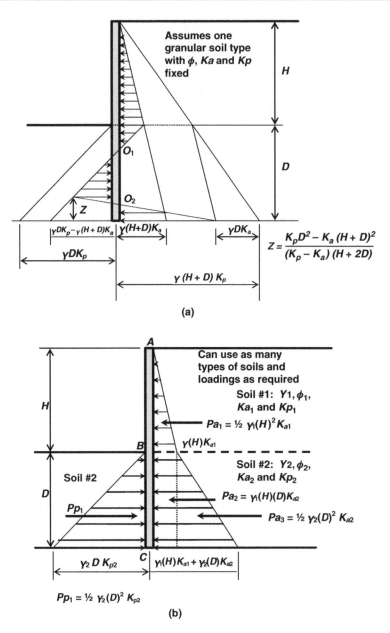

(a)

(b)

Figure 4.2.8 (a) Detailed sheet pile wall analysis – but only one soil type; (b) proposed method of analysis for sheet pile walls

I came very close to leaving you with the analysis of sheet piles required to solve summation of forces and moments for lateral forces shown in Figure 4.2.8(a). I decided to offer you an alternate method for analysis of sheet-pile walls. What made me change my mind? Just today in my office I was showing an engineer how to solve the analysis of sheet piles using the forces indicated in Figure 4.2.8(a). As soon as we started looking at this figure, his eyes glazed over. He was lost immediately. This glazing of the eyes brought back to my memory the dozens if not hundreds of engineers that I have tried to teach this sheet-pile design method. They all glazed over as soon as they saw this figure.

Five or six of my former students eventually figured out how to analyze forces in Figure 4.2.8(a) and one of them even developed a spreadsheet. Even after they figured it out, my students still had questions about how to apply this method to layered soils, surcharge loadings, and concentrated forces. The majority of engineers who never quite figured out Figure 4.2.8(a) suspended disbelief and went through the steps indicated in the US Steel Manual (US Steel Manual, 1969 and 1975) or Bowles' book (Bowles, 1996). They didn't understand how the forces were generated, so they gave up and went through a solution by rote. Darn it, that's not what engineers are supposed to do! We are supposed to understand what we are doing, and see the solution process clearly.

First thing you need to understand is that the sheet-pile wall problem is a statically indeterminate problem. There are more unknowns than equations of statics available to solve this type of problem. We don't commonly analyze the piles for stresses, strains, and deflections, we just make sure that the piles are stable by using statics and make sure that the piles don't bend over by development of a plastic hinge. I use a simplified method of analysis of sheet-pile walls that seems easier to understand because it is easier to see what lateral forces are generated; please see Figure 4.2.8(b).

Notice that I don't use the passive minus active earth pressure bounding lines in this method, and that simplification alone seems to put many engineers at ease. I draw all of the soil strata on the figure, and calculate vertical stresses at each layer boundary. Then I multiply the vertical stresses at each point just above and just below each boundary by the lateral earth pressure coefficient for each layer. After I have a figure showing the distribution of lateral active, lateral surcharge, and lateral passive stresses, I divide all of the lateral loading distributions up into rectangles and triangles so that the resultant forces act at one-half or one-third of the height, then I calculate the resultant forces on both sides of the sheet pile.

Using an iterative process, I use a trial and error method on a spread sheet to determine at what embedment depth D the summations of forces in the horizontal direction are equal to zero. After I determine the embedment depth D required for the summation of horizontal forces to equal zero, I take moments about the top of the pile of all active and passive forces. Again the summation of moments should be equal to zero or the resisting passive moment should be greater than the overturning active moments. Finally, I use the traditional method to develop a factor of safety in the pile embedment length; I increase the calculated embedment length D by 30%.

The USS manual suggests that using an embedment depth equal to $1.3D$ will give a factor of safety against wall translation or rotation of about 1.5 to 2.0. Now I have the total length of my sheet pile.

Next, starting at the top of pile, I find the point where summation of lateral forces on the sheet pile equals zero. The point of zero lateral shear is the point where the bending moment in the pile is at a maximum. Taking the summation of moments above the point of zero shear, I find the maximum bending moment. Using the maximum bending moment in the sheet pile wall (remember that when analyzing sheet piles we do all calculations on a per foot of wall basis), I calculate the maximum bending stress using the classic mechanics equation, $\sigma_{\text{maximum bending}} = Mc/I$ or M/S (where S is the section modulus of the sheet piles given in the sheet pile manual *on a per foot of wall basis*). I assume that sheet piles are embedded in soil and won't buckle, so I compare the maximum bending stress to an allowable bending stress value of $0.6F_y$, where F_y is the yield stress of the sheet piles' steel. Since most modern steel rolled in the United States has a yield strength of 50 ksi (or greater), I most frequently compare the maximum sheet pile bending stress to an allowable bending stress value of 30 ksi.

4.2.7 Design of Soldier-Pile Walls

Design of soldier-pile walls is similar to the design and analyses of sheet-pile walls except you not only have to determine pile embedment, you also have to determine appropriate pile spacing. When checking engineering design calculations for soldier-pile walls, I find one mistake over and over again. Sheet-pile walls can be designed on a per foot basis, while soldier-pile walls *cannot* be designed on a per foot basis. The method of analysis for soldier-pile walls is very similar to that used for sheet piles except you have to multiply the lateral loading by the soldier-pile spacing for the portion of pile above embedment, that is, from point A to B in Figure 4.2.9. For the portion of pile embedded below grade, sometimes the pile is driven into foundation soils and sometimes it is placed into a concreted drilled-shaft hole as shown in Figure 4.2.9. To determine the effective zone of influence of lateral pile stresses below grade, twice the width of driven piles or twice the diameter of the drilled shaft hole is used to calculate the lateral forces; again please refer to Figure 4.2.9.

4.2.7.1 Example of Soldier-Pile Wall Analysis

I'm not a big fan of giving examples of analyses, because I'm afraid that you will be tempted to just follow my example line by line without understanding what I did and why I did it. Of course my example will never be exactly like your real-world case, so there is peril afoot, "Danger Will Robinson..."

Please study this example to extract my thinking process. Ask yourself why did he do this or why did he do that. If you look at this example critically, it is highly likely

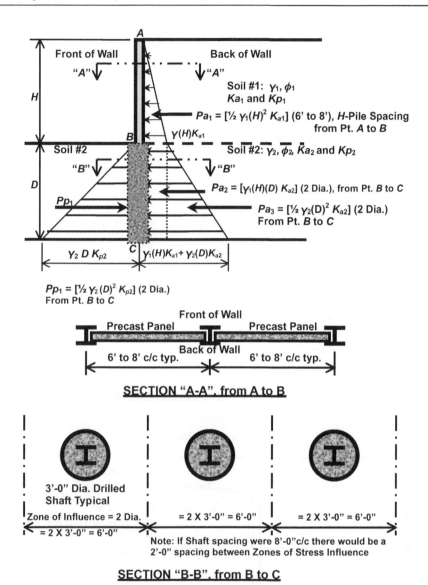

Figure 4.2.9 Analysis of a soldier-pile wall

that you may see alternative approaches and come with your own solution to this problem.

Please refer to Figures 4.2.10(a)–(d). Figure 4.2.10(a) shows vertical stresses acting on the soldier-pile wall. Figure 4.2.10(b) shows resulting lateral stresses calculated by applying the appropriate lateral earth pressure coefficients to the vertical stresses

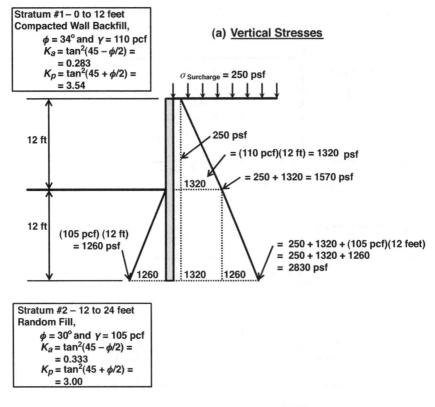

Stratum #1 – 0 to 12 feet
Compacted Wall Backfill,
$\phi = 34°$ and $\gamma = 110$ pcf
$K_a = \tan^2(45 - \phi/2) =$
$= 0.283$
$K_p = \tan^2(45 + \phi/2) =$
$= 3.54$

(a) Vertical Stresses

$\sigma_{Surcharge} = 250$ psf

12 ft

250 psf
$= (110 \text{ pcf})(12 \text{ ft}) = 1320$ psf
$= 250 + 1320 = 1570$ psf

1320

12 ft

(105 pcf) (12 ft)
$= 1260$ psf

$= 250 + 1320 + (105 \text{ pcf})(12 \text{ feet})$
$= 250 + 1320 + 1260$
$= 2830$ psf

1260   1320   1260

Stratum #2 – 12 to 24 feet
Random Fill,
$\phi = 30°$ and $\gamma = 105$ pcf
$K_a = \tan^2(45 - \phi/2) =$
$= 0.333$
$K_p = \tan^2(45 + \phi/2) =$
$= 3.00$

(b) Lateral Stresses

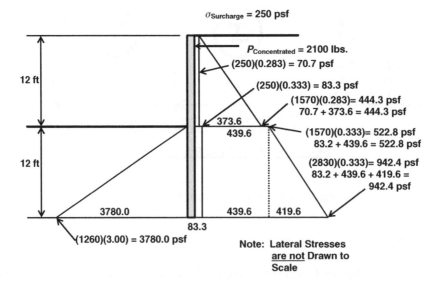

$\sigma_{Surcharge} = 250$ psf

$P_{Concentrated} = 2100$ lbs.
$(250)(0.283) = 70.7$ psf

12 ft

$(250)(0.333) = 83.3$ psf
$(1570)(0.283) = 444.3$ psf
$70.7 + 373.6 = 444.3$ psf

373.6
439.6

$(1570)(0.333) = 522.8$ psf
$83.2 + 439.6 = 522.8$ psf

12 ft

$(2830)(0.333) = 942.4$ psf
$83.2 + 439.6 + 419.6 =$
942.4 psf

3780.0   439.6   419.6

83.3

$(1260)(3.00) = 3780.0$ psf

Note: Lateral Stresses
are not Drawn to
Scale

Figure 4.2.10   (a) Soldier-pile wall example – profile with vertical stresses; (b) Soldier-pile wall example – profile with lateral stresses; (c) Soldier-pile wall example – calculating resultant forces; (d) Soldier-pile wall example – calculations

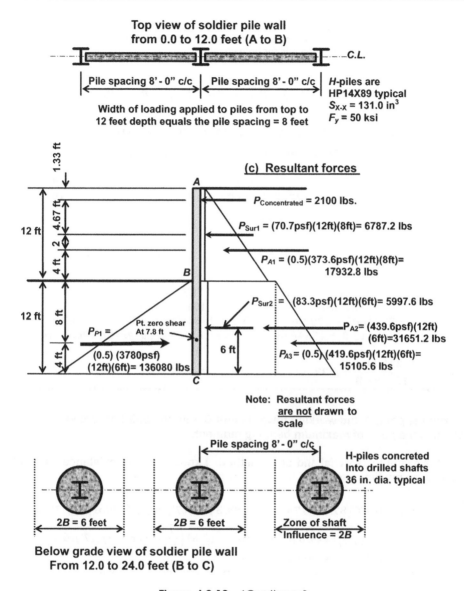

**Top view of soldier pile wall
from 0.0 to 12.0 feet (A to B)**

C.L.

Pile spacing 8' - 0" c/c | Pile spacing 8' - 0" c/c | *H*-piles are
HP14X89 typical

Width of loading applied to piles from top to
12 feet depth equals the pile spacing = 8 feet

S_{x-x} = 131.0 in^3
F_y = 50 ksi

1.33 ft

4.67 ft

12 ft

2 ft

4 ft

(c) Resultant forces

A

$P_{Concentrated}$ = 2100 lbs.

P_{Sur1} = (70.7psf)(12ft)(8ft)= 6787.2 lbs

P_{A1} = (0.5)(373.6psf)(12ft)(8ft)=
17932.8 lbs

B

P_{Sur2} = (83.3psf)(12ft)(6ft)= 5997.6 lbs

12 ft

8 ft

4 ft

P_{P1} =

Pt. zero shear
At 7.8 ft

(0.5) (3780psf)
(12ft)(6ft)= 136080 lbs

6 ft

P_{A2}= (439.6psf)(12ft)
(6ft)=31651.2 lbs

P_{A3}= (0.5)(419.6psf)(12ft)(6ft)=
15105.6 lbs

C

Note: Resultant forces
are not drawn to
scale

Pile spacing 8' - 0" c/c

H-piles concreted
Into drilled shafts
36 in. dia. typical

2B = 6 feet | 2B = 6 feet | Zone of shaft
Influence = 2B

**Below grade view of soldier pile wall
From 12.0 to 24.0 feet (B to C)**

Figure 4.2.10 (*Continued*)

(i.e., $\sigma_{horizontal} = K\,\sigma_{vertical}$). Figure 4.2.10(c) shows calculation of the resultant lateral forces and their points of application on the soldier pile. Finally in Figure 4.2.10(d), we have the analysis of lateral forces to check stability of the H-pile, to check the pile embedment depth *D*, and to check the H-pile maximum bending stress f_b.

Check stability of the soldier pile by taking summation of forces in the horizontal direction equal zero, and take summation of moments about the top of pile, point A, equal to zero. Then calculate the factor of safety of the pile for both cases.

$$\sum F_{Horiz} = 0 = P_{P1} - P_{Concentrated} - P_{sur1} - P_{A1} - P_{sur2} - P_{A2} - P_{A3}$$

$$\sum F_{Horiz} = 136080 - 2100 - 6787.2 - 17932.8 - 5997.6 - 31651.2 - 15105.6$$
$$= 56505.6, OK \ because \ resistance \ is \ greater \ than \ loading$$

$$FOS_{Horiz} = \frac{136080}{2100 + 6787.2 + 17932.8 + 5997.6 + 31651.2 + 15105.6} = \frac{136080.0}{79574.4}$$

$$= 1.71, greater \ than \ 1.50 - OK.$$

$$\sum M_A = P_{P1}(20) - P_{Concentrated}(1.33) - P_{Sur1}(6) - P_{A1}(8) - P_{Sur2}(18) - P_{A2}(18)$$
$$- P_{A3}(20)$$
$$= (136080)(20) - (2100)(1.33) - (6787.2)(6) - (17932.8)(8)$$
$$- (5997.2)(18) - (31651.2)(18) - (15105.6)(20)$$
$$= 2721600 - 2793.0 - 40723.2 - 143462.4 - 107949.6 - 569721.6$$
$$- 302112.0 = 1554838.2, OK, resistance \ greater \ than \ loading$$

$$FOS_{M_A} = \frac{2721600.0}{1166761.8} = 2.33, greater \ than \ 1.50 - OK.$$

****

Starting at point A and working toward point C, find the point of zero shear which is the point of maximum bending moment.

For this example use trial and error method. I know that the resistance starts at point B, so my first guess is 7.0 feet below point B.

$$\sum F_{horiz} = 2100 + 6787.2 + 17932.8 + (83.3 \ psf)(7')(6') + (439.6 \ psf)(7')(6')$$
$$+ (0.5)\left(\frac{7}{12}\right)(419.6 \ psf)(7')(6') - (0.5)\left(\frac{7}{12}\right)(3780 \ psf)(7')(6')$$
$$= 53921.9 - 46305 = 7616.9$$
$$Driving \ forces \ greater \ than \ resisting, increase \ depth \ below \ B.$$

(d)

Figure 4.2.10 (Continued)

Try 7.5 feet below B for the point of zero shear.

$$\sum F_{horiz} = 2100 + 6787.2 + 17932.8 + (83.3\,psf)(7.5')(6')$$
$$+ (439.6\,psf)(7.5')(6') + (0.5)\left(\frac{7.5}{12}\right)(419.6\,psf)(7.5')(6')$$
$$- (0.5)\left(\frac{7.5}{12}\right)(3780\,psf)(7.5')(6') = 56251.1 - 53156.2 = 3094.9$$

Closer but still not there, try 7.8 feet below B.

$$\sum F_{horiz} = 2100 + 6787.2 + 17932.8 + (83.3\,psf)(7.8')(6')$$
$$+ (439.6\,psf)(7.8')(6') + (0.5)\left(\frac{7.8}{12}\right)(419.6\,psf)(7.8')(6')$$
$$- (0.5)\left(\frac{7.8}{12}\right)(3780\,psf)(7.8')(6') = 57673.8 - 57493.8 = 180.0$$

This is close enough, using the point of zero shear at 7.8 below below B, calculate the maximum bending moment:

$$M_{Max} = (2100)(18.47') + (6787.2)(13.8') + (17932.8)(11.8) + (3898.4)\left(\frac{7.8}{2}\right)$$
$$+ (20573.3)\left(\frac{7.8}{2}\right) + (6382.1)\left(\frac{7.8}{3}\right) - (57493.8)\left(\frac{7.8}{3}\right)$$
$$= 456090.5 - 149483.9 = 306606.6\,ft\,lbs$$

$$f_b = \frac{(M_{Max})(12\,in/ft)}{S_{X-X}} = \frac{(306606.6\,ft\,lbs)(12\,in/ft)}{131.0\,in^3} = 28086\,\frac{lb}{in^2} < 30000\,\frac{lb}{in^2}$$

The maximum bending stress is less than the allowable bending stress, OK.

(d)

Figure 4.2.10 (Continued)

Working example

This site was backfilled 40 or 50 years ago with a silty sandy soil that was not highly compacted. I'm trying to be politically correct here. Testing of the existing fill soil indicates that it has an average moist unit weight of 105 pounds per cubic foot and a lower bound peak internal friction angle of 30 degrees. The client wants to build a retaining wall at this location to raise the site grade 12 feet.

I initially proposed using an MSE retaining wall, but the client determined that existing critical utilities would interfere with the MSE reinforcing zone.

After some consideration, the client agreed that a soldier-pile retaining wall would meet his needs and his budget. The client had plenty of existing fill material on site, so we tested existing backfilled areas and tested laboratory compacted samples and determined that the average moist unit weight of this material compacted to 95% of modified Proctor maximum density was 110 pounds per cubic. Testing of saturated, compacted samples of the fill soil by the direct shear test indicated that it had a lower bound internal friction angle of 34 degrees.

In Figures 4.2.10(a) and (b), notice the bottom of the upper fill layer Stratum #1 is at 12 feet below the top of pile. The vertical stresses at the interface between Stratum #1 and Stratum #2 are approximately the same at 11.99 feet and at 12.01 feet. Since the active lateral earth pressure coefficient in Stratum #1, K_{A1} is 0.283, and in Stratum #2 K_{A2} is 0.333, the lateral earth pressure at 11.99 feet (444.3 psf) is different than the lateral earth pressure at 12.01 feet (522.8 psf). In the physical reality of soil, I doubt that there is an abrupt change in lateral earth pressure as you cross the boundary at 12 feet, but it suits our purpose to analyze the lateral pressures with an abrupt change rather than calculate the transition of lateral stresses across the boundary of Stratum #1 to Stratum #2.

By the way, notice in Figure 4.2.10(b) that we have a 2100 pound horizontal concentrated force $P_{concentrated}$ applied 1.33 feet below top of the H-pile. This horizontal concentrated force was a surprise additional lateral force added by the owner after the wall was built! In reality, the example calculation shown in Figures 4.2.10(a)–(d) is a check of a wall that was already constructed, although the project was not quite finished and several surprises have occurred and many more surprises may yet come up.

4.2.8 What Kind of Wall Would You Use Here?

The picture shown below was a proposed site for a bridge crossing of the Rio Grande Valley in northern New Mexico. The depth of the valley at this location is nearly 1000 feet deep. What kind of retaining walls would you use at this site? The tallest gravity wall I have ever designed was 25-feet tall, although I have read about gravity walls 40-feet tall, and concrete dams are essentially gravity walls that are well over 100-feet tall. The tallest cantilevered wall I have ever designed was 45 feet tall, although I have heard about cantilevered walls 60-feet tall constructed on bedrock foundations. Tieback walls and rock-bolted shotcrete walls are required to provide lateral support on the scale needed in Figure 4.2.11. These topics are discussed in Section 4.3.

Figure 4.2.11 What kind of a retaining wall would you use at this site?

References

Bowles, J.E (1996) *Foundation Analysis and Design*, 5th edn, McGraw Hill Inc., 1024 pages.

Coulomb, C.A. (1776) Essai sur une application des règles de Maximus et Minimis à Quelques Problèmes de Statique, Relatifs à l' Architecture, Mémoires de Mathématique et de Physique, Présentés a l' Académie Royale des Sciences, par Divers Savans, et lûs dans ses Assemblées, Paris, Vol. 7, pp. 143–167.

Elias, V., Christopher, B.R. and Berg, R. R. (2001) *Mechanically Stabilized Earth Walls and Reinforced Soil Slopes Design and Construction Guidelines*, FHWA-NHI-00-043, National Highway Institute Federal Highway Administration US Department of Transportation, Washington, D.C., 394 pages.

Rankine, W.J. M. (1857) On the stability of loose earth. *Proceedings of the Royal Society, London*, **VIII**, pp. 185–187.

United States Steel Corporation (1969 and 1975) Steel Sheet Piling Design Manual, 133 pages both editions. Much of the material included in these USS manuals can be found in Wayne C. Teng's book, 1962, Foundation Design, Chapter 12, Prentice Hall, Englewood Cliffs, New Jersey, pp. 347–386.

4.3

Tieback Walls

4.3.1 Introduction to Tieback Walls

Abutments for the proposed bridge planned at the gorge site shown in Figure 4.2.11 are a good place to consider use of tieback walls and rock bolts with shotcrete facing. In fact, any retaining structure over 15 feet in height is a candidate for tiebacks. A retaining wall that uses tiebacks may be called an anchored wall, although use of the term "anchor" has a nautical flavor that seems to be fading in favor of tiebacks when construction moves inland from the ocean fronts to Midwestern United States highways. Personally I use the terms "tieback" and "tieback anchor" interchangeably as you will see below.

Why do we use tieback anchors? The primary purpose of tiebacks is to provide additional lateral support to a retaining structure because it is too tall to be cantilevered. Retaining walls with tiebacks are, as you might expect, highly statically indeterminate structures. We have to make many assumptions about the nature of the retained soil and the wall structure to design tieback walls.

I like to think of this wall design process as one of self-fulfilling predictions, or as the old saying goes, "It is a self-fulfilling prophecy." What do I mean? The idea is that you pick the magnitude and location of a retaining wall's tieback loading, and the tieback loading develops where you selected if the relative stiffnesses of the component wall parts are designed properly. Take a look at Figure 4.3.1; in this figure I propose that you can analyze a sheet-pile wall with a single level of tiebacks in two alternate ways (there are more than two ways, but we're just using this as an example).

The method on the left in Figure 4.3.1 has an active earth pressure distribution on the back side of the sheet piling and a passive earth pressure distribution in front of the piling. This distribution of lateral earth pressures is identical to that proposed for analysis of a cantilevered sheet-pile wall. The tieback loading has added a lateral force near the top of the sheet piling, but it has not changed the earth pressure distribution.

Geotechnical Problem Solving, First Edition. John C. Lommler.
© 2012 John Wiley & Sons, Ltd. Published 2012 by John Wiley & Sons, Ltd.

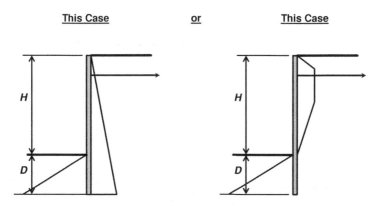

Figure 4.3.1 Are these loading cases self-fulfilling prophecies?

The method on the right in Figure 4.3.1 has an increased lateral earth pressure toward the top of the sheet piling with the same passive earth pressure acting below the toe of wall.

I've heard my students say many times, "Which case is the right answer? One of these cases has to be right and the other one is wrong, which one is it?"

The answer is that they are both correct. Which one you choose to use depends on your design assumptions and how the tieback is installed.

The wall illustrated on the left in Figure 4.3.1 has apparently deflected sufficiently for an active earth pressure distribution to develop. After the pile deflected and active earth pressure developed, the tieback anchor was installed without prestressing. All the tieback anchor does in this case is provide a greater factor of safety and prevent the top of wall from moving much further.

The wall illustrated on the right in Figure 4.3.1 has not deflected sufficiently for an active earth pressure distribution to develop. The tieback in this case was installed prior to excavating in front of the sheet pile thus preventing the pile top from moving. Alternately, the tieback could have been stressed to develop compression stresses in the retained soil. By prestressing the tieback anchor, the top of the sheet pile wall can be prevented from moving, or it may actually be pulled into the soil mass. The required depth of pile embedment D, the bending stresses generated in the sheet piling, and the tieback anchor force can be significantly different in the two cases illustrated in Figure 4.3.1. You can control the distribution of loading and the lateral stresses by changing the loading applied to the tieback anchor. Within limits of geometry and soil strength, it is a self-fulfilling prophecy!

4.3.2 Retaining Structures with One Row of Tiebacks

You have the choice to analyze a wall with a single row of tiebacks by either method illustrated in Figure 4.3.1. For years I used the method illustrated on the left, which I

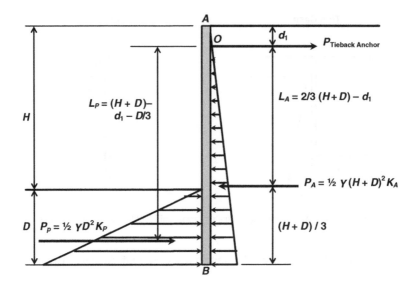

Figure 4.3.2 Analysis of one row of tiebacks – Case 1

will call Case 1. To support the top of my sheet-pile walls, I used anchor rods tied to deadmen. Today with the variety of drilled and driven anchors that are prestressed, post-tensioned and proof-loaded, I frequently use the case shown on the right side of Figure 4.3.1, which I will call Case 2. The decision on which method to use depends on your understanding and control of lateral deflections of the sheet-pile wall.

4.3.2.1 Case 1

First I will explain how to analyze a sheet-pile or soldier-pile wall for Case 1, assuming that active earth pressures are generated as shown in the case on the left in Figure 4.3.1. To further illustrate this analysis method, please refer to Figure 4.3.2.

When we start analysis of the tieback sheet-pile wall illustrated in Figure 4.3.2, we don't know what values to assign to the tieback force P and to the pile embedment length D. To isolate these two variables, take summation of moments about point O, which is the connection point of the tieback anchor to the sheet pile. Since the lever arm distance of force P about point O is zero length, the moment of force P about O is zero and we have eliminated the force P from our first calculation. Now you have one variable left in your summation of moments about point O, and that variable is D:

$$\sum M_o = P_p L_p - P_a L_a = \left(\frac{1}{2}\gamma D^2 K_p\right)\left[(H+D) - d_1 - \frac{D}{3}\right]$$

$$- \left[\frac{1}{2}\gamma(H+D)^2 K_a\right]\left[\frac{2}{3}(H+D) - d_1\right] \tag{4.3.1}$$

This looks a bit messy doesn't it? You don't know the value of D, and the variable D is included in five terms in Equation 4.3.1. I start by assuming D is $\frac{1}{2}$ H, if the resulting summation of moments is a large positive number, I reduce D and resolve Equation 4.3.1. If the result of my first estimate of D gives a summation of moments that is a large (absolute value) negative number, I increase D and resolve Equation 4.3.1. An alternate to this trial and error method is to set up a spreadsheet with D varying from 1 foot to H feet, and select the value of D that gives a value of Equation 4.3.1 closest to zero. I often use the spreadsheet method of solving for D these days because I have to include calculation of D in my calculation data package.

OK, now you have a theoretical value of the sheet-pile embedment depth D, and can calculate a value for the passive resistance P_p. Next you need to calculate a theoretical value of the tieback anchor force P. Force P is on a per foot of wall basis. To determine P, I take summation of forces in the horizontal direction:

$$P_{\text{Tieback Anchor}} = P_a - P_p \qquad (4.3.2)$$

Now we have a theoretical embedment and a theoretical tieback-anchor force (per foot of wall), and all we need to do is determine the required design values of D and P. We also need to check bending stresses in the sheet piling or select a sheet piling size. For the design value of D, I increase D by 30% as recommended in the USS Manual (US Steel Manual, 1969 and 1975), so design embedment equals 1.3D. For the required design tieback-anchor force, equal to P times anchor spacing S, the required factor of safety depends on the level of anchor testing planned for the project. If all of the anchors are proof-tested to 125% of design tieback-anchor force, I use a design tieback-anchor force of P, which is equal to the theoretical anchor force. If no anchors are proof-tested, and a few representative anchors are tested to failure, I use a design tieback anchor force equal to 50% of the failure loading, $0.5P_{\text{Failure}}$. Even if all tieback anchors are proof-tested to 125% of design loading (some specifications require 130%), I always recommend testing a few representative tieback anchors to pullout failure or a minimum of 2P, 200% of design loading. The actual numbers of failure tests depends on site variability and the size and importance of the project.

Finally we need to size the sheet piling or check bending stresses in the sheet piling suggested by the project contractor. In the old days when United States Steel and Bethlehem Steel were still in business, we always calculated the required sheet-pile section modulus and selected the next larger size of sheet piles from the steel design manual. Today, we often have to use the sheet piling that the contractor has available or the size of sheet piling that he can reasonably obtain. In this case, we don't calculate the required section modulus, we have to use the section modulus provided by the available piling. Most sheet piles these days have yield strengths of 50 kips per square inch. Don't take my word for it; always check the provided steel members' yield strength! To check bending stresses in sheet piling provided, you have to find the point of zero shear in the piling.

Start at the top of pile and working toward the bottom, take summation of forces in the horizontal direction until the value equals zero (don't stop at the location of the tieback). At the point on the piling where the shear equals zero, calculate the bending moment of the forces above the point of zero shear. Using an allowable bending stress of $0.6F_{yield}$, divide the bending moment in kip inches by the allowable bending stress in kips per square inch to find the required section modulus S_{x-x}. If you have a given pile with section modulus S_{x-x}, divide the maximum bending moment in kip inches by the section modulus S_{x-x} in inches cubed and find the bending stress in kips per square inch. Finally you compare the calculated maximum bending stress to the allowable bending stress of $0.6F_{yield}$ (30 ksi for commonly found 50 ksi yield strength steel).

4.3.2.2 Case 2

Prior to my return to highway project work in the late 1990s, I used the triangular active lateral earth pressure distribution for walls with one row of tiebacks, and I used the Terzaghi and Peck lateral earth pressure distributions (Terzaghi and Peck, 1967) for walls with multiple levels of tiebacks. I might add that Ralph Peck gave all of the credit to Karl Terzaghi for these lateral earth pressure distributions, but in modern day usage, authors of multiple texts and manuals give Ralph credit also, thus the name "Terzaghi and Peck method."

I first became aware of the Case 2 lateral earth pressure distribution while reading an FHWA manual on ground anchors and anchored systems in late 1999 (Sabatini, Pass, and Bachus, 1999). This is still a good reference and I recommend that you download it from the FHWA website. This manual is currently called "Geotechnical Engineering Circular No. 4, Ground Anchors and Anchored Systems, FHWA-IF-99-015, June 1999."

To analyze a sheet-pile or soldier-pile wall for Case 2, we assume that lateral earth pressures generated are as shown on the right side of Figure 4.3.1. I say "assume" to describe these earth pressures because they are *not* based on earth pressure theories such as Rankine's or Coulomb's theories, rather the earth pressure distribution shown in Figure 4.3.3 below comes from field testing of anchored walls. The basic method used to develop earth pressure distributions from field testing involves measuring tieback anchor loads and dividing them by a tributary area of the wall supported by each anchor.

The analysis of Case 2 starts similarly to Case 1 because we don't know what values to assign to the tieback force P and we don't know the pile embedment length D. To isolate these two variables for Case 2 take summation of moments about point O. In the equation of moments you have one unknown variable which is the depth of embedment D, see Equation 4.3.3 below:

$$\sum M_o = P_p L_p - P_a X' = \left(\frac{1}{2}\gamma D^2 K_p\right)\left[(H + D) - d_1 - \frac{D}{3}\right] - 0.65\gamma H^2 K_a X' \quad (4.3.3)$$

When solving Equation 4.3.3, you can start at 1 foot embedment and go up to a depth equal to H in a spreadsheet, then pick the value of D that gives you a summation

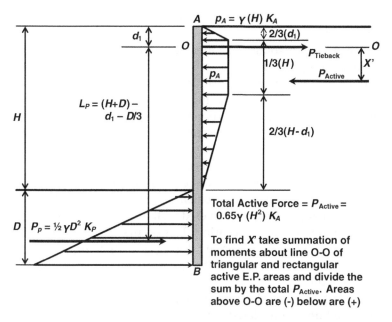

Figure 4.3.3 Analysis of one row of tiebacks – Case 2

of moments close to if not equal to zero. After you determine D, calculate a value of P_{Tieback} (again on a per foot of wall basis) by taking summation of forces in the horizontal direction equal to zero. After you have theoretical values of D and P_{Tieback} required for static equilibrium, increase the value of D by 30% and calculate the final embedment $1.3D$. Using factors of safety on the tieback anchor force that are based on your anchor-testing program (as described above or in FHWA-IF-99-015), calculate the design tieback anchor force. After you have the pile embedment and tieback force, you can calculate the point of zero shear and determine the maximum bending moment for sizing sheet piling or checking piling bending stresses.

Now you are done with the tieback wall design, right? No way; there are many more details to consider. You have to design the anchor embedment to secure your tieback. You have to make sure that the tieback length is adequate to allow development of the anchor force beyond the zone of active shear wedge development. You have to design and detail the anchor connection system to the face of your sheet pile wall or to your soldier piles. If you use waler beams, you have to design them.

There are 10 potential failure modes for tieback-wall systems listed in the FHWA-IF-99-015 manual. All 10 of these failure modes need to be considered when analyzing and designing a stable retaining wall-tieback system. These failure modes include:

1. Tension failure of the tieback tendon (i.e., the tieback rod or cable snaps!)
2. Tieback's anchor pulls out of the soil because adequate bond and passive resistance of the anchor to the soil was not generated.

3. The grouted anchor rod pulls out of the anchor's grout due to a lack of rod to grout bond strength.
4. Failure of sheet-pile or soldier-pile members in bending or excessive deflection due to high bending stresses.
5. Kick out of the toe of wall due to insufficient passive resistance of soil in front of the toe. This can be caused by installing piles in very soft, wet clay, then rapidly backfilling behind the wall causing settlement and excessive lateral shearing stresses due to excess pore water pressure, resulting in very low undrained shear strength of embedment soils. Excessive consolidation settlements in this case can also shear off the tieback anchor connection to the sheet pile wall.
6. Excessive leaning of the top of cantilevered pile prior to installing the first tiebacks. This is a construction sequencing problem that requires evaluation of the maximum allowable height of the cantilevered wall prior to installing tiebacks.
7. Failure of the piling in vertical bearing. This occurs when battered anchors and downward pile–soil friction generated during development of the active wedge apply an excess vertical loading on the piles causing them to punch into soils below the tip of pile.
8. Overturning failure of the sheet-pile wall plus anchor system acting as a whole. This type of failure is considered a global stability analysis problem, where the entire wall with tiebacks acts as a unit that slides and rotates outward.
9. Sliding block failure where the sheet piles and anchors act as a unit block. Again this type of failure is a global stability analysis problem that requires analysis of the entire wall system as a unit.
10. Rotational slide failure where a minimum factor of safety failure surface passes outside of and beneath the entire tieback-wall system. This tenth type of wall failure can affect any retaining wall system, because it is a deep-seated landslide problem that takes your retaining wall along for the ride.

4.3.3 Retaining Structures with Multiple Rows of Tiebacks

Analysis of retaining walls with multiple rows of tiebacks should never be done with a triangular active earth pressure acting on the wall. It has been known almost since the beginning of geotechnical engineering that the lateral earth pressure generated behind sheet-pile and soldier-pile walls with multiple rows of tieback anchors is a rectangular or trapezoidal stress distribution. You can find details of these stress distributions in early works by Terzaghi and Peck (Terzaghi and Peck, 1967), and in Tschebotarioff's textbook (Tschebotarioff, 1973). Discussions of retaining walls with multiple rows of tiebacks are also included in the US Army Corps of Engineers and NAVFAC publications that are available for free on the internet.

I recommend that you study and review FHWA-IF-99-015 (Sabatini, Pass, and Bachus, 1999) to help you decide which lateral earth pressure distributions to use for retaining walls with multiple rows of anchors. This reference gives updated stress distribution recommendations for sands and soft to medium clays, including the Henkel method, and it also compares and contrasts six alternate methods of determining lateral stresses in stiff to hard clays.

References

Sabatini, P.J., Pass, D.G., and Bachus, R.C. (1999) Geotechnical Engineering Circular No. 4, Ground Anchors and Anchored Systems, Report No. FHWA-IF-99-015, Office of Bridge Technology Federal Highway Administration, 281 pages.

Terzaghi, K. and Peck, R.G. (1967) *Soil Mechanics in Engineering Practice*, John Wiley & Sons, Inc., New York, N.Y., 729 pages.

Tschebotarioff, G.P. (1973) *Foundations, Retaining and Earth Structures*, 2nd edn, McGraw Hill Book Company, New York, 704 pages.

United States Steel Corporation (1969 and 1975) Steel Sheet Piling Design Manual, 133 pages both editions. Much of the material included in these USS manuals can be found in Wayne C. Teng's book, 1962, Foundation Design, Chapter 12, Prentice Hall, Englewood Cliffs, New Jersey, pp. 347–386.

5

Geotechnical LRFD

5.1

Reliability, Uncertainty, and Geo-Statistics

5.1.1 Introduction – Why Not Just Pick the Best Number?

When I started studying soil mechanics in the 1960s, we had plenty of theories and equations available to develop engineering problem solutions. The primary dilemma facing young geotechnical engineers was selection of the input "numbers" to use when solving a given problem. Fellow engineers told me at the time that with experience and a few basic field and laboratory tests, I could select the correct numbers to use to solve the client's problems. If I had any doubts about the soil parameters required for a particular problem, I could ask my mentor Neil Mason, who was a student of Karl Terzaghi, the father of soil mechanics himself!

As you might expect, I was considered to be a bit of a troublemaker by my co-workers. If I counted 12 blows per foot and the driller counted 11, which number did I use? If I observed the driller lifting the hammer 28 inches per blow rather than 30 inches per blow, what correction factor should I apply? If I ran a PI test and got 15, and the senior lab tech ran the same sample and got 18, do we each run a couple more samples and average the batch? How many PI tests should we run to get the right answer? I was continually asking questions about how much testing was enough testing.

These kinds of problems got me thinking, so I decided to pull out my college statistics text books (told you I was considered a troublemaker). The first thing that struck me in review of these math texts was the difference between integers and real numbers. What I was looking for in geotechnical practice were unique, single-valued parameters to put into my equations. But wait a minute, the only unique, single-valued "numbers" were natural, counting numbers such as 1, 2, 3, 4 or their negatives –1, –2, –3, –4, and of course zero. These unique integer numbers are

Geotechnical Problem Solving, First Edition. John C. Lommler.
© 2012 John Wiley & Sons, Ltd. Published 2012 by John Wiley & Sons, Ltd.

used for counting things, not for measuring things. When you measure something, or when you use two measurements to calculate something like the hypotenuse of a right triangle, the answer is not an integer number. It is a real number. The value of a real number can be estimated with increased precision by adding more and more decimal places, but it can never be determined exactly or uniquely because a real number is an infinite decimal number. Determination of a measured number's exact single value is impossible! No matter how hard you try, each measured number varies, depending on how you make the measurement.

What is the end result of the variation of the value of each measurement? Well, the conclusion we have to make is that there are no single-value numbers in geotechnical engineering. If you make 10 measurements of water content of a given clay sample and get the following results: 17.49%, 16.57%, 17.51%, 17.98%, 18.11%, 17.46%, 17.63%, 17.29%, 17.52%, and 17.83%, you can report the results in several ways. You could say that the water content of this clay varies from 16.57 to 18.11%, but then you would have to add that this was a variation of measurement and not of water content in the field. You could say that the water content of this clay has an average value of 17.54%. But then you might prefer to give the average water content value and include a description of the variation of the 10-value set. This is where a degree of complication can set into your analysis. How do you describe variation of measurements in a set? Of course there are several ways to describe the variation of a set of measurements, but a common method used involves calculation of the standard deviation. For the set of moisture content measurements given above the standard deviation is 0.425. I used my calculator to determine this value, but the equation that is used for determining the standard deviation is:

$$s_{N-1} = \sqrt{\frac{1}{N-1} \sum_{i=1}^{N} (x_i - \bar{x})^2}$$

You probably heard of the mean and standard deviation in your statistics class in college. Oh, you didn't take statistics in college. Did your advisor forget to tell you that you needed statistics? What, it wasn't required? You are not alone if you missed statistics in college. It seems that many engineers missed taking a statistics class. How did I know to take statistics before it became so darned important to engineers? I have to admit, it was just dumb luck. A mathematics minor was required to finish my masters degree. The only graduate math classes available during the summer were a series of statistics classes. It was my luck that the head of the mathematics department was a statistics professor, so he considered statistics as an appropriate study to complete an engineer's required math minor. Many engineering and mathematics professors at the time looked down on the study of statistics. They don't take statistics lightly these days, especially in geotechnical engineering!

Don't worry; we will be able to study the implications of uncertainty in geotechnical engineering without resorting to equations with multiple integrals and equations with dozens of terms. Understanding uncertainty in our field data, our laboratory data and our analysis methods will help you *see* the nature of complexity in the study and practice of geotechnical engineering.

5.1.2 How Do we Know that Our Designs Are Safe?

Recall back in Section 1.4, I discussed mistakes and errors, and pointed out that no matter how hard you try, no geotechnical determination will result in exactly the same number every time. I used the example of shooting an air rifle at a target. It was my intention to put all of the B-Bs through the center of the target, but reality was a pattern of holes that fit inside a circle about 1 3/8" in diameter, please refer to Figure 1.4.3. Let's assume that the horizontal scale on Figure 1.4.3 represents ultimate bearing capacity of a 4-foot square spread footing in kips per square foot, as shown in Figure 5.1.1 below. Also assume that each of the air rifle shots represent calculated estimates of ultimate bearing capacity derived from SPT blow count data from 11 test borings (I thought I took 10 shots and when I recounted just now I actually took 11 shots). Table 5.1.1 gives a list of results of the 11 ultimate bearing capacity estimates (i.e. shots) illustrated in Figure 5.1.1.

For this hypothetical case, assume that the structural engineer gave us a service column loading of 40 kips and an expressed desire to have an allowable bearing capacity of 2500 pounds per square foot. Given the request for an allowable bearing capacity of 2.5 ksf, what is our footing's factor of safety? Take a look at the third column in Table 5.1.1. If we use the average or median value based on the sample of

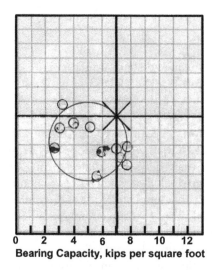

Figure 5.1.1   Eleven estimates of 4 × 4 footing bearing capacity

Table 5.1.1   Ultimate bearing capacities from Figure 5.1.1

Point Number	Calculated nominal bearing capacity, ksf	Factor of safety for a footing bearing pressure of 2.5 ksf
1	2.50	1.00
2	3.00	1.20
3	3.25	1.30
4	4.00	1.60
5	5.25	2.10
6	5.50	2.20
7	6.00	2.40
8	7.00	2.80
9	7.25	2.90
10	7.75	3.10
11	7.75	3.10
Average	5.39	2.16*
Std.Dev.	1.95	–
Median	5.50	2.20**

*FOS for 5.39 ksf; **FOS for 5.50 ksf.

11 tests, we calculate a factor of safety of about 2.2, which is less than 3.0 (as often used). A factor of safety of 2.2 sounds pretty good. You might be tempted to go with a recommended allowable bearing pressure of 2.5 ksf as the structural engineer desires. Make him happy, why not?

What if every one of the values given in Table 5.1.1 represents an actual ultimate bearing pressure somewhere on your site? If the 40 kip column and its footing end up over soil represented by points 8, 9, 10, or 11, everything will be just fine. If the 40 kip column and footing end up over soil represented by points 4, 5, 6, or 7, it may experience excessive settlement or even some punching shear action for soils at point 4. Finally, what might happen if the 40 kip column and footing end up over soil represented by points 1, 2, or 3? In the case of point 1, a bearing capacity failure is indicated, and points 2 and 3 will likely experience some punching shear and excessive settlement. We have a 1 in 11 chance of a bearing capacity failure, that's 9.1% chance of failure if the 11 data points in Figure 5.1.1 and in Table 5.1.1 are representative of soils on the tested building site. Another way of expressing this chance of failure is to say that the probability of failure is 9.1%.

But wait a minute, does that one service loading of 40 kips represent the complete range of potential loadings that may be experienced by this column during the life of the building? Sometimes the column loading may be less, but then again sometimes it may be more than 40 kips.

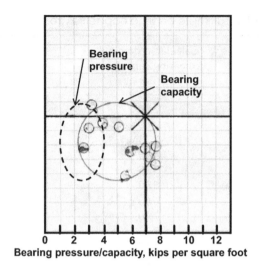

Bearing pressure/capacity, kips per square foot

Figure 5.1.2   Column 4 × 4 footing bearing pressure and capacity

For discussion purposes, assume that the area inside the dashed oval in Figure 5.1.2 represents the complete range of footing bearing pressures due to actual column loading cases that will be experienced during the building's lifetime. Rather than use 11 point estimates, we will assume that the area inside the circle in Figure 5.1.2 represents all of the possible bearing capacity values that a 4 × 4 footing may achieve on the proposed building site. Use of the oval shape represents the idea that the building column loadings are better known (or are less uncertain) and are focused around 2500 psf. The circular shape of bearing capacity values represents more variability, suggesting that the subsurface soil conditions and the method used to calculate the bearing capacity result in higher uncertainty.

Notice in Figure 5.1.2 that the oval representing column bearing pressures (i.e. soil loading) and the circle representing bearing capacity (i.e. soil resistance to loading) intersect. The area of intersection between the dashed oval and the solid circle in Figure 5.1.2 represents the combinations of footing bearing pressures and footing ultimate bearing capacities that result in foundation failures. The area of the solid circle to the right of the area of intersection represents cases where the ultimate bearing capacity exceeds all of the potential loading cases. The area of the dashed oval to the left of the area of intersection represents cases where footing bearing pressures are less than the minimum potential bearing capacity.

If all of the loading and resistance cases represented by the solid circle and the dashed oval have equal probability of occurrence, then the probability of foundation failure represented by the area of intersection between the two shapes in Figure 5.1.2 is calculated by dividing the area of the intersection by the sum of areas of the oval and circle. I roughly estimate that the probability of failure in this case is 0.352 square inches divided by 1.84 square inches or 1 in 5.23 chance, or 19.1%. It is intuitive to me from looking at Figure 5.1.2 that the probability of failure is greater when the loading cases and the ultimate bearing capacities are represented by two sets of values (represented by the oval and circular areas) rather than by a single given service loading bearing pressure value and a single ultimate bearing pressure represented by the average or median of 11 data points. Looking at the probability of failure in this case and realizing that it may be as high as 19% makes me much more nervous about the design of my 4 × 4 footing than the somewhat comforting factor of safety value of 2.20 based on an average of results shown in Table 5.1.1. It makes me think that we have used bearing capacity factors of safety of 3.0, 3.5, or 4.0 for good reason in the past! I am also reminded of footings that have settled excessively even though they were designed with a bearing capacity factor of safety of 3.0.

You as the geotechnical or structural engineer will have a better visualization and feel for the safety and reliability of your foundation system if you can represent foundation loadings and soil resistances as distributions of values and use simple statistical tools. You may not feel comfortable with statistics terms or statistical calculations. I recommend that you take a statistics course at your local university or attend a short course at an ASCE conference. When you are ready for the really meaty material, I recommend you purchase and study a copy of Fenton and Griffiths' book (Fenton and Griffiths, 2008).

5.1.3 What is Reliability and How is it Used in Design?

The word reliability is a common word. Most English speaking people would tell you that something is reliable when it does what it is supposed to do. Reliability is commonly considered to be synonymous with dependability, consistency, certainty, and trustworthiness. As you know well, common meanings of words and technical meanings of words are often different. As used in geotechnical engineering the word reliability is a measure of the probability of success or of *not* having a failure. In other words, reliability is the probability that a geotechnical design (or prediction) will perform as determined by geotechnical tests and analyses.

Reliability is related to the probability of success. If the probability of all outcomes is equal to 1.00 then reliability is equal to 1.00 minus the probability of failure. Please review Figure 5.1.3 from Fenton and Griffiths' book (Fenton and Griffiths, 2008), where the curve on the left represents the distribution of all loads and the curve on the right represents the geotechnical resistance. The small area of intersection between the two curves represents the probability of failure, and the area under the resistance curve to the right of the intersection of curves is a representation of the probability of success.

The peaks of these load and resistance curves are not equal to the mean value of the load and resistance because these distributions are not symmetrical normal distributions. Mean values of load and resistance, μ_L and μ_R are shown on

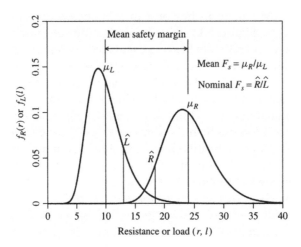

Figure 5.1.3 Load and resistance distributions (from Figure 7.5 of Fenton and Griffiths, 2008). With permission from John Wiley & Sons

Figure 5.1.3. The distance between the mean load and the mean resistance is a measure of the mean safety margin. The further apart the load and resistance distributions are and the smaller the intersection of the distributions, the safer the design will likely be.

The values L and R (with the top hats ^) in Figure 5.1.3 are quite interesting. The value L is the unfactored sum of conservative estimates of loads, similar to what we call the service loading. R is the resistance calculated by geotechnical formulas using conservative estimates of soil properties. The position of L and R in Figure 5.1.3 is an indication of the factor of safety obtained in current allowable stress design (ASD) practice.

The theory behind reliability-based design is that a design criterion can be developed to provide an acceptably low probability of failure.

Reliability-based design may sound like a rather simple concept, but mathematically the calculations required to develop a consistent method of maintaining an acceptably low probability of failure is quite complex. This mathematical complexity was resolved by development of the reliability index to simplify probabilistic design (Cornell, 1969).

When two structures or two components, like a column and its foundation, have the same reliability index, they have the same probability of failure. The reliability index β (Kulhawy and Phoon, 2002) is defined as:

$$\beta = \frac{\log_e \left(m_{FS} \sqrt{\frac{1+\text{COV}_F^2}{1+\text{COV}_Q^2}} \right)}{\sqrt{\log_e \left[\left(1+\text{COV}_F^2\right) \left(1+\text{COV}_Q^2\right) \right]}}$$

Table 5.1.2 Relationship between reliability index and probability of failure

Reliability Index, β	Probability of Failure, $p_f = \Phi(-\beta)$
1.0	0.16
1.5	0.07
2.0	0.023
2.5	0.006
3.0	0.001
4.0	0.00003
5.0	0.0000003

$\Phi(-) = $ standard normal probability distribution function.

where,

$\beta = $ reliability index (Note: β is a non-linear function of the probability of failure p_f, see Table 5.1.2).
$m_{FS} = $ mean factor of safety
$m_Q = $ mean of capacity Q
$m_F = $ mean of load F
$COV_Q = (s_Q)/(m_Q) = $ coefficient of variation (COV) of capacity (resistance)
$COV_F = (s_F)/(m_F) = $ COV of load
s_Q and $s_F = $ standard deviation of capacity and load.

The reliability index is the basis for development of the load and resistance factor design(LRFD) method discussed in Section 5.2.

5.1.4 Certainty and Uncertainty

When we're certain about something, we are sure of its value. When asked if the sun came up today, you can answer with 100% certainty, yes! If it didn't come up, I'm sure you would know, if you were still alive. If you weren't still alive how could anyone ask you the question? So certainty is knowing something for sure. Other examples of certainty are flipping a coin or throwing a dice. If you flip a coin, you are 100% certain that the coin will land heads or tails. If you throw a dice, you are 100% certain that the dice will land with a value of 1, 2, 3, 4, 5, or 6 on the up-facing side.

Uncertainty is not knowing for sure. When you flip a fair coin, you are uncertain as to the outcome, but you do know that the probability of heads is 50%. When you throw a dice, again you are uncertain as to the outcome, but you do know that the probability of getting a 3 is a 1 in 6 chance or 16.6%.

The geologic map shows that your site is in a glacial till area. When you drill a test boring at this site, you are quite certain that the test borings will encounter brown over gray, sandy, silty, clay glacial till soil. You are quite certain, but are you 100% certain that all test borings on your site will encounter glacial till. No, you are not 100% certain. Borings on your site could encounter a lens of gravel or a deposit of peat. You could be very unlucky, and the site may have been used as a disposal site for excess fill soil from adjacent properties, or it may even be an undocumented waste disposal site! So how certain are you that the site has undisturbed glacial till soil? Your certainty increases and your uncertainty decreases as your knowledge of the site increases. Unlike the coin flip or the dice throw, you may never know if an unknown pocket of fill or refuse exists on your site. Adding more and more test borings and soil tests decreases uncertainty, but it never eliminates it.

Statistical tools and the reliability theory are used to quantify uncertainty of structural loadings and foundation resistances. When uncertainty is quantified, we have numbers to use to develop tools and procedures that balance the probability of failure. Since engineers don't like to scare the general public, we don't express our designs based on a probability of failure. We use the probability of success or the reliability of a structure to communicate the positive aspects of our designs with clients and the public. We are in the process of transitioning between uses of factors of safety to the use of reliability index. The public knows safety is a good thing, but they are not sure what reliability is all about. Some engineers also have doubts about reliability analyses.

5.1.5 Factors of Safety and Reliability

Reliability analyses are characterized by some geotechnical engineers as involving complex theory and unfamiliar terms. A soon-to-be-classic (in my opinion) geotechnical paper in the ASCE *Journal of Geotechnical and Geoenvironmental Engineering* by Mike Duncan (Duncan, 2000) addresses how reliability analyses can be used in standard geotechnical practice. I suggest that you download this paper from ASCE's library (which will cost you a few dollars) or that you go to your local library and check out a copy of the April 2000 journal (if they don't have it, they can get it by interlibrary loan).

A summary of this paper is as follows:

- In standard geotechnical practice, the same value of factor of safety is applied to cases that involve varying degrees of uncertainty. This violates the notion that designs should have consistent reliability (i.e. consistent probability of failure).
- The concept of reliability can be applied simply, without additional test borings, laboratory testing, and with a bit of additional calculation time and effort.
- Methods are presented for estimating the parameters required for reliability analysis from standard geotechnical data. The primary method presented is the Taylor series method.

- Likely values of the soil parameters are used to calculate most likely values of the factor of safety.
- Standard deviation of the soil parameters used in calculation of the FOS are estimated by methods given in the paper.
- Using a spreadsheet, values of the FOS are recalculated using soil parameters increased by one standard deviation and then calculated using parameters decreased by one standard deviation. Soil parameters are varied one at a time, while keeping the other parameters at their most likely values.
- Taking the maximum and minimum factors of safety calculated by varying soil parameters one at a time, the range (FOS_{max} minus FOS_{min}) of each parameter's FOS is caclulated. Based on the range of FOS for each parameter the paper shows how to calculate the standard deviation and coefficient of variation of FOS for each parameter.
- Using the value of the most likely FOS and the value of the COV of factor of safety, the paper provides a table for determination of the probability of failure. Knowing the probability of failure helps the design engineer evaluate the uncertainty embedded in his geotechnical analysis. Knowing the size of the range in FOSs for each parameter gives an indication of the sensitivity of the analysis to each soil parameter.

Let's move on to Section 5.2 and see how reliability, uncertainty, and probability of failure are used to perform load and resistance factor designs (LRFD).

References

Cornell, C.A. (1969) A probability-based structural code. *Journal of the American Concrete Institute*, **66**(12), 974–985.

Duncan, J.M. (2000) Factors of safety and reliability in geotechnical engineering, American Society of Civil Engineers. *Journal of Geotechnical and Geoenvironmental Engineering*, **126**(4), 307–316.

Fenton, G.A. and Griffiths, D.V. (2008) *Risk Assessment in Geotechnical Engineering*, John Wiley & Sons, 461 pages.

Kulhawy, F.H. and Phoon, K.K. (2002) Observations on geotechnical reliability-based design development in North America. International Workshop on Foundation Design Codes and Soil Investigation in view of International Harmonization and Performance Based Design, Balkema, Netherlands, pp. 31–48.

5.2

Geotechnical Load and Resistance Factor Design

5.2.1 Limit State Design – General

Starting in 2007 the Federal Highway Administration (FHWA) required that all bridges built with federal funds (or partially funded by federal funds) had to be designed by the LRFD design procedures included in the LRFD Bridge Design Specifications (currently AASHTO, 2010). This included geotechnical LRFD design procedures. To my knowledge this is the first time that geotechnical LRFD was required to be used in the United States.

AASHTO's LRFD design approach used reliability theory (as introduced in Section 5.1) to quantify uncertainty in bridge loadings and resistances. In the AASHTO code there are many load combinations requiring analysis, and load factors are given to increase or decrease these loadings as required by reliability analyses.

The four general limit states considered by AASHTO are described as service, strength, extreme, and fatigue limit states.

Geotechnical/foundation designs generally involve the first three limit states. The service limit state is an evaluation of load-settlement performance of foundations.

Geotechnical strength design evaluates when a structure's loading exceeds the foundation soil's strength. Unlike settlement and deflection analyses, which consider soil movements and strains under service loading (i.e., everyday loading cases), strength analysis considers loadings that are highly unlikely, but are statistically possible at the upper end of loading cases.

Extreme loadings are once in a structure's life-time events such as the design earthquake, design hurricane winds, or extremely overloaded illegal trucks crossing your bridge.

Geotechnical Problem Solving, First Edition. John C. Lommler.
© 2012 John Wiley & Sons, Ltd. Published 2012 by John Wiley & Sons, Ltd.

Some clients like to say that engineers are pessimists, because they are always imagining bad things happening. Real pessimists always believe that they are being realistic. Engineers are not necessarily being pessimistic when considering strength and extreme events, rather they are considering the events that have a low or a very low probability of occurrence, but are not impossible. The reliability method used to generate factors for analyses of service, strength, and extreme loading events is called the load factor method, where we multiply predicted loadings by a number equal to or greater than one.

What soil shear strength do we use when the strength or extreme loading events occur? This can be a rather complex consideration because soils do not have constant shear strength. Some soils, such as soft saturated clays have a minimum strength when loaded, which is often called the "construction case." Other soils, such as stiff fissured clays or clay-shales, have a minimum strength many years after loading due to "soil softening or progressive failure." It is the geotechnical engineer's job to determine when critical loading-soil strength cases occur and analyze these cases. Such analyses often consider several cases with a range of loadings and strengths considered for each case. For each soil strength or soil resistance value there is a range of probable values and selection of an appropriate minimum value is necessary in analysis of a structure's performance. The resistance factor method reduces the calculated strength by multiplying predicted strength by a number less than one. Load and resistance factors and their use in geotechnical load and resistance factor design, that is, geotechnical LRFD is discussed in the following subsections.

5.2.2 Let's Stop and Think about this for a Moment

For years geotechnical engineers have been performing analyses and providing recommendations to structural engineers, architects and owners based on allowable stress design or ASD. Besides all of the high-minded talk about reliability and probability of failure, are there any down to earth and practical benefits to be had by changing from geotechnical ASD to geotechnical LRFD?

Let's stop and think about this for a moment. Earlier in this book I described the importance of foundation settlement and lateral deflection on the performance of a structure. I believe that the large majority of structures that fail to perform to the client's expectation have settled or deflected too much. Foundations that settled excessively have not experienced a strength failure; rather they have experienced a service failure.

In the past, using the ASD method, we applied a factor of safety of 3.0 to the ultimate footing bearing capacity calculated from the Terzaghi or Meyerhof equations to determine the allowable footing bearing pressure. We designed footings sized for the allowable bearing pressure and were done. What about the service loading settlement and the design earthquake's design bearing pressure? Geotechnical analyses of foundations bearing pressures did not include everyday loadings and once in a life-time loadings unless they were specifically requested to be considered by the structural engineer. When foundation load settlement analyses were requested during allowable stress design, there was always some confusion about what loadings to use in the analyses. Similarly when extreme, events such as the design earthquake, were analysed for foundation bearing stresses, there was some confusion about what factor of safety to use. Do we use an FOS of 3.0 or do we reduce the FOS to 2.0 because the extreme loading case is very highly unlikely to occur during the life of the structure. If we do reduce the FOS to 2.0 from 3.0, why didn't we reduce it to 2.25 or 1.75? What is the basis for our selection of a reduced factor of safety?

The benefits of using geotechnical LRFD analyses as required by the AASHTO Design Manual is that a system of factoring load cases and reducing soil/foundation resistances is clearly outlined. You need to check the foundation settlement for service loadings. You also need to check the factored strength and extreme case loadings against factored resistances. This system directs the structural and geotechnical engineers to work together to evaluate all of the applicable limit loading cases and it gives factors for use in the required analyses. If you believe that the specified resistance factors are excessively conservative, the LRFD method allows you to specify field load testing to reduce uncertainty and increase the resistance factor. Using the geotechnical LRFD method, you can also increase the number of supporting elements, thus providing alternative load paths and increased redundancy in structural support. The AASHTO code has provisions for increasing the resistance factor for increased foundation redundancy.

5.2.3 Geotechnical LRFD Design

The LRFD design equation compares factored loadings to factored soil resistances, see Equation 5.2.1 for a simplified form of the LRFD equation, considering live and dead loads, used in transportation design in the United States (AASHTO, 2010). In the 2010 AASHTO code there are 13 load combination limit states given, four service limit states, five strength limit states, two extreme events, and two fatigue limit states. Rather than copy all of the loading cases here, I suggest that you review AASHTO Section 3.3.2

$$(\gamma_{LL} * LL) + (\gamma_{DL} * DL) < \phi R_n \tag{5.2.1}$$

This equation basically says that if we factor (increase) the loads to include improbable events and factor (reduce) the nominal resistance R_n to cover uncertainty in soil resistance, then probability of a failure, that is, when the loads equal or exceed the resistance, is very low. Before I leave you with the impression that the load factor γ is always greater than one, there are cases where it is less than one.

How can a load factor have a value less than one? Examples of dead loads that both help resistance and at the same time have to be fully supported by the factored soil resistance are dead loads applied to a footing. A dead load helps increase sliding resistance, requiring use of a γ of 0.9 in sliding calculations, while the same dead load must be fully supported by vertical bearing pressure, requiring a γ of 1.25 in bearing capacity calculations.

Why do we refer to the ultimate resistance as the nominal resistance when using the LRFD method? This is a case of what do we know and how do we know it. We need to define the "nominal resistance," R_n. If you calculate the maximum or ultimate bearing pressure for a square footing using the Terzaghi bearing capacity equation, you have calculated the maximum loading that a real footing can support before experiencing failure . . . right? No, not really. What you have done with the Terzaghi bearing capacity equation is estimate the maximum bearing capacity of a real footing. The only way you can be 100% certain what a real footing can support is to construct a full-sized prototype footing on your site and load it to failure. Will the calculated Terzaghi bearing capacity value match the real load test capacity? It is not likely. In fact, Bowles' 3rd edition text (Bowles, 1982) on page 141 reports eight tests where several bearing capacity equations were compared to tested results. The Terzaghi equation's predicted ultimate bearing capacity varied from 56 to 114% of the tested values. Since there is uncertainty in the calculated maximum bearing capacity value, we call it the nominal bearing capacity, not the ultimate bearing capacity. The same goes for pile and drilled-shaft nominal resistance values, they are predicted values determined by analysis methods. Since each analysis method has unique characteristics, each method requires its own resistance factor ϕ.

References

AASHTO (2010) *AASHTO LRFD Bridge Design Specifications*, 5th edn, **2** volumes, American Association of State Highway and Transportation Officials, Washington, D.C.

Bowles, J.E. (1982) *Foundation Analysis and Design*, 3rd edn, McGraw-Hill Book Company, New York, 816 pages.

5.3

LRFD Spread Footings

5.3.1 LRFD and ASD Spread Footing Analyses – An Overview

Geotechnical LRFD analyses of spread footings are basically analyses of ultimate bearing capacity for the strength limit state and footing settlement for the service limit state. As we discussed earlier in Section 5.2, three primary loading conditions need to be analyzed to satisfy life-cycle loading analyses: service loadings for every day conditions, strength loadings for unlikely, but possible, overloading conditions that may result in structural failure, and extreme loadings representing once in a life-time events.

Current spread footing design practice is a mixture of both allowable stress design (ASD) and strength design (LRFD). Structural engineers estimate a range of wall and column loadings and give these loadings to the geotechnical engineer in the document requesting a geotechnical investigation. If building or bridge loadings are not given in the request for services document, the geotechnical engineer calls the structural engineer and asks for them. Many geotechnical engineers don't seem to be aware that the loadings received from structural engineers are service loadings. If the structural engineer is a young rather inexperienced engineer, he or she often gives the geotechnical engineer factored loadings when requested to provide building loadings. Using factored loadings to analyse footings and then applying a factor of safety (ASD) to the design, results in designing a footing with compounded conservatism.

I recall many times in the past asking my young geotechnical engineers whether they received service loadings or factored loadings from the structural engineer and they did not know. Upon investigation by calling a senior structural engineer at the structural firm, I often find that the young structural engineer gave my young geotechnical engineer factored loadings.

Continuing with the current footing design method description, by some established methods using soil information from field and laboratory testing and service

Geotechnical Problem Solving, First Edition. John C. Lommler.
© 2012 John Wiley & Sons, Ltd. Published 2012 by John Wiley & Sons, Ltd.

loadings provided by the structural engineer, the geotechnical engineer calculates an allowable bearing pressure that results in reasonably sized footings. If the estimated footing sizes are about 10-feet wide or greater, the geotechnical engineer calls the structural engineer and discusses the possibility of using piles or drilled shafts for building support. Assuming the probable footing sizes are reasonable, the geotechnical engineer calculates estimated footing settlements using the probable footing sizes and allowable bearing pressure. If the predicted footing settlements are all less than 1 inch or the predicted differential settlements are less than about $\frac{3}{4}$ inch, the geotechnical engineer is satisfied that the recommended spread footing allowable bearing pressure is suitable and gives his results and recommendations to the structural engineer in the geotechnical report. Whether the geotechnical engineer is aware of it or not, everything that he or she has done to this point is part of an allowable stress design (ASD) because he or she has essentially analyzed service loadings for settlement.

Upon receipt of the geotechnical report, the structural engineer looks up the recommended allowable bearing pressure and if it is reasonable, he or she can start spread footing design as soon as column and wall loadings are available. If the geotechnical engineer's recommended allowable bearing pressure is not reasonable to the structural engineer, he or she often calls the geotechnical engineer to complain about their conservatism and many times they convince the geotechnical engineer to "up their recommendation."

Using service loadings and the recommended allowable bearing pressure the structural engineer calculates the required area for column footings and footing width for wall footings. Up to this point in the design and sizing of spread footings, the structural engineer is also participating in allowable stress design (ASD). Next the structural engineer takes his factored loadings and the calculated spread footing size and computes a factored bearing pressure for use in his or her spread footing concrete design. At the point where the structural engineer calculates a factored soil bearing pressure, they have transitioned into strength design to be compatible with LRFD concrete design procedures. From this point forward in structural design all of the analyses are done in the LRFD format.

Given this explanation, you can see why I say that the current spread footing design process is a mixture of ASD and LRFD design procedures. Assuming that service loadings are used in geotechnical analyses and factored spread footing bearing pressures are used in concrete footing designs, current practice is a mixture of ASD and LRFD, whether the participating engineers are aware of it or not.

If you are interested in reviewing the procedures used by structural engineers to design concrete spread footings, I am aware of two good references published by the Portland Cement Association (PCA). The first one is *Notes on ACI 318-05 Building Code Requirements for Structural Concrete with Design Applications* (Kamara and Rabbat, 2005). Chapter 22 of this book discusses design of footings and gives detailed examples of the concrete footing design process, including a clear description of calculating factored soil bearing pressure for use in LRFD concrete design. I might add that the older I get the more I turn into an editor. On the bottom of page 22–25 in this reference when describing the area A_2, the text should read *2 vertical to 1 horizontal* not 1 vertical to 2 horizontal.

The second book from PCA that clearly describes the factoring of allowable soil bearing pressure for concrete footing design is titled *Simplified Design: Reinforced Concrete Buildings of Moderate Size and Height* (Alsamsam and Kamara, 2004). I told you that design was becoming more and more complex as time goes on, and structural engineers like the rest of us are always on the outlook for references and methods that "simplify" their lives. Chapter 7 of this book is devoted to simplified design of footings. I expect that geotechnical engineers reading the footing design description in Chapter 7 will bristle at the material included in Section 7.3 which suggests that structural engineers may simplify footing design by foregoing geotechnical investigations and using local building code values for allowable soil pressures. Forgoing a geotechnical investigation may make footing design simpler, but it will also fatten the wallets of attorneys suing structural engineers for failed structures built over undetected waste dump sites!

5.3.2 A Spread Footing LRFD Design Approach

Requirements for geotechnical LRFD spread footing analyses are included in the 2010 AASHTO Bridge Design Specifications Section 10.6. When describing LRFD design of spread footings, the 2010 AASHTO manual states, "Spread footings shall be proportioned and designed such that the supporting soil or rock provides adequate nominal resistance, considering both the potential for adequate bearing strength and the potential for settlement, under all applicable limit states in accordance with the provisions of this Section" (i.e., Section 10.6). I'm tempted to do some editing of my own on this wordy, run-on sentence, but then it is a quote from a government document!

AASHTO requires, and as a practicing geotechnical engineer I prefer using an effective footing area to account for overturning moments and eccentrically applied foundation loadings in accordance with the Meyerhof method. When using the effective footing area, B' wide by L' long, the applied loadings are assumed to act at the centroid of the reduced effective area. Recall from Section 4.2, Equation 4.2.2 that the reduced footing dimensions are defined as:

$$B' = B - 2e_B \quad \text{and} \quad L' = L - 2e_L \qquad (5.3.1)$$

where,

e_B = eccentricity in the B direction
e_L = eccentricity in the L direction.

When calculating footing settlements for service limit states and bearing resistance for strength limit states, the reduced footing area based on B' and L' is used.

AASHTO commentary 10.6.2.1 suggests that spread footing design is frequently controlled by service limit state settlements. They also suggest that it is "advantageous" to size footings for service limit states and then check the strength and extreme limit states. I agree. This approach of checking service loadings roughly follows the footing design approach previously followed by geotechnical and structural engineers for decades, as described above in Section 5.3.1.

As I have mentioned before and will mention again, geotechnical and structural engineers must work together to assure that both geotechnical and structural footing design concerns are resolved.

Starting with the service loading analyses first, the geotechnical engineer should develop load versus settlement charts for service loadings for a range of footing sizes. These charts are used by the structural engineer to size footings for each column, wall, abutment, and so on. With the preliminary footing sizes based on service loadings and settlement limitations, the structural engineer calculates factored loadings for each applicable strength limit state and calculates the footing bearing pressure for each case. Given the preliminary footing sizes and factored bearing pressures for strength limit states, the geotechnical engineer calculates the nominal and factored bearing pressure resistance. Comparing the factored loading footing bearing pressure with the factored bearing pressure resistance, the preliminary sizes of footings are checked to see if the footings are large enough to satisfy strength limit state loading cases. If the footings are large enough so that their factored bearing resistances are greater than the factored bearing pressures, footing concrete and reinforcing are designed and the footings are checked for extreme loading cases.

5.3.3 Development of Spread Footing Load-Settlement Curves

I would like to give my friends at the Arizona Department of Transportation's Geotechnical Design Section a tip of the hat. They have issued an LRFD design policy document SF-1 (ADOT, 2010a) for design of spread footings that is very helpful in showing consulting geotechnical engineers how to prepare LRFD foundation design charts for use by bridge designers. I highly recommend that you go on the internet and find a copy of this document or contact the Geotechnical Design Section at ADOT and request a copy. I might add that they have several design policy documents for spread footings and for drilled-shaft LRFD designs that illustrate how geotechnical engineers can report the results of their analyses in formats that are beneficial to bridge design engineers. Figure 5.3.1 is a copy of the example "Factored Bearing Resistance Chart" for spread footing design from ADOT's Geotechnical Design Policy SF-1.

Figure 5.3.1 includes a lot of information in one small package. The horizontal axis is effective footing width in feet, applying the Meyerhof method, as given above in

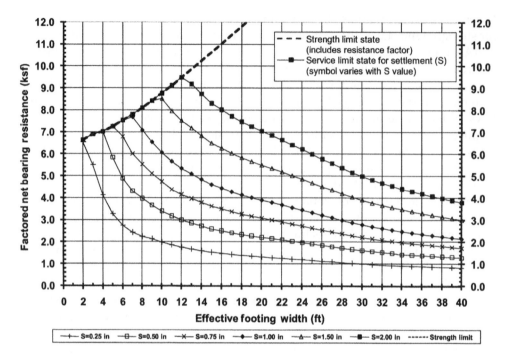

Figure 5.3.1 Spread footing factored bearing resistance chart example (ADOT, LRFD Geotechnical Design Policy Document SF-1, 2010)

Equation 5.3.1. The vertical axis is factored net bearing resistance in kips per square foot. By net bearing resistance, we mean that the nominal bearing resistance has been reduced by subtracting the overburden stress at the base of footing level. The factored net bearing resistance is the net bearing resistance times the appropriate resistance factor $\varphi_{bearing}$.

The rising bold dashed line in the upper left corner of Figure 5.3.1 is the factored net bearing resistance for the strength limit state using a resistance factor φ_b of 0.45 because this example is for SPT data in sandy soil. The family of falling curves running from the upper left to the lower right side of Figure 5.3.1 represents the service limit state where the factored nominal net bearing resistance is given for selected settlements. In this example the service load settlements range from 0.25 to 2.0 inches. I might add that these factored nominal net bearing resistances for service loadings equal their nominal net bearing resistances because the resistance factor for the spread footing service limit state, φ_b equals 1.0.

You might now be asking yourself, how can we plot strength and service loading cases on the same plot in Figure 5.3.1? The strength and service limit states are different, but we can benefit by plotting them on the same figure so that service and strength cases can be checked by the project structural engineer at the same time.

Settlement values given for the family of curves shown in Figure 5.3.1 are immediate settlements. How do we calculate settlement values used to plot service limit state curves like those shown in Figure 5.3.1? This is where difficulties and complexities start to enter the spread footing analysis process.

We know from earlier discussions that footing settlements may develop as immediate/elastic settlements, consolidation settlements, or secondary compression (i.e., creep) settlements. To decide what settlements to compute, you need to have an idea of when in the life of a structure footing settlements are important to structural function. If settlements don't cause structural distress, and if people don't see them, who cares? Oh, you would be surprised who cares.

After you have determined what types of settlements are important, you need to roughly determine how accurately you need to estimate footing settlements.

5.3.4 Development of a Spread Footing Service and Strength Resistance Chart

To develop a spread footing service limit state chart like that given in Figure 5.3.1, you first have to decide what calculation method to use in estimating footing settlements. After you decide on a settlement calculation method, you have to solve the settlement equations backwards, in other words, you select a settlement for a given footing width, and back calculate the bearing pressure that is required to generate the selected settlement. Doesn't that sound relatively simple?

AASHTO 2010 manual section 10.6.2.4.2 indicates that you can calculate immediate settlements in sands by use of elastic theory or by an empirical method proposed by Hough (Hough, 1959). I expect that many engineers would pick one of these two methods and go with it. Why not? These methods are specified by AASHTO so their use is in compliance with government requirements. You might believe that this use of AASHTO-specified methods of calculating spread footing settlements is prescriptive design requiring compliance by the geotechnical engineer, but it is not.

Both AASHTO 2010 and FHWA 2006 (Samtani and Nowatzki, 2006) suggest that empirical methods such as the Hough method may over-estimate spread footing settlements by a factor of two or more, resulting in use of costly deep foundations when spread footings may have performed satisfactorily. First AASHTO 2010 specifies a settlement calculation method to be used, and then they tell you that it is likely too conservative. What is this all about? To explain, let me first share ADOT's approach to the conservatism issue, then let me point out the small print in AASHTO 2010.

ADOT spread footing LRFD design guidance outlined in SF-1 repeats the wording included in FHWA 2006 about conservatism of the Hough method, and then they propose using the Schmertmann method for calculating immediate settlements of spread footings as outlined in Section 8.5.1 of FHWA 2006. Schmertmann's method is generally considered to be more precise than the Hough SPT chart method, although as I mentioned earlier even the Schmertmann method has significant limitations when correlated CPT values are used (among other limitations). Schmertmann originally proposed his method of calculating settlements of spread footings on sands in 1970. In 1978 Schmertmann revised his method to increase the soil modulus calculated from cone penetration data. ADOT recommends using the 1978 Schmertmann method as described in FHWA 2006, although they recommend using SPT-correlated values of soil modulus. In Section 3.2 above I have given you a rather detailed description of the 1978 Schmertmann method for calculating settlements in sandy soils.

Although I am aware that CPT equipment is rather scarce in many parts of the United States and that SPT equipment is commonly used in geotechnical investigation work, I still believe that you should use CPT equipment whenever possible to achieve better settlement estimates with the Schmertmann method.

I might also add that for LRFD service loading cases FHWA 2006 recommends that you limit use of the time factor to 0.1 year for immediate settlements and 1.0 year for secondary compression or creep settlements. They also recommend that you use a time factor C_2 equal to 1.0 if you are also calculating consolidation settlements.

Based on my review of FHWA 2006 and ADOT LRFD spread footing design guidance, it is apparent that they suggest a refinement of spread footing settlement calculations for bridge structures over the Hough method to improve accuracy of spread footing settlement estimates. They suggest that the Hough method may be more suitable for calculating embankment fill immediate settlements.

In AASHTO 2010 LRFD manual Section C10.6.2.4.2, they mention that refined settlement calculations may be required, and they give several additional references for other settlement calculation methods. AASHTO 2010 also recommends in Section 10.4.6.3 that "where evaluation of elastic settlements is critical to the design of the foundation ... in-situ methods such as PMT (i.e., the pressuremeter test) or DMT (i.e., the dilatometer test) for evaluating the modulus of the stratum should be used." The only reason that I can imagine for performing pressuremeter or dilatometer tests is to obtain improved estimates of soil moduli to improve calculated spread footing settlements.

Again when considering methods of computing spread footing settlements, we are presented with an application of the graded approach. If empirical methods provide suitable prediction accuracy for our design purposes they are appropriate to use. If advanced tests and analysis procedures are required to fulfill our design purposes then they are more appropriate to use.

5.3.4.1 Calculation of the Strength Limit State for Spread Footings

To plot a strength limit state line on the factored bearing resistance chart shown in Figure 5.3.1, you have to calculate the nominal bearing resistance for a series of footing widths. Calculation of the strength limit state for spread footings requires an equation for calculating of the nominal bearing resistance (what we used to call the ultimate bearing capacity). Like we have done for decades, the nominal bearing resistance of spread footings is calculated using the bearing capacity equation (what some folks call the 3 Ns equation), see Equation 5.3.2 below.

$$q_{\text{nominal}} = c N_c s_c i_c + q N_q s_q d_q i_q C_{wq} + \frac{1}{2} \gamma B N_\gamma s_\gamma i_\gamma C_{w\gamma} \qquad (5.3.2)$$

where,

c = cohesion or undrained shear strength (kips per square foot, ksf)

N_c = cohesive bearing capacity factor (dimensionless)

N_q = surcharge bearing capacity factor (dimensionless)

N_γ = unit weight (of soil below footing) bearing capacity factor (dimensionless)

$q = \gamma D_f$, q is the surcharge pressure of soil above the bearing surface of the footing, if there is no applicable surcharge loading on the ground surface, then q is calculated from γ the moist unit weight of soil (kips per cubic foot) above the bearing level times D_f the depth of footing in feet (q is expressed in units, ksf)

γ = total moist unit weight of soil below the footing bearing surface in the third term of the bearing capacity equation (kips per cubic foot, kcf)

B = footing width (feet)

$C = C_{wq}$ and $C_{w\gamma}$ ground water table correction factors (dimensionless)

$s = s_c, s_q$, and s_γ are shape correction factors, since the bearing capacity equation was originally formulated for strip footings not square or rectangular footings (dimensionless)

d_q = depth correction factor since the shearing resistance of the soil between the bottom of footing and ground surface was not originally included in the bearing capacity equation (dimensionless)

$I = i_c, i_q$, and i_γ are inclination factors to correct for loadings that are not vertically oriented or do not act perpendicular to the footing bearing surface (dimensionless).

If you intend to determine the nominal bearing resistance of spread footings for bridge or transportation-related structures, I recommend that you refer to the AASHTO 2010 manual or to the FHWA 2001 shallow foundations manual (Munfakh *et al.*, 2001). If you are designing a spread footing for a building using geotechnical LRFD, you are free to use whichever bearing capacity equation you deem fit. Personally I am partial to the Meyerhof version of the bearing capacity equation.

After you calculate an appropriate value for the nominal bearing resistance you need to select a resistance factor that is appropriate for the soil type, field testing type, laboratory testing, and the bearing capacity equation that you used. Current AASHTO 2010 guidance for selection of a spread footing strength resistance factor is given in their Table 10.5.5.2.2–1. In this AASHTO table for the spread footing strength limit state, they give resistance factors that range from 0.45 to 0.50. AASHTO also indicates that you can use a resistance factor of 0.55 when a plate load test is used to determine the bearing resistance. I wonder ... what would the resistance factor be if you did a full-scale footing load test to failure? If we conducted, properly designed, monitored, full-scale footing load tests, I would lobby for an increased resistance factor similar to increased factors allowed when you perform full-scale pile or drilled-shaft load tests.

Of course AASHTO gives you guidance on selecting spread footing bearing resistance factors, φ_b, and that's fine, but I recommend that you periodically check geotechnical journals for current papers on this topic because a significant amount of research is currently ongoing in this field.

Wait a moment, what did I say in the first sentence two paragraphs above? "After you calculate an *appropriate value* of the nominal bearing resistance ..." Are you going to let me get away with a comment like that? What "appropriate value" are we talking about? Very good question, I'm glad you asked.

The nominal bearing resistance or ultimate bearing capacity calculated by the 3 Ns bearing capacity equation, Equation 5.3.2, requires values of the soil cohesion, soil internal friction angle and soil moist unit weight. I hope you noticed that we have two values of the moist unit weight in play in this equation. The second term, called the surcharge term, uses the moist unit weight of soil above the footing's bearing elevation to calculate the surcharge q, and the third term, called the gamma term, uses the moist unit weight of soil below the footing's bearing elevation. Now what about the soil shear strength parameters c and ϕ?

Standard practice indicates that for sandy, cohesionless soils, we should use effective stress or drained soil shear strength parameters in the bearing capacity equation. Laboratory tests that give drained shear strength parameters c' and ϕ' would be the direct shear test and the CD (consolidated drained) triaxial test. There are also correlated effective stress values of friction angle reported for CPT and SPT tests that you might consider using to find ϕ'.

For spread footings on clayey, cohesive soils, standard practice indicates that a total stress analysis with undrained shear strength parameters should be used in the bearing capacity equation. Tests used to find the undrained shear strength of clayey soils includes the unconfined compression test and the UU (unconsolidated undrained) triaxial test (i.e., $s_u = c_u = q_u/2$).

This is all fine, but please don't forget that the drainage condition and stress path of the state of stresses that you are analyzing controls the actual shear strength of your foundation soils. For example, if your foundation is supported by clayey soil and is loaded slowly enough to avoid generation of excess pore water pressures, you should use effective stress parameters c' and ϕ' in your bearing capacity calculations. If on the other hand your foundation is supported on saturated silty, fine sand and the anticipated loading is an extremely quick dynamic loading, you should use undrained shear strength parameters in your bearing capacity calculations. One more case, what if your foundation soil is clay shale or stiff, fissured clay that looses strength with time due to wetting and dissipation of negative pore water pressures (i.e., loss of suction and dry strength)? For this case, you should use softened, effective shear strength parameters in your bearing capacity calculations, that is, soil shear strength parameters which model the soil's softened stress state at a future time. The three telescope spread footing foundations shown in Figure 5.3.2 bear on select compacted granular fill.

Figure 5.3.2   Three large telescope spread footing foundations

5.3.5 Other Spread Footing LRFD Considerations – Eccentricity and Sliding

Although analyses of spread footing settlement and bearing resistance are of primary importance when doing LRFD design of spread footings, we still have to consider foundation stability, which includes considerations of overturning moments and lateral sliding forces.

5.3.5.1 Overturning Moments – Limiting Eccentricity

Back when allowable stress design was king, we used to have a rule of thumb that the resultant force on a footing should act within the middle third of the footing base. The middle third was defined as $B/3$ where B was equal to the footing width. The idea behind this middle third concept was simple. If the resultant force on a footing acted outside the middle third of its base, the edge of the footing would start to lift off the ground. Somewhere along the line, engineers in charge decided that use of the phrase "resultant in the middle-third" was confusing. In its place they declared that the eccentricity of the loading acting on a footing should not exceed $B/6$, where eccentricity was defined as the overturning moment acting on the footing divided by the vertical loading acting on the footing. I guess this was simple enough. They were just applying the rule of statics that says that two statically equivalent systems have the same forces and moments acting. But wait a moment, the allowable eccentricity $B/6$ acts on both sides of the center of the footing resulting in a width of permissible eccentricity of two times $B/6$ which is equal to $B/3$. Use of the eccentricity definition resulted in the same criterion as the earlier middle third rule.

So long as we were using ASD with service loadings, the middle third or the allowable eccentricity of $B/6$ was the limiting criterion for application of overturning moments (i.e., $M/V \leq B/6$) to a spread footing. Now we are factoring applied footing loadings for the strength limit case in LRFD design, what limiting eccentricity do we use for factored loadings?

AASHTO 2010 in Section 10.6.3.3 says that the eccentricity of loading for the strength limit state with factored loads should not exceed $B/4$. Use of an eccentricity of $B/4$ would be the same thing as using the middle-half for earlier limiting eccentricity criteria. Physics hasn't changed; won't an eccentricity of $B/4$ result in footing edge lift. The answer is yes. This makes me wonder if AASHTO is anticipating a limited amount of footing edge lift or are they considering the limited probability that fully factored loadings will actually occur. In the commentary to AASHTO 2010 Section 10.6.3.3, they have the following statement, "A comprehensive parametric study was conducted for cantilevered retaining walls of various heights and soil conditions. The base widths obtained using the LRFD load factors and eccentricity of $B/4$ were comparable to those of ASD with an eccentricity of $B/6$." I'm not sure. I have some doubts about this commentary statement. This sounds like an economic decision to me.

This $B/4$ versus $B/6$ eccentricity situation piqued my curiosity about what the Arizona Department of Transportation might be doing. Sure enough, ADOT has issued a geotechnical design policy document SF-2 titled "Limiting Eccentricity Criteria for Spread Footings based on Load and Resistance Factor Design (LRFD) Methodology." ADOT took a look at AASHTO's $B/4$ eccentricity limitation and checked it against their retaining wall standards and concluded that the $B/4$ limitation would result in an increase in footing widths of about 15 to 20% for standard walls that previously met the ASD footing width limitations. I didn't call my friends at ADOT to hear their reasoning, but it doesn't take much imagination to see that increasing the footing width of all wall and bridge footings in the State of Arizona that had eccentric loadings by 20% would cost a lot of money.

In ADOT document SF-2 (ADOT, 2010b) they proposed their own limited eccentricity criteria based on calibration of past ASD wall design practice using LRFD load factors in Section 3 of AASHTO 2010. Again, I suggest that you obtain a copy of ADOT's document SF-2, but below I have included a sample of their proposed limiting eccentricity criteria:

"The eccentricity, e, of loading at the strength limit state, evaluated based on factored loads shall meet the following limits:

- For footings on soils: e $\leq$ B $[(1/3) - (\beta/320)]$
- For footings on rocks: e $\leq$ B $[(3/7) - (\beta/500)]$

where,

B = the footing dimension (width or length) in which the eccentricity is being evaluated,

β = the backslope inclination angle of the soil retained behind the wall in degrees with respect to the horizontal. The maximum limit on β is 26.56 degrees (i.e., 2H : 1V slope; H = Horizontal, V = Vertical). In addition, the slope shall satisfy the minimum slope stability requirements for the project.

The eccentricity, e, computed by the above equations has the same units as B."

ADOT has several limitations on the application of their proposed eccentricity criteria, such as granular backfill only, active earth pressures only, and so on, which I will not list here, but again I suggest you look up their SF-2 document.

5.3.5.2 Spread Footing Sliding Failure

Failure of a footing by sliding in LRFD methodology is considered to be a strength limit case. Factored lateral loadings are compared with factored sliding resistances developed from sliding friction resistance and passive sliding resistance. As I mentioned earlier in Section 4.2 when discussing sliding resistance of gravity and cantilevered retaining walls, the passive soil resistance must be present throughout the life of the footing to be included in sliding resistance.

Apparently the reliability of base frictional resistance is considered to be much higher than passive soil resistance to sliding, because resistance factors given in AASHTO Table 10.5.5.2.2-1 range from 0.8 to 0.9 for frictional sliding resistance and are given as 0.5 for passive soil resistance. Passive resistance may not be highly reliable, but you also have to remember that large lateral deformations are required to mobilize full passive resistance, while relatively small lateral deformations are required to mobilize frictional resistance to sliding. Without doing strain-compatible calculations, I would expect a greater reduction in passive resistance when it is added to frictional resistance.

References

AASHTO (2010) *AASHTO LRFD Bridge Design Specifications*, 5th edn, **2** volumes, American Association of State Highway and Transportation Officials, Washington, D.C.

ADOT Geotechnical Design Section (2010a) *Geotechnical Design Policy SF-1, LRFD, Development of Factored Bearing Resistance Chart by a Geotechnical Engineer for Use by a Bridge Engineer to Size Spread Footings on Soils based on Load and Resistance Factor Design (LRFD) Methodology*, December 1, 2010, 15 pages including the cover letter.

ADOT Geotechnical Design Section (2010b) *Geotechnical Design Policy SF-2, LRFD, Limiting Eccentricity Criteria for Spread Footings based on Load and Resistance Factor Design (LRFD) Methodology*, December 1, 2010, 7 pages including the cover letter.

Alsamsam, I.M. and Kamara, M.E. (2004) *Simplified Design, Reinforced Concrete Buildings of Moderate Size and Height, Chapter 7 Simplified Design for Footings*, 3rd edn, Portland Cement Association, Skokie, Illinois, 19 pages.

Hough, B.K. (1959) Compressibility as the basis for soil bearing value, Paper No. 2135. *Journal of the Soil Mechanics and Foundations Division, American Society of Civil Engineers*, **85**(SM4).

Kamara, M.E. and Rabbat, B.G. (2005) *Notes on ACI 318-05 Building Code Requirements for Structural Concrete with Design Applications, Chapter 22 Footings*, Portland Cement Association, Skokie, Illinois, 22 pages.

Munfakh, G., Arman, A., Collin, J.G., *et al.* (2001) *Shallow Foundations Reference Manual*, FHWA-NHI-01-023, Federal Highway Administration, United States Department of Transportation, Washington, D.C.

Samtani, N.C. and Nowatzki, E.A. (2006) *Soils and Foundations Reference Manual, Volumes 1 and 2*, FHWA-NHI–06-089, **2** volumes, National Highway Institute, Federal Highway Administration, Washington, D.C., 1056 pages.

5.4

LRFD Pile Foundations

5.4.1 LRFD Piles – Overview

Unlike drilled shafts which are basically large drilled piles, the topic of pile foundation design and construction includes drilled piles, driven piles, rammed concrete and aggregate piles, and dozens of patented pile-installation systems.

Both geotechnical ASD (allowable stress design) and LRFD (load and resistance factor design) designs of piles includes analyses of ultimate load capacity and settlement of individual piles and pile groups. Frequently piles are load tested by static and dynamic methods to verify load-settlement characteristics, which reduces uncertainty and allows use of smaller factors of safety or larger resistance factors.

Differences between geotechnical ASD (Bowles, 1982 and 1996) and LRFD design and analysis requirements for piles are partly terminology and partly analysis methods used. As we discussed above in Section 5.2, three primary loading conditions need to be analyzed: service loadings, strength loadings, and extreme loadings. Pile foundations are designed for ultimate pile capacity (called nominal load capacity in LRFD) for strength loadings and checked for service and extreme loadings. An illustration of pile driving using a vibratory hammer is given in Figure 5.4.1 below.

In geotechnical LRFD design the ultimate load capacity of a pile is called the nominal load resistance of the pile. To calculate the nominal load resistance of a pile, we calculate the sum of pile side friction (skin friction) resistance and the pile tip resistance, as we discussed in Section 3.5, see Equations 3.5.1, 3.5.1a, and 3.5.1b. For allowable stress design, we divided the total pile nominal resistance by a factor of safety of three or less, depending on the extent of pile load testing performed. For geotechnical LRFD design of piles, we have to apply separate factors to the pile side friction and tip resistances. These factors are called resistance factors, $\varphi_{side\ friction}$ and φ_{tip}. Use of separate resistance factors for side friction and tip resistances allows us to tailor our resistance values to account for installation techniques, soil types and many

Geotechnical Problem Solving, First Edition. John C. Lommler.
© 2012 John Wiley & Sons, Ltd. Published 2012 by John Wiley & Sons, Ltd.

Figure 5.4.1 Checking plumb before driving, vibratory hammer shown

factors contributing to uncertainty. If resistance factors are selected appropriately and are based on statistically significant data, the result should be increased economy of pile installations with increased certainty of proper pile performance.

5.4.2 Geotechnical LRFD Codes for Piles and Drilled Shafts

As discussed in Section 5.2.3 geotechnical LRFD design in the United States is primarily driven by requirements of the Federal Highway Administration (FHWA) for design of bridges. To the best of my knowledge there are currently no comprehensive geotechnical LRFD codes for buildings and other structures in the United States (although I understand that geotechnical LRFD building codes are present in Canada and Europe).

The current applicable FHWA code for geotechnical LRFD is included in the 2010 American Association of State Highway and Transportation Officials LRFD Bridge Design Specifications, 5th Edition, which I will call AASHTO 2010 (AASHTO, 2010). Piles, drilled shafts and micropiles are covered in Sections 10.7, 10.8 and 10.9. It is clear to me after review of AASHTO 2010 that they give a full measure of coverage to driven piles in Section 10.7, good coverage to drilled shafts in Section 10.8, and they give the "short end of the stick" to micropiles in Section 10.9. I am not sure why micropiles receive brief coverage, but I imagine it is because micropiles are not often used in large transportation projects, and because many micropile types are patented commercial products.

When reviewing the AASHTO 2010 coverage of LRFD pile foundation design, my first question was where is the driven-pile section? It's in AASHTO 2010 Section 10.7, check. My second question was where is the drilled-pile section? Oh my, I didn't see a separate section of AASHTO 2010 for drilled piles or augercast piles. Not seeing a separate AASHTO 2010 section for drilled piles, I turned to the drilled-shaft section 10.8. In Section 10.8.1.1, in the third paragraph, AASHTO 2010 says that, ". . . provisions of this Section shall not be taken as applicable to drilled piles, for example, augercast piles . . . ," so it is clear that AASHTO 2010 does not cover drilled piles in their section 10.8.

Review of the micropile Section 10.9 of AASHTO 2010 reveals that drilled and augercast piles are defined as Type A and Type B micropiles (even though augercast piles up to 60 inches in diameter don't seem to fit the description "micro"). Drilled piles and augercast piles are given resistance factors in AASHTO 2010's Table 10.5.5.2.5-1 and you are directed to portions of the driven pile Section 10.7 for most analyses required for drilled and augercast piles.

Given the lack of specific LRFD material for drilled piles, I will concentrate on driven piles here in Section 5.4 and on drilled shafts in Section 5.5.

5.4.3 Development of a Driven-Pile Axial Strength Resistance Chart

Development of an axial strength resistance chart for driven piles involves calculating the driven-pile side resistance and end bearing (or tip) resistance, selecting resistance factors based on the amount of field load testing anticipated, factoring side and end bearing resistances, and plotting their sum versus depth. Soil types encountered by each driven pile dictate the permissible methods for calculating side friction and end bearing resistance. Details of calculation methods for determining side friction and tip resistance are given in Section 3.5.2 of this book (Coyle and Castello, 1981, Hannigan et al., 2006). AASHTO 2010 guidance for selecting methods to use for calculating side and tip resistances are outlined below in Table 5.4.1.

To prepare an axial strength resistance chart, you first have to prepare a representative soil profile from your available test boring and CPT tests. Then, using available laboratory data and correlated values from corrected SPT and CPT data, you determine soil shear strength parameters for each of the soil layers in your soil profile. Using methods described in Section 3.5.2, you calculate the side resistance and end bearing resistance at the center of each soil layer. Referring to the field sampling method, laboratory testing method, soil type, analysis method used, and likely pile load testing to be performed, you next determine the appropriate load resistance factors to use for side friction resistance and end bearing for each soil layer; please refer to

Table 5.4.1   Methods for determining driven-pile nominal bearing resistance

Analysis or Test Type	Soil Type	Comments
Static load test	All types	Best method for predicting pile load-settlement and pile nominal axial resistance
Dynamic testing	All types	Economical method for testing many piles during construction, can determine pile driving criteria, and can check pile setup and driving damage.
Wave equation analysis	All types	Used to predict pile driving performance with selected hammer type.
Dynamic formulas	All types	Easy to use but not highly reliable
α-Method	Clays	Used to calculate side adhesion resistance from undrained shear strength (S_u) of clayey soils.
β-Method	Clays	Effective stress method for predicting nominal side resistance in normally consolidated clays
γ-Method	Clays	Used to calculate side resistance in cohesive soils
Nordlund method (Nordlund, 1963, 1979)	Sands	Effective stress method for calculating side and tip resistance of piles driven into sands and non-plastic silts
Tip resistance equations	Sands and clays	For total stress analyses of clayey soils the nominal tip resistance is taken as $9S_u$ and for mixed soils bearing capacity equations or correlated values to SPT or CPT are used.

AASHTO 2010 Table 10.5.5.2.3-1 for driven pile resistance factors. After multiplying side friction and tip resistances by the appropriate load resistance factors, you have determined the factored nominal resistances for your given driven pile. Finally, you plot factored nominal axial resistance versus depth below top of pile for use in pile foundation design by the structural engineer.

But wait, what if you are required to give the structural engineer three or four or more alternate sizes of piles for use in his foundation design? Then you need to repeat this process for each of the pile sizes recommended for the project so that he or she can choose between the alternate pile sizes.

5.4.4 Development of a Driven-Pile Axial Service Resistance Chart

Analysis of the service limit state of driven-pile foundations includes evaluation of pile vertical load-settlement, pile group vertical load-settlement, pile lateral load-deflection, pile group lateral load-deflection, and pile group stability (it could be located on or near a slope with stability issues). For LRFD design of piles sized for the strength limit state described above, they also need to be checked for vertical deflections under service loadings and stability under extreme loadings.

Analysis of an individual pile's load-settlement performance under service loadings is a relatively simple matter if you have a sufficient number of full-scale pile load tests on your site. Why? Because the plot of pile load versus deflection that you develop during pile load testing can be used directly to estimate pile deflections under service loadings. As a rule of thumb, I look for a service load pile deflection of about 0.25 inches and a nominal pile load deflection of about 1 inch when designing individual driven piles.

Great you say, but how often do we use individual piles to support our structure's foundations? My answer: not often do we use individual piles, we normally use pile groups to support our structures. So then how do you calculate the settlements of pile groups to develop vertical loading service resistance charts? Personally I'm not happy with the profession's current answer to this question. AASHTO 2010 directs geotechnical engineers to use an equivalent footing analogy to model pile group load-settlement behavior. Why don't we have some nifty computer programs for calculating pile group load-settlement behavior? We do have computer programs like GROUP that model deflections of pile groups, and we can use finite element and finite difference computer programs, but these are not mentioned in AASHTO 2010 so we don't have a service limit state resistance factor for their use. I might suggest that use of these computer programs with high-quality field and laboratory data, checked by a senior or principal geotechnical engineer should provide better pile group settlement estimates than the equivalent footing analogy, *so* if you use them you should have a good case for using a service limit state resistance factor of 1.0.

5.4.4.1 Pile Group Settlements Using the Equivalent Footing Analogy

We have completed our LRFD pile group design for the strength limit state so we have the size of the group and the depth of the pile tips. Now we are ready to check the service limit state deflections of our pile group using permanently applied service loadings. AASHTO 2010 makes a point that we do *not* apply the Meyerhof reductions in footing size (i.e., $B' = B - 2e$) when performing an equivalent footing analogy to solve for pile group settlements.

The equivalent footing method used to calculate pile group settlements assumes that at some depth below the top of piles the pile group acts like a large block spread footing applying its loading (transferred from the pile cap) on the underlying soils. This method has been used for years to check the pile group efficiency against the capacity of a single pile. The idea is that piles and soil between the piles act together as a large block with a perimeter equal to the outside perimeter of the pile group. If the center-to-center pile group dimensions are 20-feet wide by 30-feet long and the piles are 2 feet in diameter, the equivalent footing has dimensions of 20 feet plus one pile diameter or 22-feet wide by 30-feet plus one pile diameter, or 32-feet long. If the pile group is an end bearing pile group driven on or into the top of bedrock, dense granular soil or very hard clay like glacial till, then the settlement of the group is considered negligible and no settlement calculation is attempted. If the pile group is driven through a soft layer into a firm layer of soil (sand or clay), the bearing of the equivalent block footing is assumed to be at 2/3 the embedment into the firm layer. If the piles are friction piles with little or no end bearing capacity, the bearing of the equivalent block footing is assumed to be 2/3 the length of the pile (that is 2/3 of the zone that generates the pile resistance capacity).

After you have determined the length and width of the equivalent footing and the depth where the bottom of the equivalent footing bears, you calculate the settlements just as you would for a large spread footing bearing at that depth. You don't take friction of the sides of the big block footing into account in the settlement calculations, rather you apply the total pile group loading to the bottom of the equivalent footing block. If the soils beneath the equivalent footing are saturated clayey soils you calculate consolidation settlement, if the soils are granular soils or unsaturated sandy, silty, clayey soils you calculate immediate settlements using elastic methods, Hough's method, Schmertmann's method, and so on. In AASHTO 2010, they give an elastic method (Meyerhof, 1976) for estimating settlements of pile group equivalent footings on sandy soils which looks surprisingly similar to the settlement methods proposed earlier by Janbu and Schmertmann. I guess physics is physics no matter who comes up with the equation.

One more thing about the equivalent footing analogy method that is very important for you as a practicing structural or geotechnical engineer. The equivalent footing analogy method is very useful in helping you avoid a very costly, embarrassing blunder! This blunder has cost some unlucky geotechnical engineers hundreds of thousands if not millions of dollars if they make it. I might add that this blunder occurs in spectacular fashion at least once a decade in the United States.

Have I got your attention? "What is it?" you ask!? Take a look at Figure 5.4.2. Notice that a single driven pile develops tip resistance in the top of a hard crust that overlies a thick layer of soft clay. If you do a series of pile load tests on individual piles in the large pile group, they will all have plenty of load capacity because they have plenty of tip resistance bearing in the hard crust layer. Now look at the large pile group as a large footing bearing on top of the hard soil crust. This large equivalent footing transmits significant loading to the soft clay layer to initiate consolidation settlement, lots of consolidation settlement. In some cases the pile tips are driven too far into the hard soil crust and the entire equivalent footing (i.e., pile group) punches through the crust into the underlying soft clay. To avoid such problems, always consider the presence of a soft soil layer beneath a hard soil crust being used as a foundation bearing layer as a red flag! I am aware of a huge foundation problem caused by placing very large spread footings too close to the bottom of a soil crust over a soft

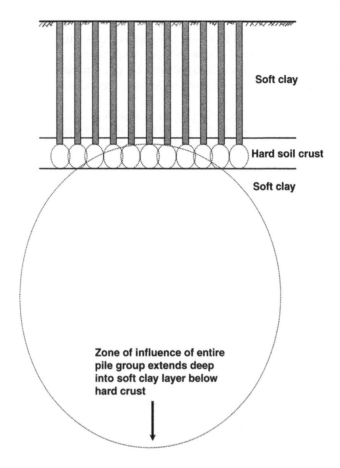

Figure 5.4.2   Pile group blunder – beware!

clay layer. Whether it is pile groups or large deep spread footings beware of soft clay beneath a crust.

References

AASHTO (2010) *AASHTO LRFD Bridge Design Specifications*, **2** volumes, 5th edn, American Association of State Highway and Transportation Officials, Washington, D.C.

Bowles, J.E. (1982 and 1996) *Foundation Analysis and Design*, 5th edn, McGraw Hill Inc., 816 and 1024 pages.

Coyle, H.M. and Castello, R.R. (1981) New design correlations for piles in sand. *Journal of the Geotechnical Engineering Division*, The American Society of Civil Engineers, **107**(GT 7), 965–986.

Hannigan, P.J., Goble, G.G., Likins, G.E., and Rausche, F. (2006) *Design and Construction of Driven Pile Foundations* – Volumes I and II, Report Number FHWA-HI-05-042, U.S. Department of Transportation, Federal Highway Administration, 968 pages.

Meyerhof, G.G. (1976) Bearing capacity and settlement of pile foundations, The Eleventh Terzaghi Lecture. *Journal of Geotechnical Engineering Division*, American Society of Civil Engineers, **102**(GT3), 195–228.

Nordlund, R.L. (1963) Bearing capacity of piles in cohesionless soils, The American Society of Civil Engineers. *Journal of the Soil Mechanics and Foundations Division*, **89** (SM3), 1–35.

Nordlund, R.L. (1979) *Point Bearing and Shaft Friction of Piles in Sand*, University of Missouri-Rolla, 5th Annual Short Course on the Fundamentals of Deep Foundation Design.

5.5

LRFD Drilled-Shaft Foundations

5.5.1 LRFD Drilled Shafts – Overview

Geotechnical LRFD analyses of drilled shafts are basically analyses of drilled shafts for ultimate load capacity and settlement (Brown *et al.*, 2010). The differences between what geotechnical engineers did for decades (Bowles, 1982 and 1996; Reese and O'Neil, 1999; Chen and Kulhawy, 2002) and LRFD design and analysis requirements for drilled shafts is partly terminology and partly the analysis methods used. As we discussed above in Section 5.2, three primary loading conditions need to be analyzed: service loadings, strength loadings, and extreme loadings. Generally drilled shafts are designed for strength loadings and checked for service and extreme loading cases.

During the design process, geotechnical and structural/bridge engineers work together to assure that both geotechnical and structural drilled-shaft design concerns are resolved. The geotechnical engineer first develops a drilled shaft axial resistance chart for use by the structural engineer in design of foundations. The structural engineer needs to check factored structural strength limit state loadings to be sure that they are equal to or less than factored foundation resistances. This first geotechnical chart is a plot of factored drilled-shaft axial strength resistance versus depth of embedment for a range of shaft diameters that will likely cover the range of anticipated structural factored loadings.

Then to check the foundation's service load performance against structural requirements, a second series of charts/graphs is developed by the geotechnical engineer. A drilled shaft diameter and embedment length is selected from consideration of factored strength loadings, this second set of charts is used to check if the anticipated drilled-shaft settlement is within acceptable limits for service loadings. The second set of charts plot service axial resistance versus shaft embedment for a range of shaft

Geotechnical Problem Solving, First Edition. John C. Lommler.
© 2012 John Wiley & Sons, Ltd. Published 2012 by John Wiley & Sons, Ltd.

diameters for a set value of vertical displacement of the shaft top. For example, you would prepare a service axial resistance versus embedment depth graph for four shaft diameters for a top of shaft displacement of 0.25 inches. Then you would prepare another chart for 0.50 inches, another for 0.75 inches, and so on, until you had a set of graphs covering the range of shaft top deflections considered reasonable for your project. Given this set of service loading charts, the structural engineer can estimate the anticipated settlement of drilled shafts selected for the project.

Using the service load-settlement charts, a third type of design chart can be developed. After the structural engineer selects a shaft diameter and a shaft embedment length, an axial load versus displacement chart can be developed for use in evaluating shaft displacements for a range of likely loadings.

5.5.2 Development of a Drilled-Shaft Axial Strength Resistance Chart

The basics behind developing an axial strength resistance chart involves calculating the drilled-shaft side resistance and end bearing (or tip) resistance, factoring side and end bearing resistances, and plotting their sum versus depth, as shown in the example from the Arizona Department of Transportation (ADOT) in Figure 5.5.1 (ADOT, 2010).

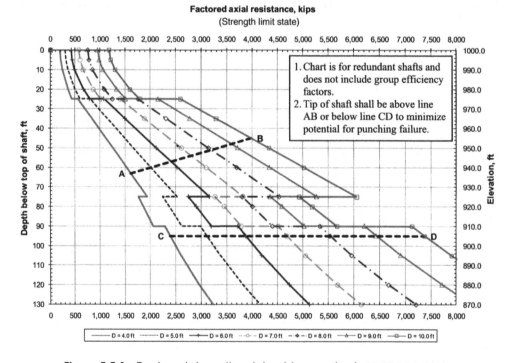

Figure 5.5.1   Factored strength axial resistance chart (ADOT DS-1, 2010)

Depending on the soil types encountered by each drilled shaft, various methods for calculating side friction and end bearing resistance may be required, as discussed in Section 3.5.

There are several decisions that you have to make to accomplish preparation of a chart like the ADOT example shown in Figure 5.5.1. For example, it is likely that your soil profile will include several soil layers. If you are doing the calculation by hand or if you just prefer to use layers in your spreadsheets, you can calculate the side resistance and end bearing resistance at the center of each layer and plot these values versus depth. My suggestion if you use layers is to subdivide thick layers into sublayers that are not thicker than 4 feet. Personally I don't use soil layers. I analyze CPT or SPT boring data on a foot-by-foot basis, assigning each spreadsheet row as a soil sublayer interval 1-foot thick.

After you decide how to divide soil layers for shaft analysis, you have to decide how to characterize each layer, that is, as sand, clay, or an intermediate geomaterial. Then you have to decide which methods of analysis for side and end bearing resistance to use in your calculations. The field testing, laboratory testing, and calculation methods used in your analyses affect the selection of resistance factors φ; for resistance factors please see AASHTO 2010 Design Manual's Table 10.5.5.2.4-1 (AASHTO, 2010).

Let's briefly review how to prepare a factored axial resistance design chart (or graph, I use these terms interchangeably although charts and graphs are actually somewhat different). Starting at the top of your shaft, neglect resistance from the upper 5-feet of drilled shaft or two diameters, whichever is greater from calculation of the nominal side resistance. Starting at 5-feet depth, calculate the side resistance of the shaft by multiplying the nominal side resistance times the perimeter of the shaft and times 1-foot length (or the applicable length for your stratum layer thickness). For example, if the soil is a silty sand with an N_{60} value of 25 from 5 to 10 feet, the unit side resistance may be calculated using the Beta method (Equation 3.5.14) for each foot depth from 5 to 10 feet, and the end bearing resistance is calculated using Equation 3.5.20. Calculating the nominal shaft resistance at each depth, multiply the side resistance times the perimeter area and the unit tip resistance by the end area of the drilled shaft.

To calculate the total factored axial resistance, multiply side resistance values by their resistance factor φ_{qs} and multiply the end bearing resistance by their tip resistance factor φ_{qt} and add side and tip values to get the total factored axial resistance. Repeating this process from 5-feet depth to the estimated total depth of shaft for a series of shaft diameters gives you the data needed to develop a drilled shaft axial strength resistance chart like that in Figure 5.5.1.

Take a close look at Figure 5.5.1, there are a few more issues to consider. Notice the notes in the box in the upper right hand corner. Note #1 indicates that this axial strength resistance chart is for redundant, widely spaced shafts. What does this mean? First it says these drilled shafts are redundant, that means there are two or more shafts supporting the same structure. For example, these multiple shafts might be supporting the same bridge pier or abutment. If an entire bridge pier is supported by one drilled shaft, it has to be a good one because it has no backup, that is, no alternate load path is available. Using only one non-redundant support element (i.e., one drilled shaft) to support a bridge pier introduces uncertainty into the design process, and so the use of one non-redundant drilled shaft requires that shaft resistance factors be reduced by 20% (AASHTO 2010 Article 10.5.5.2.4, second paragraph).

The second part of Note #1 says that the axial resistance chart values do not include group efficiency factors. In other words, all shafts are far enough (more than four diameters in AASHTO 2010; earlier versions required spacings up to six diameters) apart to avoid shaft vertical stress interactions that could reduce their load capacity.

Note #2 directs the structural engineer to avoid specifying shaft tip elevations between lines AB and CD (the dashed lines on Figure 5.5.1). Why? Notice the drop in shaft load resistance between about 75 and 90 feet depth; this drop in shaft capacity is likely due to the presence a relatively softer clay layer that is encountered between 75 and 90 feet. Line AB is drawn three shaft diameters above the clay layer and line CD is drawn five feet below the bottom of the clay layer to prevent loss of shaft tip resistance by bearing on or near the soft clay layer. Some engineers suggest that 1.5 shaft diameters above the clay layer is adequate; personally I have been using three diameters like ADOTs example chart in Figure 5.5.1. It's your choice, based on your understanding of subsurface conditions below the shaft tip. The notion of avoiding bearing of shafts tips in or near soft clay layers has been a practice of experienced geotechnical engineers for years, and I'm pleased to see it applied in this LRFD design example.

One more thing I might mention. When plotting an axial resistance chart like that shown in Figure 5.5.1, you can smooth the graph or plot it as a rough line. Personally I like the rough line because it visually appears similar to "real data" to me. I think smoothed plots look unrealistic, but then again it's up to you to select. Next we will develop service load axial resistance charts.

5.5.3 Development of a Drilled-Shaft Axial Service Resistance Chart

Figure 5.5.2 is an ADOT (ADOT, 2010) example of a service load resistance chart for shaft head deflection of 0.25 inches. Again, as with Figure 5.5.1, this chart is for redundant shafts with capacities that do not include a group reduction factor. The service load resistance factor is given in AASHTO 2010 as $\varphi_r = 1.0$.

To develop a service load resistance chart like the ADOT example given in Figure 5.5.2, you need to solve the settlement analysis problem backwards, that is, you need to select fixed values of settlement and shaft diameter, then, varying the shaft length, calculate the shaft service loading that generates the selected top of shaft settlement. Keep calculating service loadings for various sizes of shafts at different depths until you generate a table of data suitable for preparation of a Figure 5.5.2 for your shaft foundation design.

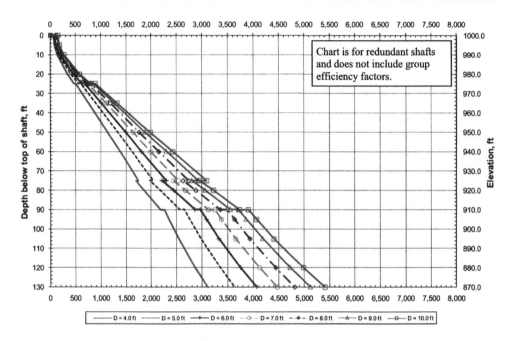

Figure 5.5.2 Service axial resistance chart for 0.25" shaft head settlement (ADOT DS-1, 2010)

If you are performing a shaft settlement analysis for a building structure, you can select your own method of shaft settlement analysis. If you are performing a shaft settlement analysis using the AASHTO 2010 code (or a similar AAHSTO code), you have to use an appropriate normalized load transfer curve for side and tip resistance for sands, clays or intermediate geomaterials as shown in AASHTO 2010 Article 10.8.2.2.2.

I don't know about you, but the first time I saw the AASHTO drilled-shaft normalized load transfer curves, I was very suspicious. I mean, what's up with taking a few drilled shafts in standard sand or clay soils, performing load testing, normalizing shaft settlement and ultimate resistances, and then using these few load tests to calculate shaft settlements on my project site. Why am I using these few load tests, which are not on my project site, and probably not in my State, to estimate shaft settlements? I was not pleased with the requirement that I use homogenized load test data! Why not do project specific load testing on my site? Yes, I hear you, smaller projects cannot afford drilled shaft load testing, and besides, these service load-settlement estimates are used by the structural/bridge engineer in his or her design well before a construction award is made and well before construction starts.

Having completed several shaft settlement analyses using the normalized curves, I have changed my mind (a bit). Use of the AASHTO curves seems to do a pretty good job of establishing shaft–soil strain compatibility by weighting side friction and end bearing components of shaft settlement. In my experience with fully instrumented (36" to 96" diameter) drilled shafts in various types of soils, tip resistance develops much slower than shaft side resistance. Many shafts develop full side friction resistance at shaft top deflections of about $1/4$ to $1/2$ inch, while full tip resistance requires shaft settlements of 2 to more than 4 inches. My hope is that we get more and better data from full-scale drilled-shaft load tests in the future. We need to develop a much broader-based library of site appropriate load-settlement data for use in drilled-shaft service load calculations.

To continue with shaft service limit state analysis, what about the shaft shortening (often called elastic shortening although I'm not sure it is really elastic), that is the load-settlement caused by settlement of the concrete shaft itself? Doesn't shaft shortening affect a drilled shaft's load-settlement properties? Standard practice suggests that the general rule when considering shaft shortening in service load analyses is that long shafts require this analysis. AASHTO 2010 article C10.8.2.2.2 says that elastic shortening is not included in their load-settlement curves, but should be considered for adjustment of settlements for long shafts. AASHTO 2010 refers you to the AASHTO 1999 Drilled-Shaft Manual (Reese and O'Neil, 1999) for a discussion of elastic shortening of long drilled shafts. General guidance for estimating elastic shortening of long shafts is given in Appendix C of the 1999 Drilled-Shaft Manual (Reese and O'Neil, 1999) as summarized below:

1. Elastic shortening of drilled shaft concrete is estimated by use of a modified form of the column deflection equation, $S_{column} = PL/AE$, where P is the column axial loading, L is the column length, A is the cross-sectional area of the column, and E is Young's modulus of the column material.
2. Adoption of the column deflection equation PL/AE for a column (i.e., drilled shaft) embedded in the ground has some complications. First of all, the sides of a building column don't have frictional resistance pushing up along the column side like a drilled shaft. So we need to correct the column deflection equation for shaft side friction. Secondly, our drilled shaft is constructed of reinforced concrete with steel and concrete that has difference values of Young's modulus. So we will have to calculate an effective value of Young's modulus that considers the presence of both reinforcing steel and concrete.
3. The resulting equation for estimating long drilled-shaft elastic shortening given in Appendix C of the AASHTO 1999 Drilled Shaft Manual (Reese and O'Neil, 1999), considering the adjustments to the column deflection equation given in item 2 above is given below.

$$\text{Shaft compression, } w_{concrete} = (P - 0.5P_{ms})\left(\frac{L}{AE}\right) \qquad (5.5.1)$$

where,

$w_{concrete}$ = elastic compression of the drilled shaft, inches
P = load applied to the top of drilled shaft, lbs
P_{ms} = estimated mobilized shaft side resistance, lbs
L = length of the drilled shaft, inches.
A = cross-sectional area of drilled shaft, square inches
E = effective Young's modulus of drilled shaft, psi
  = $(E_{concrete})(A_{concrete} + nA_{steel})$
$E_{concrete}$ = Young's modulus of drilled shaft concrete, psi
$A_{concrete}$ = cross-sectional area of shaft concrete, sq. inches
n = $E_{steel} / E_{concrete}$, dimensionless
E_{steel} = Young's modulus of reinforcing steel, psi
A_{steel} = cross-sectional area reinforcing steel, sq. inches.

In addition to giving guidance for calculating elastic shortening of long drilled shafts, Appendix C of the AASHTO 1999 Drilled Shaft Manual (Reese and O'Neil, 1999) gives a good summary of several alternate techniques for calculating total drilled shaft settlements including side, tip, and concrete shaft settlement. I recommend you check it out if you are interested. I checked and you can find copies of this reference on the internet for downloading to your computer (for free, free is good).

Let's finish up our discussion of service load resistance calculation. When calculating the total service load resistance of a selected shaft, you pick a settlement value for a given shaft diameter with a given shaft length, you adjust the settlement value for elastic shortening then sum the side resistance for each soil layer from head to tip of shaft and add the tip resistance. The weight of the concrete shaft is *not* subtracted from the shaft resistance value because we assume that the weight of soil removed from the shaft excavation is approximately equal to the weight of the shaft concrete.

If you don't agree with this assumption, you will have to estimate the weight of soil removed, and the weight of concrete and reinforcing steel placed, and compute the net increase in shaft weight in the actual shaft volume (don't forget that actual installed shaft volumes can be 10 to 50% greater than nominal volumes indicated by design shaft diameters used in calculations). The net increase in weight is then subtracted from the service shaft resistance. Most geotechnical engineers don't think that this calculation is worth the effort. Some structural engineers disagree because they never knowingly neglect any structural loading. My position on this matter is rooted in the graded approach. If I am working on a bridge or commercial building, I neglect the net increased loading due to shaft concrete weight. If I am working on a telescope foundation or on a foundation for high-tech scientific instrumentation, I work hard to include all identifiable loadings in my calculations.

5.5.4 Drilled-Shaft Load-Settlement Curves

During my discussion of spread footings and the bearing capacity 3 Ns equation, I presented material suggesting that load-settlement curves were more useful and appropriate for analysis of footing performance than an allowable bearing capacity value. The same point can be made for service resistance analyses of drilled shafts. Assuming that you have developed a series of shaft service settlement charts, as shown in Figure 5.5.2, you have the data needed to develop a specific shaft load-settlement chart, as shown in Figure 5.5.3.

Figure 5.5.3 is another example from ADOTs Drilled Shaft Design Policy DS-1 (ADOT, 2010). To prepare a chart for your project like Figure 5.5.3, select a shaft diameter and length, in this example case in Figure 5.5.3 it was a 7-foot diameter, 95-foot long shaft. Using Figure 5.5.2 service axial resistance charts for 0.10", 0.25", 0.50", 0.75", 1.00", and 2.00", for example, pick values of service axial resistance for the selected shaft and plot them as shown in Figure 5.5.3. This process takes considerable work effort that can be reduced somewhat by building spreadsheets for each of your typical drilled shafts. It is worth the effort to have load-settlement curves for each selected drilled shaft on your project. The structural/bridge engineer can make good use of these load-settlement curves when designing structural elements for

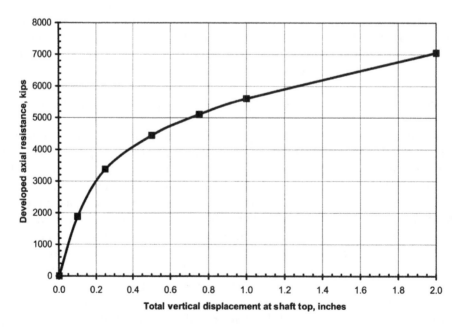

Figure 5.5.3 Example of an axial resistance versus vertical shaft top displacement curve for 7-foot diameter shaft, 95 feet long (ADOT DS-1, 2010)

uniform settlement, thus avoiding unanticipated load redistribution in the supported structure.

References

AASHTO (2010) *AASHTO LRFD Bridge Design Specifications*, **2** volumes, 5th edn, American Association of State Highway and Transportation Officials, Washington, D.C.

ADOT Geotechnical Design Section (2010) *Geotechnical Design Policy DS-1, LRFD, Development of Drilled Shaft Axial Resistance Charts for Use by Bridge Engineers based on Load and Resistance Factor Design (LRFD) Methodology*, 17 pages including the cover letter.

Bowles, J.E. (1982 and 1996) *Foundation Analysis and Design*, 5th edn, McGraw Hill Inc., 816 and 1024 pages.

Brown, D.A., Turner, J.P., and Castelli, R.J. (2010) *Drilled Shafts: Construction Procedures and LRFD Design Methods*, NHI Course No. 132014, FHWA NHI-10-016, Geotechnical Engineering Circular No. 10, 970 pages.

Chen, Y.J. and Kulhawy, F.H. (2002) *Evaluation of Drained Axial Capacity for Drilled Shafts*, American Society of Civil Engineers, Geo-Institute, Geotechnical Special Publication No. 116, Deep Foundations 2002, M. W. O'Neill and F. C. Townsend, Editors, Reston, VA, pages. 1200–1214.

Reese, L.C. and O'Neill, M.W., 1999, *Drilled Shafts: Construction Procedures and Design Methods*, Publication No. FHWA-IF-99-025, Federal Highway Administration, Washington, D.C., 758 pages.

5.6

LRFD Slope Stability

5.6.1 Introduction

In my opinion slopes are the most difficult topic in geotechnical engineering. Why? Because slopes are highly statically indeterminate.

You may ask, "What does that mean? Structural professors talk about statically indeterminate structures, in fact, I recall having a college class by that name. But what in the world is a statically indeterminate slope?"

OK, I'm going to have to work a bit to give you a better explanation. Of course I could give you the excuse that the topic of slope stability is extensive and is covered by several thick books! But then I imagine that you will not accept this excuse.

I recall that my friend Ralph Peck used to give his students the assignment of writing a project geotechnical report with analyses and recommendations on a single sheet of $8\frac{1}{2}$ by 11 inch paper. Ralph used to say that concise, clear writing is an indication of clear thinking. But, I'm a story teller and I like to be a bit long winded (some might say more than a bit). How can I argue with Ralph's "clear thinking" philosophy? I guess I can't argue. I give up. Here is my explanation of the complexities of landslides as clear and concise as I can put it. But first, let me illustrate my point by one short story.

> **Working example**
>
> A client of my called one morning and asked me the famous landslide question, "Why is my computed factor of safety for our slope different than yours? We used the same geometry, the same soil parameters, everything was the same, so how in the world can our factor of safety answers be different!?"

Geotechnical Problem Solving, First Edition. John C. Lommler.
© 2012 John Wiley & Sons, Ltd. Published 2012 by John Wiley & Sons, Ltd.

Good question, but not an easy one to answer. We used different computer programs and selected different methods of analysis. He selected the Bishop Circle method (also called the Swedish Circle method) and I selected the Morgenstern–Price method.

"Ah ha," the client said, "one of our computer programs is wrong!"

Since the slope-stability problem has more unknowns than equations of statics, each slope-stability analysis method has to make assumptions about the nature of the slide analyzed.

"What assumptions," he replied, "I don't recall making any assumptions. All I did was input the soil profile, select values for c, ϕ, and γ_{moist} and push the 'start analysis' button."

The assumption he made was selecting or allowing the program to select the Bishop Circle method, which suggests he agreed that it was appropriate for this slope problem. By selecting this method, he has to take the internal assumptions included in the method. The Bishop Circle method includes the following assumptions:

1. The slide mass is a rigid body. Internal side mass forces are not considered.
2. The slip surface is a circular surface or arc portion of a circle.
3. One equation of statics, summation of moments about the circle's center is used. Summation of forces in the horizontal and vertical directions is ignored.
4. The friction angle of the soil is assumed equal to zero, and all shearing resistance is generated by cohesion.
5. Since internal friction is assumed to be zero, the shear strength of the soil is independent of depth.
6. Slices can be used with this method to assist in finding the center of gravity of the slide mass, but slices are not required to be used for the analysis.

Then I gave him a copy of John Krahn's 2003 paper, *The Limits of Limit Equilibrium Analyses*, and a copy of my class notes based on Duncan and Wright's text (Duncan and Wright, 2005) that I used to teach several slope-stability classes. After we closely reviewed Krahn's paper (Krahn, 2003), he saw that identical slope-stability analysis methods give different answers for different soil conditions and geometry.

OK, that's my slope-stability story. I do recommend Krahn's 2003 paper from the Canadian Geotechnical Journal and Duncan and Wright's 2005 Wiley book to help you understand the complexity of slope-stability analyses. Now here is my explanation of statically indeterminate slopes and the impacts of assumptions made to solve slope-stability problems. I didn't limit myself to one page!

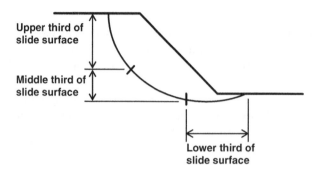

Figure 5.6.1 Slope with slide (slip) surface

5.6.2 Slope Stability by the Beam Analogy Method

Like many scientists and engineers before me, I like to simplify complex problems by using a simple model called an analogy. An *Encarta Dictionary* definition of analogy is "a comparison between two things that are similar in some way, often used to help explain something or make it easier to understand." To explain the effect of assumptions on the solution of a statically indeterminate slope-stability problem, I'm going to introduce a beam analogy of slope stability.

In Figure 5.6.1 we have a uniform clay slope with a defined slide surface. Soils in a slope provide both driving forces and resisting forces. The portion of slope soil above the slip surface generates driving forces, while the soil along the bottom of the slip surface generates the resisting forces. For our beam analogy, we will call the soil above the slip surface our loading, and we will call the soil below the slip surface our reactions. To simplify our beam model, we will represent the soil above the slip surface as a uniform loading, and the soil below the slip surface as discrete reactions, as shown in Figure 5.6.2.

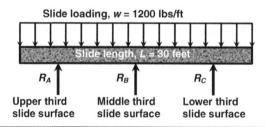

Assumption No. 1 – Cohesive slide resistance is constant from top to bottom of slide surface. $R_A + R_B + R_C$ = Total resistance
Conclusion: $R_A = R_B = R_C = \frac{1}{3}$ (30 ft) (1200 lbs/ft) = 12,000 lbs.

Figure 5.6.2 Slope-beam analogy for assumption No. 1

To start, we will divide the slide surface into three equal length sections: the upper, the middle, and the lower sections. All three sections of the slip surface provide support for the slide mass. In Figure 5.6.2., the analogous beam is supported at three points: the left support R_A represents the upper portion of the slide, the center support R_B represents the middle portion of the slide, and the right support R_C represents the lower portion of slide. For this example, we assume that the slide slip surface is 30-feet long, and that the analogous beam is 30-feet long.

To solve the slide problem in Figure 5.6.1 for shearing stresses along the upper, middle and lower thirds of the slip or slide surface, we solve the analogous beam in Figure 5.6.2 for reactions R_A, R_B, and R_C. Since each beam reaction in Figure 5.6.2 represents 10 feet of the slide slip surface, we calculate the shearing resistance for each portion of the slide by dividing its reaction force by 10 square feet. Why are we dividing by 10 square feet? The length of each third of the slide surface is 10 feet, and we are assuming that this slide analysis is for a 1-foot thick slice of the slide perpendicular to the image plane in Figures 5.6.1 and 5.6.2.

To calculate the factor of safety of our example clay slope, we will be comparing the shearing stresses generated by the slope loading to the shearing strength of the clay along the slide surface. Let's assume that the undrained shear strength of the uniform slope clay soil is 1500 pounds per square foot.

For our first slope-stability analysis, please refer to Figure 5.6.2. We have assumed that the beam is uniformly loaded along its entire 30-foot length. To find the beam reactions, we assume that the cohesive soil resistance is uniformly applied along the slide surface (similar to the assumption used in the Bishop Circle method). With this assumption, each beam reaction resists 10 feet of the uniform loading, which is given as 1200 pounds per foot, so $R_A = R_B = R_C = $ (10 feet) (1200 pounds per foot) = 12 000 pounds. Since the shearing resistance along the slide surface is uniform, the shearing stress generated along the slide surface in also uniform and equal to 12 000 pounds divided by 10 square feet or 1200 pounds per square foot. Given that the clay soil shearing strength is 1500 pounds per square foot (psf), the factor of safety of the slope based on assumption number 1 is (1500 psf)/(1200 psf) = 1.25.

Our second slope assumption is illustrated in Figure 5.6.3. For assumption 2, we assume that the slide resistance contributed by the middle third of the slip surface is greater than the resistances generated by the upper and lower thirds of the slide.

Using the beam analogy, we assume that the resistances are distributed to the re-actions like a three support beam shown in Figure 5.6.3. By using this three support beam assumption, we are assuming that the soil slide mass can act like a beam, by redistributing loadings to the three portions of the slide surface through soil stiff-ness generated by internal shearing stresses in the sliding soil mass. Using equations given in beam tables, with a uniform loading of 1200 pounds per foot and two 15-foot long spans, the reactions at the ends of the analogous beam, $R_A = R_C = $ 6750 pounds and the center reaction $R_B = $ 22 500 pounds. Calculating shearing stresses, the upper and lower thirds of the slide surface have (6750 pounds)/(10 square feet) = 675 pounds per square foot, and the center third has (22 500 pounds)/(10 square

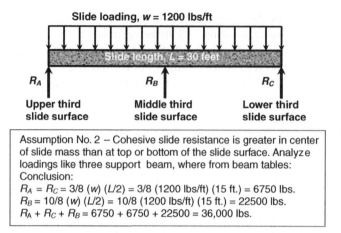

Figure 5.6.3 Slope-beam analogy, assumption 2

feet) $= 2250$ pounds per square foot shearing stress. The center third of the slide has calculated shearing stresses that exceed the shearing strength of the soil, that is, 2250 psf is greater than 1500 psf. Now we have to make additional assumptions. We can calculate the factor of safety for each third of the slide surface, but how do we calculate the overall slide factor of safety knowing that the middle third of the slide surface has failed?

Let's try a couple of approaches. First calculate the factor of safety of each slide segment: FOS of upper third $=$ FOS of lower third $= (1500\text{ psf})/(675\text{ psf}) = 2.22$, and FOS of the middle third $= (1500\text{ psf})/(2250\text{ psf}) = 0.67$. We could calculate a weighted overall factor of safety: FOS entire slide surface $= [(2)(2.22) + (1)(0.67)]/3 = (5.11)/(3) = 1.70$. This weighted factor of safety result doesn't seem reasonable since our first case had a FOS of 1.25 and none of its segments failed.

We know that the middle slide segment cannot generate a shearing resistance greater than the shearing strength, so it makes more sense to limit the center reaction R_B to an upper bound value of resistance equal to $(1500\text{ psf})(10\text{ square feet}) = 15\,000$ pounds. The reaction at B would be even lower than 15 000 pounds if we assumed that the soil has peak strength of 1500 pounds per square foot and a lower residual value at higher strains. Using the peak strength assumption for R_B, the calculated factor of safety of the slope would be equal to sum of the generated resistance reactions divided by the total loading of the slope: $(6750\text{ pounds} + 15\,000\text{ pounds} + 6750\text{ pounds})/36\,000\text{ pounds} = 0.79$.

Notice that the slope geometry and soil shear strength and weight parameters have not changed, but the calculated factor of safety has changed, based on differing assumptions.

What if we introduce a crack at the ground surface of the slide, as shown in Figure 5.6.4? This will be our third slide case: Assumption 3, a surface crack 5-foot

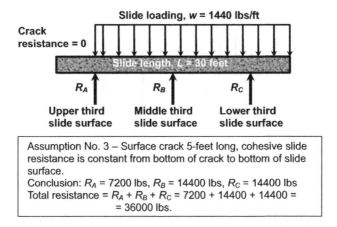

Figure 5.6.4 Slope-beam analogy, assumption 3

long reduces the resistance of the upper third of the slide and transfers weight to the remaining slide surface.

The transferred loading increases our beam loading on the remaining 25 feet of slide surface to 1440 pounds per foot, as shown in Figure 5.6.4. Assuming that R_A resists 5 feet of loading and that R_B and R_C each resist 10 feet of loading, the reactions are 7200 pounds at R_A and 14 400 pounds at reactions R_B and R_C. Shearing stresses at the three segments are R_A 1440 psf (the resistance is over 5 feet due to the crack), R_B 1440 psf, and R_C 1440 psf. So the factor of safety for all three side segments is equal to 1.04 and we can say that the overall slide factor of safety is 1.04.

So far we have made three different slide assumptions to solve the given simple slide problem and we have come up with three different answers, 1.25 (stable), 0.79 (failed), and 1.04 (marginally stable).

In Figure 5.6.5, we have an example of a slide analysis problem that is closer to what we actually solve when we use one of the advanced slope-stability analysis methods. Loadings vary from slice to slice because they have varying heights and varying soil unit weights. Resistances on the bottom of slices vary because multiple layers of soil are used and most often soils are modeled with cohesive and frictional strength components. The slide mass is not modeled as a rigid body, requiring calculation of interslice forces and redistribution of forces in the slide mass.

Calculation of the reactions for the beam shown in Figure 5.6.5 would require an extensive trial-and-error iterative solution approach, which I will not attempt.

My point in presenting these beam-analogy examples is that varying slope-stability analysis assumptions generate varying calculated factors of safety. When you use a complex slope-stability analysis model, it is not easy to identify the assumptions used and their affects on the calculated factor of safety.

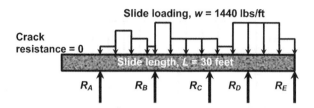

Assumption no. 4 – Surface crack 5-feet long. Both cohesive slide resistance and loadings are <u>not</u> constant. Load redistribution between slide blocks is an unknown function of interslice forces.
Conclusion: Reaction forces along the side surface unknown. Total resistance = unknown. Use the trial and error method.

Figure 5.6.5 Slope-beam analogy, assumption 4

5.6.3 Slope Stability – ASD and LRFD Analysis Methods

Methods of slope-stability analysis that you learned in college and those that are covered by nearly all available geotechnical texts are allowable stress analyses. They use factor of safety values to characterize the stability of a slope.

LRFD methods of slope-stability analysis are intended to provide consistent reliability for all slope analysis limit states, similar to analyses we discussed in earlier sections for footings, piles, and drilled shafts. My study of geotechnical LRFD for slopes uncovered two LRFD slope-stability approaches. The first approach to geotechnical LRFD slope analysis uses statistical analyses of soil parameters used in standard slope-stability analysis procedures, such as the Bishop, OMS, Spenser, Morgenstern-Price and GLE methods.

The second approach to geotechnical LRFD slope analysis uses statistical analyses of the entire slope and determines the probability of failure without reference to previous slope analysis methods. This second type of LRFD analysis attempts to follow the path of least resistance through the slope's soil mass.

The first type of LRFD slope analysis is covered by an FHWA research paper RI03.030 (Report No. OR 06-010) by the University of Missouri-Columbia titled *Procedures for Design of Earth Slopes using LRFD* (Loehr, Finley, and Huaco, 2006). Parameters used in this method's slope-stability analysis include soil total unit weight γ, undrained shear strength s_u, effective stress parameters c' and ϕ', and pore water pressure u.

Loehr, Finley and Huaco consider pore water pressures as "deterministic" values and use a resistance factor of 1.0 for pore pressure u. All of the other parameters used in this first type of LRFD analysis are factored nominal resistance values obtained from field tests, laboratory tests, and empirical geotechnical correlations.

The second type of LRFD slope analysis is presented by Fenton and Griffiths in Chapter 13 of their *Geotechnical Risk Analysis* book (Fenton and Griffiths, 2008). They use a unique approach that incorporates a rigorous probabilistic geotechnical analysis. Fenton and Griffiths use a non-linear finite-element method combined with a random-field generation technique. They call this method the random finite-element method or RFEM. They promote this method as a way to allow the slope-stability solution to seek a path of minimum shear strength resistance through the soil mass, thus leading to higher probabilities of failure than could be generated by using statistical methods of shear strength characterization for LRFD designs using ordinary slope-stability methods.

5.6.4 Three Basic Slope-stability Problem Types

When most engineers and geologists hear the words "slope stability", they imagine a natural slope. In many cases we are talking about natural slopes when we use the term "slope stability", but there are at least two other types of stability analyses commonly required on projects: (1) embankment slope stability, which involves calculating a stable slope angle for a fill slope planned to be built on the project, and (2) trench slope stability, where the side walls of a utility trench excavation require a stability evaluation.

5.6.5 Closing Thoughts on LRFD Slope-Stability Analyses

I consider the material given here as an introduction to geotechnical LRFD slope-stability analysis. My feeling is that much additional work needs to be completed by researchers in the field of geotechnical LRFD slope-stability analyses to present a complete picture of LRFD slope analysis. Personally I feel that the complexity of slope-stability problems has not been adequately studied to provide appropriate resistance factors for all of the variables involved in slope-stability problems. So at present for my consulting practice, I will analyze natural slopes, fill slopes, and trench excavations using traditional ASD methods (I prefer the Morgenstern–Price method of analysis). I will continue to study and try out proposed geotechnical LRFD slope-stability methods to see how they work on real-world problems. When the time is right, and my review of geotechnical LRFD slope-stability material converges on a comprehensive approach, I propose that an entire book be written to cover this topic.

References

Duncan, J.M. and Wright, S.G. (2005) *Soil Strength and Slope Stability*, John Wiley & Sons, 297 pages.

Krahn, J. (2003) The 2001 R.M. Hardy Lecture: The limits of limit equilibrium analyses. *Canadian Geotechnical Journal*, **40**, 643–660.

Fenton, G.A. and Griffiths, D.V. (2008) *Risk Assessment in Geotechnical Engineering, Chapter 13 – Slope Stability*, pages 381–400.

Loehr, J.E., Finley, C.A., and Huaco, D.R., Report Date December (2006) FHWA issue date January 2006, Procedures for Design of Earth Slopes Using LRFD, Organizational Results Research Report RI03.030, Prepared by University of Missouri-Columbia and Missouri Department of Transportation, 80 pages.

6

Closing

6.1

The Big Picture

6.1.1 How Do Geotechnical Engineers Miss the Big Picture?

About three years ago I was involved in a post mortem geotechnical investigation of extensive soil damage to a large expensive home, see Section 1.3, Figures 1.3.1 and 1.3.2. During my first visit to the site, I followed the Terzaghi Method and explored the geology of the area around the home. My first pass involved walking a 1500-foot radius circle around the damaged home. From the vantage point of distance, I noticed that the site had steeply dipping bedrock with alternating bands of sandstone and expansive clay shale crossing beneath the home site. This situation set the home up for differential foundation support conditions. Reviewing the geotechnical report provided for this home, I saw evidence that they did not identify the differential support conditions, nor did they identify the presence of expansive clay shales. My question is: why did the geotechnical engineers miss an obvious geologic condition? During a lunch meeting with our semi-retired, senior geologist, Mr. Jerry Lindsey, I asked him to consider his 60+ years of experience in the geo-industry and write down his "explanation" for why geotechnical engineers often miss obvious geologic hazards. The following material with some of my editing is Jerry's answer.

6.1.2 The Big Picture – What a Soils Engineer Should Know about the Geologic Setting before Going to the Job Site

Geotechnical engineers should keep in mind that the "geo" in geotechnical means that foundations interface with a geologic strata that is part of a landform created by a geomorphic process. This applies not just to the bearing surface and what is directly

Geotechnical Problem Solving, First Edition. John C. Lommler.
© 2012 John Wiley & Sons, Ltd. Published 2012 by John Wiley & Sons, Ltd.

below, but the big picture, that is, the geological environment that created the present conditions. In these times of large engineering companies that share manpower across the continent and around the globe, geotechnical engineers can find themselves in unfamiliar territory.

The physical characteristics of the material immediately underlying the site may be only part of the problem. Rather than just evaluating the site by analyzing the boring logs and getting a verbal report from the field engineer, I suggest that the geotechnical engineer first find out where the site is in relation to the surrounding geologic features. What kind of deposits should you expect on the site? Will the lithology (the physical characteristics of a soil or rock) change with depth or laterally? Are there adjacent conditions that could affect the project site? Are the soils derived from marine shale? The soil represents a history book of that depositional energy that surely involved water flow: fine-grained sediments are the result of low energy flow, such as ponding or flood-plain backwater, and coarse-grained deposits result from higher energy, such as found in the axis of streams. Many soils are the result of wind deposition that gets reworked by runoff. What processes are at work now? If the site is well drained then erosional processes may be at work. The type of deposits implies what will lie below or adjacent to the site.

Rapidly changing soil conditions occur because of changing energy of deposition. Lateral changes in lithology, are called facies changes. In a river valley environment a steep-sided cut bank that is subsequently backfilled by river (fluvial) sediments, or overburden alluvial fan deposits, becomes an inset, and may result in an abrupt change in lithology and potential bearing capacity. A layer of cobbles that represents strong current (axial) river deposits may not be continuous, but could change laterally into a wet clay stratum. Again differential settlement and variable bearing capacity is a concern.

In recent years finding information on the computer has become increasingly easy (you older gentlemen may like to know that Jerry is 82 years old). The entire set of topographic maps for your state can be (and should be) on your computer. GPS bearings can be obtained from those maps just by movement of the cursor. This allows the geologist and engineer to locate possible problem areas (drainages, flood potential) and plan a drill-boring program. Site plans provided by the client very seldom show the adjacent topography. Regional topographic maps and aerial photographs are available from a web site called Terra Server that uses United States Geological Survey resources. Another web site that gives you road maps with topographic features, but also has aerial photographs is Live Search. The most used source for aerial photographs is probably Google Earth. The engineer should try all of these resources to get the best photographic resolution. In some states the Bureau of Land Management (BLM) and State Land Offices have their own maps and photos. Geologic maps for the state and region should be somewhere on the office wall so that the younger engineers and geologists can become familiar with geologic formations that are potential problems.

6.1.3 Bedrock

Examples of formations to watch out for are deposits of Cretaceous and Jurassic age. These formations crop out in a broad area of the central United States. These marine- and delta-type deposits respectively consist of thick deposits of shale that include the Mancos Shale (or equivalent) and the Morrison Formation in the mountain states. Developments placed on outcrops of these formations, or on alluvial soils derived from these formations can encounter expansive soil conditions; salt deposits harmful to concrete; collapsible soils; dip slope slides; rock slides; erosion problems; unusable waste material from excavation cuts; and other "plastic clay" problems.

The alternating shale-limestone strata found in Paleozoic and Mesozoic strata (Mississippian, Pennsylvanian, Permian, Triassic, Jurassic, and Cretaceous) that occur in a broad swath across the mid-western United States, offer their own unique problems. Dipping beds of alternating shale and limestone are problems, especially if the limestone is highly jointed. These joints can result in blocks of limestone migrating down dip with very little encouragement. The message is: watch out for shale!

Limestone is typically jointed and will allow infiltration of moisture to areas that were not exposed before site grading. That is not a good thing if the beds are underlain by shale and the beds are dipping significantly. In areas with thick strata of limestone, cavernous conditions tend to develop especially where the formations are jointed and faulted. Active karst (sinkhole) conditions can be expected in areas with closely spaced joints adjacent to river valleys or where ground water is within 300 feet of the surface. Topographic maps that reveal alignments of depressions or springs, or shows a grain pattern, that is having ridges and drainages with a common trend, may have Karst or cavernous problems.

6.1.4 Structural Problems

Besides the concern of faulted and jointed bedrock discussed above, folded bedrock strata present other problems. Interbedded strata that dip (even a few degrees) indicates that dip slope hills may be prone to slides, spring flow, and changing lithology below leveled building pads. Settings in mountains and valleys (basins and ranges) mean that the regional geomorphology is controlled by faults, folds, or by rivers.

If you are west of the Rocky Mountains it is likely that faults lie at the boundary of every range and basin. Even though there hasn't been an earthquake in your lifetime don't rule out future regulations that recognize Quaternary age faults and call out setbacks for structures from these faults that should have been recognized. If you are in country of low relief, those wide river bottoms are full of cut-off meanders (easily seen in aerial photos), flood plain clay deposits, organic sediments, and other surprises.

6.1.5 Previous Land Usage

Cities with rapidly expanding suburbs, such as Las Vegas, NV, Austin, TX, Denver, CO, Indianapolis, IN, Louisville, KY, Albuquerque, NM, or Wichita, KA, have concerns with previous land usage that may not show up on environmental assessments which are focused on landfills or service stations. Where are the old gravel pits that got filled in? Is the site on a man-made terrace or an old railroad grade?

Environmental survey studies often consider the value of old aerial photographs in evaluating sites for previous uses. The same thing applies to geotechnical engineering projects to examine the big picture. Be very wary when the site topography is already level, or that vegetation is strangely younger than that on adjacent terrain, or has recently been cleared of all vegetation. The developer may think they are making the land more attractive to a buyer by dressing up the site. There's a good chance that uneven terrain has been smoothed over or that evidence of previous use has been obliterated. What surprises underlie the site? It's not likely you can answer all your concerns just by drilling. Look up surveyors in the phone book and find the regional aerial photographers who have a history in that area. Develop a database of aerial photographs that cover your area. Those pasture/farm/desert lands could be soon covered with subdivisions and the natural features that reveal the geology will be lost forever.

6.1.6 Paleo Channels

Paleo channels can occur in a wide variety of circumstances. These are former stream channels that became incised into the underlying material, later become abandoned by shifting of flow patterns and subsequently became filled with alluvium. Do the circumstances of the Leaning Tower of Pizza ring a bell? There, tidal inlets became abandoned and subsequently filled as the land uplifted (the sea retreated) causing highly variable soil conditions below a relatively small building footprint. A paleo channel is not only usually well hidden, but can contain fill deposits completely different than the surficial material. Then, for builders, and for engineers, it gets worse – these old channel deposits that are typically more permeable than the eroded sediments, can become saturated after construction and allow differential wetting of foundations soils, or even carry perched ground water.

6.1.7 Jerry's Closing Comment and a Thought from Ralph Peck

I would like to close this section by saying that Jerry Lindsey is a true gentleman. Jerry's final comment to me was that many engineering failures occur because of surprises in subsurface conditions. He could have pointed out that geotechnical engineers often are surprised because they ignore study of site and surrounding geology due to lack of project budget, lack of time, or dare I say lack of interest.

Ralph Peck told me that engineers' who miss the opportunity to look into fresh excavations, check road cuts, and study geologic references are missing out. Ralph suggested that they should get out from behind their computers and see what is in the real world.

6.2

V and V and Balance

6.2.1 Have Hand Calculations Died?

A few years ago, I finished a retaining-wall design, and sent my recommendations to the project structural engineer. The structural engineer finished his calculations and construction drawings and submitted the design to his client, a United States National Laboratory (to remain un-named). Nearly six months after finishing this rather ordinary retaining-wall project, I received a phone call from the structural engineer's office manager. The office manager wanted proof that the geotechnical computer programs that I used had been V and V'd. I asked innocently, "What is V and V?" He replied, verification and validation to NQA-1 quality standards. I told him that our company has quality plans for commercial and federal projects including federal work done to NQA-1 quality standards, *but* the retaining wall in question was designed to commercial quality standards, as agreed to in our contract. I reminded him that we submitted a received approval of a work plan for the retaining-wall design, and it was clearly not totally NQA-1 compliant.

One of the computer programs I used on the retaining-wall project was an Excel spreadsheet. No problem, Excel had an exemption from the rigorous V and V because it was a commonly used math program. The other program used was a standard geotechnical computer program (to remain un-named) developed by a well-respected group of university professors working on federally funded research. This geotechnical program had been used successfully on hundreds, if not thousands, of State and Federal projects over a period of more than 30 years. It did not matter to the office manager. He insisted that I needed to submit quality assurance documentation for this computer program so that he could submit it to his client.

I called the company that markets this computer program and told them about our quality assurance "problem."

Geotechnical Problem Solving, First Edition. John C. Lommler.
© 2012 John Wiley & Sons, Ltd. Published 2012 by John Wiley & Sons, Ltd.

"No problem," they said, "we've had dozens of similar requests and we have prepared a document that satisfies these quality assurance requests."

I received a copy of the quality document and submitted it to my client. Problem solved, or so I thought. A few weeks later, the flood gates of questions and quality audits broke upon us. I won't bore you with the gory details, but the V and V requirements for checking the commercial computer program used to design a simple retaining wall were more costly than the entire fee that we originally received for designing the retaining wall. We concluded after much discussion with the quality auditors that it was simpler, easier, and cheaper to re-do the entire set of calculations for the retaining wall by *hand*! What happened to the graded approach? We could have continued to argue with the auditors, but it was a losing battle. We were not designing a nuclear reactor structure; we were just designing a 10- to 12-foot high retaining wall. It did not matter.

For single calculations on small projects, consultants can't afford to have their commercial computer programs verified and validated. It is easier to do hand calculations, list your references, explain why these are appropriate references, and then do your calculation by hand, as shown in Figure 6.2.1, explaining where all data came from and why you used it. The bottom line is we have come full circle. We used to do hand calculations, then we transitioned to computer-generated calculations and now again we need to do detailed hand calculations much like we did prior to the development of commercial geotechnical computer codes! The only difference is that today we have to justify all methods used, and justify input data used in our calculations. Often these justifications require writing one or two "white papers," that is, technical papers that accompany the calculations. I guess that I am longing for the "good old days" when engineers were experts and our clients believed what we recommended.

6.2.2 What about the Graded Approach and Balance?

Please don't misunderstand me, I know that we need to have quality control and quality-assurance plans to make sure that our engineering calculations are properly checked and peer reviewed. I am also in full agreement that important, critical government projects require strict quality standards, such as those included in NQA-1 standards.

My point is that we need balance in our engineering projects such that the level of QA/QC required should match the importance of the project. Development of the graded approach where the level of effort matches the requirements of the project was started (to the best of my understanding) by government engineers and scientists who were trying to make sure public funds were being used wisely and appropriately.

I would also like to comment on the importance of balance in your engineering career. When I take too much of my time, not hours, but days and weeks at a time talking and presenting, and marketing our firm's consulting services, and sitting in endless meetings, I start to feel that I'm losing my way. I start to feel like I'm losing my

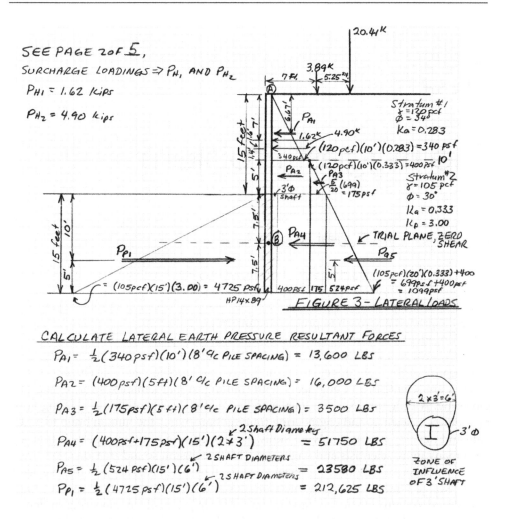

Figure 6.2.1 Example of a hand calculation required by V and V

engineering identity. Who am I? The antidote for these feelings is doing engineering work. I feel great when I go out into the field on a project, or when I sit down with a young engineer to review his or her spreadsheet or hand calculations. I have to admit it. I love solving problems. When I sit down to do hand calculations or a detailed spreadsheet calculation that takes me a day or two to focus on the problem without talking to others, I feel great!!

6.3

The Biggest Problem

6.3.1 What is the Biggest Problem?

We have discussed the process of problem solving and the nature of complexity in this book. Geotechnical engineering problems can be relatively simple, or difficult, or complex, or compound-complex. We have discussed developing a problem plan and using the graded approach to problem solution.

US Presidents, historians, and commenters in newspapers and on television all seem to agree that maintenance of competitive advantage for the United States and European countries requires creative, innovative, and problem-solving citizens. These citizens are scientists and engineers with training in mathematics, science, and engineering. A program I saw on television had a science contributor explaining how mathematics, science, and engineering training hard-wires the human brain to innovate and develop problem-solving skills. The conclusion of these experts is that we need many more scientists and engineers. This conclusion sounds reasonable to me.

Many politicians, engineering societies, and citizen engineers suggest that we need to encourage students to study mathematics and science in elementary and high school, and to contribute to scholarship funds to help promising engineering students. These are wonderful pursuits, but we are still falling far behind the reported numbers of engineers required.

I have attempted to take the "big picture" approach to defining this biggest of all engineering problems. My approach is to look at the life cycle of an engineering career as a flow of time. Young students are trained, mentored, and advanced into the engineering profession. Mid-level engineers work to solve problems and learn the engineering profession on the job. These working engineers are required to continually upgrade their skills by personal study, by attending seminars and classes, and by acquiring enough professional development hours (pdh's) to renew their professional engineering licenses. Mid-level engineers work with and help mentor young

Geotechnical Problem Solving, First Edition. John C. Lommler.
© 2012 John Wiley & Sons, Ltd. Published 2012 by John Wiley & Sons, Ltd.

engineers in the problem-solving process. Senior and principal engineers direct teams of young and mid-level engineers working on client projects. Senior and principal engineers are required to keep abreast of advances in technology, to participate in life-long learning, to mentor and encourage younger and mid-level engineers, and to solve complex problems associated with client projects.

How do we maintain, improve and increase the flow of engineering talent for the benefit of society? This is the biggest engineering problem.

6.3.2 How do We Solve the Biggest Problem?

My view of the engineering career life cycle is that we need to get young engineers into the profession, retain and upgrade skills of working engineers, and encourage older engineers to forgo early retirement and mentor younger engineers.

How do we accomplish maintaining and improving the engineering profession for the benefit of society? The easy answer is that I don't know. Why am I copping out and admitting that I don't have the answer to this question? My reason is relatively simple. The solution to this problem cannot be defined by one person or one group of people. The solution to this problem requires that society agrees to improve the status of the engineering profession. Society has to hold up this profession as a valuable pursuit that is required for the better good of all. Students looking at the hard work and educational expense of an engineering education, working engineers coping with family and professional needs, and older engineers looking for validation and respect: they all want the engineering profession to provide more money, more training, more benefits and more respect. Engineers want to be special. In the end, most people want an answer to one question, what's in it for me? Please don't judge them as selfish; it is self-interest to want an interesting, rewarding, well-paid career that is fun.

6.4

Topics Left for Later

6.4.1 Geotechnical Engineering Topics are Endless

In New Mexico, we have a monthly geotechnical group meeting that we call the Albuquerque Soil Mechanics Series, ASMS. Most groups have a presenter or a lecturer to entertain the members with a slide presentation. Our ASMS group has a different philosophy; we normally have short presentations and then have open discussions of the topic by all of the members present. About every fourth or fifth meeting, we have an open discussion topic for the group without having a speaker. Today was our monthly meeting, and we chose the open topic of compacted fills to discuss without a speaker. All geotechnical and geological engineers have an opinion about how fill soils should be compacted and tested, and most have plenty of horror stories to share about compacted fill projects that went sour. I was amazed at the depth and breadth of our discussion today. We discussed how many compaction tests to perform, what kind of tests to perform (nuclear densometer, sand cone, SPT, or CPT), the role of the geotechnical engineer, and so on. One hot topic was the difference between testing, inspection, and testing with inspection services. One engineer quoted Ralph Peck, who said that the most important inspection tool was a knowledgeable engineer with eyes to see and a mind to evaluate what he or she was seeing. At the end of today's meeting, all in attendance agreed that we needed more than an hour and a half to discuss compacted fills, testing of fills etc., so we scheduled a second meeting to discuss this apparently simple, but not actually simple geotechnical topic.

On the way home today, I couldn't help but be amazed at the complex details involved in every aspect of geotechnical engineering. A couple of years ago, I challenged the ASMS group at our monthly meeting to discuss the complexities of performing the water content test and interpreting its results for use in a geotechnical report. I anticipated that the group would stare at me with nothing to say about the lowly

Geotechnical Problem Solving, First Edition. John C. Lommler.
© 2012 John Wiley & Sons, Ltd. Published 2012 by John Wiley & Sons, Ltd.

water content test. No way. We had to cut short the water content discussion after 45 minutes to finish other business.

In case you are wondering, we discussed gypsum soils that should not be heated above 60 degrees Centigrade to avoid driving off structural water, Smectite clays that should be heated and weighed in increments up to 200 degrees Centigrade to evaluate inner layer water, the presence of small amounts of organic matter in water content samples, how a silty sand could have an *in situ* moisture content of 1%, among other topics.

With all of the infinite possibilities of geotechnical problem-solving topics for us to discuss in this book series, I would like to propose that we continue on in later volumes with detailed discussions of stability analyses of natural slopes, cut slopes, trenches, and embankment slopes. I would also like to cover compaction and testing of fills for buildings, landfill covers, levees, dams, roadway and airport pavements, retaining walls, and MSE retaining walls. Speaking of MSE walls, I would like to discuss MSE walls in general and MSE stress relief walls in particular. On the topic of problem soils, I propose to discuss design, construction, and QA/QC testing of gypsum soils, identification of dispersive soils, more details on identification and treatment of collapsible soils, and how to build on expansive clays.

6.4.2 Closing

We have discussed the process of problem solving and the nature of complexity in this book. Geotechnical engineering problems can be relatively simple, or difficult, or complex, or compound-complex. Geotechnical problems can even evolve into wicked problems if too many "stake-holders" with conflicting positions weigh in on the issue.

We have discussed developing a problem plan and using the graded approach to problem solving. I have shared my motto, "You have to see it to solve it," and I have urged you to seek out several alternate references when studying a geotechnical problem.

I would also like to pass on Ralph Peck's advice to me which was to attend national geotechnical conferences and between sessions make sure to approach and talk to senior engineers. Ralph was the most eminent engineer I ever knew, and he was absolutely approachable. His greatest joy was talking with young engineers. Do not be afraid, we were all young like you at one time, and we would love to share our geotechnical knowledge with *you*.

Presidents, historians, and economic experts agree that maintenance of a competitive advantage for the United States, European, South American, African, and Asian countries requires creative, innovative, and problem-solving citizens. These citizens are scientists and engineers with training in mathematics, science and engineering. They are or should be *you*.

Index

Geotechnical Problem Solving, First Edition. John C. Lommler.
© 2012 John Wiley & Sons, Ltd. Published 2012 by John Wiley & Sons, Ltd.

Printed and bound by CPI Group (UK) Ltd, Croydon, CR0 4YY

16/04/2025

14658385-0004